IMPACT OF PESTICIDE USE ON HEALTH IN DEVELOPING COUNTRIES

IMPACT OF PESTICIDE USE ON HEALTH IN DEVELOPING COUNTRIES

Proceedings of a symposium
held in Ottawa, Canada,
17–20 September 1990

Editors:
G. Forget, T. Goodman, and A. de Villiers

INTERNATIONAL DEVELOPMENT RESEARCH CENTRE
Ottawa • Cairo • Dakar • Johannesburg • Montevideo • Nairobi
New Delhi • Singapore

©International Development Research Centre 1993
PO Box 8500, Ottawa, Ontario, Canada K1G 3H9

Forget, G.
Goodman, T.
de Villiers, A.

Impact of pesticide use on health in developing countries : proceedings of a symposium held in Ottawa, Canada, 17–20 September 1990. Ottawa, Ont., IDRC, 1993. x + 335 p. : ill.

/Pesticides/, /public health/, /health hazards/, /poisoning/, /developing countries/ — /epidemiology/, /research results/, /agricultural workers/, /product safety/, /occupational hygiene/, /information services/, /health services/, /pest control/, /disease control/, /conference reports/, references.

UDC: 632.95.02:614   ISBN: 0-88936-560-1 √

Technical editors: G.C.R. Croome, S. Garland

A microfiche edition is available.

Abstract — The introduction of modern inputs, such as pesticides and chemical fertilizers, has substantially increased agricultural productivity. Pesticides have also revolutionized the fight against endemic diseases in developing countries. Unfortunately, there is reverse side to this coin: the World Health Organization estimates that these agents cause 2 million cases of poisoning a year, of which 20 000 result in death. This book contains papers presented at a symposium to review the impact of pesticide use on health in developing countries. Part I presents a brief survey of the global situation and the results of 12 epidemiologic studies carried out by researchers in Africa, Asia, Latin America, and the Middle East. These focus on poisonings resulting from organophosphates, herbicides, and pyrethroids. Part II illustrates how the processes of development and production of pesticides, spraying techniques, and legislation can affect the health of workers. Part III consists of discussion of the benefits and ways of obtaining pertinent information for the prevention of pesticide poisonings. Finally, consideration is given to the advantages and disadvantages of certain alternatives to the use of synthetic pesticides in agriculture and public health, such as botanical pesticides and integrated pest management strategies.

Résumé — C'est avec l'introduction des intrants modernes tels les pesticides et les engrais chimiques que l'agriculture a atteint son niveau de productivité actuel. De même, les pesticides ont révolutionné la lutte contre les grandes endémies dans les pays en développement. Malheureusement, il y a un revers à la médaille : l'Organisation mondiale de la santé estime à 2 millions par année les cas d'empoisonnement dus à ces agents, dont 20 000 sont mortels. Cette publication contient les communications présentées au colloque sur l'effet des pesticides sur la santé dans le tiers monde. La première partie donne un aperçu global de la situation et les constatations de douze études épidémiologiques faites par des chercheurs d'Afrique, d'Asie, d'Amérique latine et du Moyen-Orient. Ces études ont porté sur les empoisonnements dus aux organophosphatés, aux herbicides et aux pyréthrinoïdes. La deuxième partie illustre comment la création et la production des pesticides, les techniques de pulvérisation et la législation peuvent influer sur la santé des travailleurs. La troisième partie traite des avantages que présente l'existence d'informations pertinentes sur les pesticides pour la prévention des empoisonnements et des modalités d'accès à ces informations. La dernière partie porte sur les avantages et les inconvénients, en agriculture et en santé publique, de moyens de lutte autres que les pesticides synthétiques comme les pesticides d'origine végétale et les stratégies de défense intégrées des cultures.

Resumen — Gracias a la introducción de insumos modernos tales como pesticidas y fertilizantes químicos la agricultura ha aumentado notablemente su productividad. Los pesticidas han revolucionado a lucha contra las enfermedades endémicas en los países en desarrollo. Lamentablemente, la medalla tiene también un reverso: La Organización Mundial de la Salud considera que estos agentes causan 2 millones de casos en envenenamiento por año, de los cuales 20 000 se traducen en muertes. En este libro se recogen las ponencias presentadas en un simposio organizado para examinar el impacto del uso de pesticidas en la salud de los habitantes de los países en desarrollo. La Parte I presenta un breve estudio de la situación global y los resultados de 12 estudios epidemiológicos realizados por investigadores en Africa, Asia, América Latina y en el Medio Oriente. Estos se centran en envenenamientos causados por fosfatos orgánicos, herbicidas y piretroides. La Parte II ilustra cómo los procesos de desarrollo y producción de pesticidas, técnicas de rociado por aspersión, y la legislación pueden afectar la salud de los trabajadores. La Parte III del volumen examina los beneficios y modos de obtener la información pertinentes para la prevención de envenenamientos por pesticidas. Finalmente, se consideran las ventajas y desventajas de ciertas alternativas hacia el uso de pesticidas sintéticos tanto en agricultura como en salud pública, tales como pesticidas botánicos y estrategias integradas para control de plagas.

Contents

v

Part V: Appendices

Acknowledgments

The editors thank James Mullin, Vice President, International Development Research Centre (IDRC), William F. Tordoir, Secretary/Treasurer, Scientific Committee on Pesticides, International Commission on Occupational Health, and Deogratias Sekimpi, Secretary/Treasurer, Scientific Committee on Occupational Health Services in Developing Countries, International Commission on Occupational Health, for their welcoming remarks at the symposium. Appreciation is also extended to William Durham and Deogratias Sekimpi for their valuable summary contributions and comments, which helped make the symposium a success. The presentations made by participants elicited a great deal of interest and discussion, both on the floor during question-and-answer periods, and outside the formal sessions. Unfortunately, not all of these comments were recorded and, in the interests of uniformity, we have not included them in this review.

Foreword

From their first appearance, pesticides were recognized by farmers and by public-health officials as valuable adjuncts to food production and disease prevention. New families of pesticides have since been developed, and the search for others continues. As research progresses, older pesticides — such as the organochlorinated compounds — have been banned or restricted because of their adverse effects on the ecosphere. The development of resistance in pest species has resulted in increased dosages or the application of new compounds. Finally, we now realize that pesticide poisoning of humans, especially in the Third World, is a serious problem.

In recognition of this health hazard, two scientific committees of the International Commission of Occupational Health (ICOH) — the Committee on Pesticides and the Committee on Occupational Health in Developing Countries — and the International Development Research Centre (IDRC) of Canada sponsored a symposium entitled "The Impact of Pesticide Use on Health in Developing Countries." The symposium was held in Ottawa, on 17–21 September 1990, to precede the 23rd International Congress of ICOH held in Montreal the following week. This book is a collection of expanded and edited versions of papers that were presented at the symposium.

Soon after the first comprehensive study, commissioned by the World Health Organization (WHO) in the early 1980s in Sri Lanka, IDRC provided financial support to four groups of researchers in Southeast Asia to look at occupational intoxication by pesticides. Their studies confirmed that pesticides do pose a serious health risk in many Asian countries. The findings are discussed, in the context of our present knowledge of the risks pesticides pose to public health, in a monograph by Professor J. Jeyaratnam who spearheaded the WHO study as well as serving as the regional coordinator of the IDRC project, carried-out in Indonesia, Malaysia, Sri Lanka, and Thailand.

Results from this project made several points clear: agricultural workers had little knowledge, if any, of the toxicity of pesticides or of their proper use; what knowledge that did exist was often erroneous; personal spraying equipment was more often than not in a sad state of disrepair and operators seldom used appropriate protective equipment and clothing; and pesticide intoxication in Southeast Asia had become a serious problem.

More recently, IDRC has supported a number of research projects on pesticide intoxication in different regions of the Third World, investigating these and other issues that have surfaced since the first investigation. Most studies have concentrated on agricultural workers, and these are reported in this publication.

There are alternatives to the use of pesticides in agriculture and public health. IDRC has also been involved in the research effort to develop other methods, which can be as simple as crop rotation or as highly technical as genetic engineering of pest-resistant crops. Effort is also being expended on the identification, development, and field-testing of botanical pesticides. These topics were discussed during the symposium, as was integrated pest management and how it can afford a measure of pest control with a reduction of costly and sometimes hazardous inputs.

The symposium was an opportunity for lively and constructive exchange of information, presentation of new data, and thought-provoking discussions by scientists from academia, industry, government, and international agencies. Scientists from 18 developing countries and 7 industrialized countries, as well as representatives of WHO and the Pan American Health Organization (PAHO) participated in the symposium. Whenever possible, papers in this proceedings were edited to reflect some of the discussion that they elicited.

We hope that this symposium has increased understanding of the issues and has helped create links between scientists from the South and their colleagues from the North, and that it may lead to interventions that will promote safer living and working environments for the people of developing countries.

Gilles Forget, Tracey Goodman, and Arnold de Villiers
Health Sciences Division
International Development Research Centre

PART I

EPIDEMIOLOGY OF
PESTICIDE POISONING —
PRACTICAL RESEARCH EXPERIENCE
IN DEVELOPING COUNTRIES

Balancing the need for pesticides with the risk to human health

Gilles Forget

Health Sciences Division,
International Development Research Centre,
Ottawa, Ontario, Canada

Pesticides have been used successfully in the control of a number of diseases such as typhus and malaria. It is more difficult to assess the benefits of pesticides in agricultural practice because their introduction has often coincided with that of chemical fertilizers and resistant crop varieties. The development of resistance by target organisms is now eroding past successes. Farmers and public-health workers must often apply increasing concentrations of pesticides, thereby increasing the risks of occupational and accidental intoxications. New compounds are developed that, although often safer to users and to the environment, can be considerably more costly. Surveys conducted in the tropics suggest that a large proportion of noncriminal, nonsuicidal incidents of pesticide poisoning is due to lack of knowledge and unsafe practices in the storage, handling, and spraying of these chemicals. Concern is growing that cases of poisoning and death could be grossly underreported because of lack of diagnostic expertise on the part of treating health practitioners. The key to the future appears to lie not in discarding pesticides but in their integration into sound pest-management practices: proper training of users, continued development of environmentally safer compounds, initiation and maintenance of sound storage and transport practices, and integrated pest management, such as intercropping and biological control.

Pesticides are used worldwide to control pests that destroy crops and transmit diseases to people and animals. Developing countries are steadily increasing their demand for imported chemicals, many of which are used in agriculture. Between 1970 and 1980, the value of pesticide purchases in Third World countries increased 6.5-fold in constant dollars (World Resources Institute 1986).

Although pesticides are used both in agricultural and public-health programs, the methods differ and the parameters for assessing their efficacy or cost-effectiveness vary greatly. However, the development of pest resistance to pesticides and their effects on nontarget organisms, including humans, have a similar impact and are of equal significance. Moreover, the use of pesticides in one sector effects the other. It has been surmised that large-scale spraying of dichlorodiphenyltrichloroethane (DDT) on cotton in Central America was responsible for the development of resistance to the pesticide in *Anopheles* spp., which are malaria vectors (Chapin and Wasserstrom 1981).

In the past, pesticides have been very successful in controlling and eradicating several diseases from endemic areas. Regretfully, these past successes in health are now being eroded because of the appearance of resistant vectors. Similarly, although agricultural experiments have demonstrated the benefits of pesticides for crop protection, the complexity of intervening factors has made it difficult to measure their real impact on world agricultural production.

The use of toxic chemicals is fraught with risks both to users and to the environment in which they are released. The number of reports of human poisonings and mortality due to pesticides is growing (Foo 1985), although it is not clear whether this represents a rising trend or better reporting because of increased awareness. Nevertheless, pesticides will remain a necessary part of our arsenal for crop protection and disease prevention. In this paper, issues related to their efficacious and safe use are reviewed.

Uses of pesticides

Public health

As Metcalf (1970) wrote, "There are many examples of insect pests and of vector-borne diseases that were never controlled successfully until the advent of modern insecticides." The first documented instance of disease control by pesticides was that of typhus accomplished by the large-scale delousing of people in Algiers in 1943 and Naples in 1944 using DDT (Metcalf 1970). Residual spraying with DDT was also responsible for the eradication of malaria on the island of Sardinia in the 1940s (Metcalf 1970). Since 1950, there has been no evidence of malaria transmission on the island.

Considerable success was also obtained in India where a malaria-control program was launched in 1953. Total reported cases were reduced from 75 million annually in 1947 to 100 000 in 1965, while number of deaths due to malaria dropped from 800 000 per year to none (Reuben 1989). In Sri Lanka, similar results were achieved. The crude death rate fell from 25 per 1 000 in 1921 to 8 per 1 000 in 1971; 23% of the decrease is believed to be due to the reduced incidence of malaria (Ault 1989). Deaths from malaria were reduced by 36% the year immediately after the inception of residual DDT spraying in 1935 (Metcalf 1970).

Another example of successful disease control by pesticides is the onchocerciasis control program carried out by the World Health Organization (WHO) in West Africa. In 1970, an estimated 20 million people in the region suffered from onchocerciasis (Metcalf 1970). In the seven West African countries where the program was implemented, it was believed that 1.5 million people harboured the parasite, which caused severe vision impairment or blindness in 70 thousand of them. For example, in Burkina Faso, 400 000 of the 4.5 million inhabitants were affected by onchocerciasis (Remme and Zongo 1989). The onchocerciasis control program was launched in 1974, by killing larvae with temephos, an organophosphate insecticide. After 12 years, only 9 new cases were reported from 184 indicator villages, compared with the expected 800, and the microfilarial load had declined by 95%. The program is considered a success by many (LeBaerre et al. 1989).

A major setback in the use of pesticides for public health has been the development of resistance in many disease vectors. For example, body lice, which are vectors for typhus, are now resistant to DDT and other organochlorine insecticides such as lindane and to organophosphates such as malathion (PAHO/WHO 1973). The development of insecticide resistance is escalating. The number of resistant arthropods grew from 2 in 1946 to 150 in 1980, while resistant mosquito species, which numbered 7 in 1957, rose to 93 in 1980 (WHO 1984). This has had severe consequences for malaria-eradication programs. Although resistance was observed in the course of the onchocerciasis-control program, rotation of five insecticides and application of mixtures allowed for efficient management of this problem (WHO 1982).

Social resistance to the use of pesticides has also hampered their usefulness. Often residual spraying in homes for malaria control is less popular when people acquire better living quarters (Reuben 1989). Notwithstanding these problems, the consensus is that chemical pesticides will probably continue to be the principal component of most vector-control programs, at least in the foreseeable future (WHO 1988).

Agriculture

Assessing the benefits of pesticides in agriculture is difficult, partly because their introduction (Table 1) has often coincided with the use of fertilizers, mechanization, and high-yield crop varieties. For example, between 1952 and 1968 pesticide use in the United States increased by 269%, fertilizers by 292%, and machinery by 130%, while agricultural labour was reduced to 53% (Carlson and Castle 1972).

Although it has been established that pesticides kill pests, there is controversy surrounding their efficacy in actually increasing world food production. Field trials in the USA showed substantially increased production in test plots

4

Table 1. Introduction of pesticides in the tropics for common agricultural use.

Family	Year of introduction
Organochlorines	1940
Organophosphates	1960
Herbicides	1960
Carbamates	1970
Pyrethroids	1975
Bacillus thuringiensis	1980

Source: World Resources (1986).

treated with fungicides compared with control plots (Zweig and Aspelin 1983). The use of pesticides can increase rice yield by 50% compared to an untreated test plot (Zweig and Aspelin 1983), and treating tomatoes with pesticides increased production by 141% compared with untreated plants (Decker 1974). Similar treatment with fertilizers alone increased the yield by 181%, while combining fertilizers with pesticides increased the yield by 253% (McNews 1967).

The common wisdom is that pests (insects, weeds, and pathogens) are responsible for a loss of one-third of preharvest yields (Cramer 1967). Surprisingly, this ratio changed little between 1942–45 (31.4%) and 1974 (33.0%) when pesticide use was common (Fig. 1). In fact, losses due to insects increased steadily from 7.1% to 13.0% during this period. The loss of cotton crops to insects increased from 10% in 1900 to 19% in the 1950s. For maize, the loss increased from 8% to 12% over the same period (Plucknett and Smith 1986). In 1986, the percentage of all US crops lost to insects was twice that of 1940 (Metcalf 1980). This has led many to conclude that crop damage from insect infestation has not decreased as a result of pesticide usage, at least in the USA, and that it has actually increased (Pimentel 1976; Odhiambo 1984; Plucknett and Smith 1986).

Overall, pesticides have had a positive effect on crop yields that is not offset by the poor record of certain insecticides. Herbicides, for example, have become central to many types of agriculture and their use has been instrumental in decreasing preharvest losses in the USA from 13.8% in 1942–51 to 8% in 1974 (Cramer 1967). Headly (1968) estimated that each additional $1 spent on pesticides in 1968 would increase the value of protected crops by $4. However, he stressed that these values were based on field trials and did not account for geographic, climatic, and other real-life variables such as variations in pest infestations.

Modern, high-yield cultivars depend on higher levels of inputs that must be applied at the right time. However, the monocultures that are now the mainstay of world agricultural production have made this system unstable. This may explain why the use of pesticides has not increased world food production to the levels predicted in experimental trials.

5

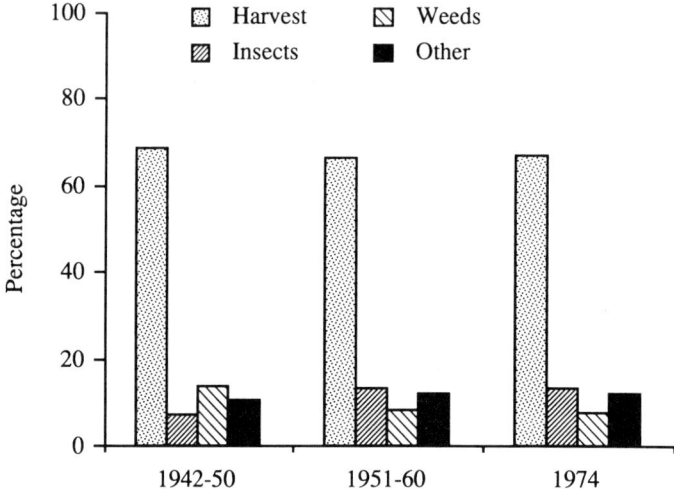

Fig. 1. Preharvest yields and losses to insects, weeds, and other causes, over the period when insecticides became widely used (Pimentel 1976).

Intoxication, morbidity, and mortality

Acute poisoning

Pesticides are toxic chemicals by design and, as such, they represent risks to users. In developing countries, where users are often illiterate, ill-trained, and do not possess appropriate protective equipment, the risks are magnified. Furthermore, comprehensive bodies of legislation to regulate the use and distribution of pesticides often do not yet exist. When legislation is enacted, it is often difficult to enforce given existing structures and budgets. For these reasons, nonoccupational poisonings constitute an important public-health problem in the Third World.

From first suggesting in 1972 that 500 000 cases of pesticide poisoning were occurring annually, WHO has now increased its estimate to 1 million annual poisonings with 20 000 resulting in death (WHO 1986). These estimates were based on area mortality reports submitted to WHO, area morbidity surveys, national mortality statistics for accidental pesticide poisoning as reported to WHO, and preliminary analyses of mortality statistics for accidental pesticide poisoning as reported to WHO. Estimates vary from 1 111 000 cases per year reported in area mortality reports to 834 000 cases determined through analysis of mortality statistics for accidental poisoning reported to WHO. Death estimates range from 20 000 in area mortality reports to 3 000 recorded from mortality statistics for accidental poisoning. The actual figures may be much higher (Chapin and Wasserstrom 1981) as estimates have fluctuated widely.

The first large-scale survey of pesticide poisoning was carried out by WHO in Sri Lanka (Jeyaratnam et al. 1982). The results revealed the seriousness of the problem. The International Development Research Centre (IDRC) funded a similar survey in Indonesia, Malaysia, Sri Lanka, and Thailand in 1983 (Jeyaratnam et al. 1986). Again, the results confirmed the significance of pesticide poisoning in Asia (Table 2). Professor Jeyaratnam presents an updated account of these results and the conditions in several Asia countries and the need for a common research agenda are described by Dr Lum and his colleagues in this volume.

A more recent survey conducted in Sri Lanka by the National Poison Information Centre (NPIC) indicated that, in 1986, 47.3% of all poisonings for which victims were admitted to state hospitals were due to pesticides, principally acetylcholinesterase inhibitors (Fernando 1990). Of these, 10.1% (1 452) resulted in death. These figures do not include another estimated 1 500–2 000 deaths by poisoning from all causes that were never reported to hospitals (Fernando, this volume). In this volume, the results of research conducted in several developing countries are presented (Liesivuori, Sansur et al., Condarco et al., and Castaneda) and reveal the seriousness of the problem in these countries.

Clearly, all these cases of poisoning were not occupational. In the first Sri Lankan study, the investigators reported that a large proportion (73%) of cases were suicidal in nature (Jeyaratnam et al. 1982). Using the NPIC registry, Fernando (1990) reported that 53% of all inquiries about poisonings were related to self-inflicted incidents. In Latin America, however, it appears that most cases of pesticide poisoning may be occupational. In Costa Rica, for example, 67.8% were work-related compared with 6.4% that were suicidal (PAHO 1986).

Herbicides are not highly toxic, but surprisingly they are responsible for many cases of poisoning. Paraquat, in particular, is implicated in many cases of poisoning in developing countries. In Fiji, Papua New Guinea, and Western Samoa, for example, it is responsible for the majority of human poisonings (Mowbray 1986). In Malaysia, Mahatevan (1987) reported that out of 569

Table 2. Extent of pesticide poisoning in Indonesia, Malaysia, Sri Lanka, and Thailand.

| Country | Agricultural workers (%)[a] | |
	Ever poisoned	With AChE between 50 and 75%
Malaysia	13.3 (4 351)	37.4 (821)
Thailand	8.1 (4 971)	45.7 (318)
Sri Lanka	4.6 (3 439)	17.2 (144)
Indonesia	4.1 (1 192)	1.5 (99)

Source: Jeyaratnam et al. (1986, 1987).
Note: AChE = acetylcholinesterase.
[a] Sample size is given in parentheses.

deaths caused by pesticide poisoning, 310 were due to this herbicide. In Colombia, Dr Arroyave conducted an epidemiological survey of paraquat poisoning (Arroyave, this volume).

Loevinsohn (1987) reported a 27% higher nontraumatic mortality rate in rural men compared with an urban population. The increase was ascribed mainly to cardiovascular and cerebrovascular incidents that may be confused with symptoms of poisoning by organochlorides. Such misdiagnoses are not unlikely in rural, primary health-care settings of the Third World. Loevinsohn has convincingly argued for an association between this high mortality rate and pesticide poisoning.

Even pesticides that are purported to be relatively safe for humans can cause health problems when mishandled. Severe intoxications have been associated with pyrethroid insecticides in China (He et al. 1988; He et al., this volume).

In the pesticide formulating and packaging industry of Egypt, Amr and his coworkers found an astounding number of workers suffering from pesticide-induced symptoms. This work may well serve as a model for more studies in the pesticide-manufacturing process in developing countries (Amr et al., this volume).

Chronic and long-term effects

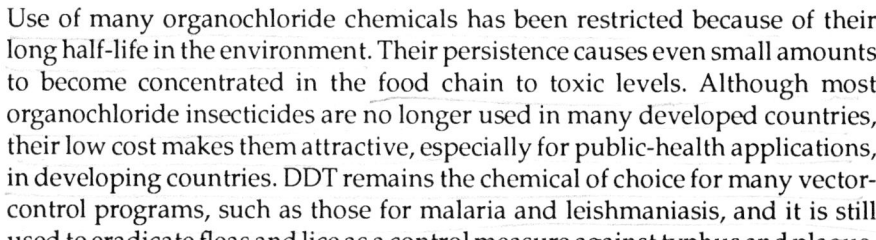

Use of many organochloride chemicals has been restricted because of their long half-life in the environment. Their persistence causes even small amounts to become concentrated in the food chain to toxic levels. Although most organochloride insecticides are no longer used in many developed countries, their low cost makes them attractive, especially for public-health applications, in developing countries. DDT remains the chemical of choice for many vector-control programs, such as those for malaria and leishmaniasis, and it is still used to eradicate fleas and lice as a control measure against typhus and plague.

Organochlorine residues have been reported in various foodstuffs in developing countries: red meat, poultry, game, and vegetables in Nigeria (Atuma 1985), eggs in Kenya (Mugambi et al. 1989), and potatoes in Egypt (El Lakwah et al. 1989). Organochlorine residues are also found repeatedly in human milk (for an excellent review, see Jensen 1983) in all parts of the world, including countries where they have been banned for many years. Levels measured in Third World countries suggest that nursing infants are often ingesting residues at levels many times greater than the acceptable daily intakes proposed by the Food and Agriculture Organization (FAO/WHO 1988).

Many organochlorinated pesticides have now been banned from all but a few public-health uses, because their resistance to environmental degradation would allow them to accumulate with dire consequences to the biota. This volume would have been incomplete without some discussion of the effects of pesticides on nontarget organisms. Drs Mineau and Keith have studied the impact of pesticides on Canadian wildlife; their paper will be of considerable interest to all who care for the conservation of nature.

The molecular mode of action of organophosphates and carbamates (acetylcholinesterase inhibitors) may help explain why they are the cause of most pesticide intoxication reported. The effects of acute poisoning are rapidly felt by the victim, and quickly lead him or her to seek health care. However, several delayed effects of organophosphate intoxication have now been described (Senanayake and Johnson 1982) and their full significance has not yet been established in relation to community health.

Acetylcholinesterase inhibitors may pose a greater risk to human health in the Third World where hunger is often a daily reality. Laboratory studies have shown that the hepatic enzyme activity of animals suffering from protein deficiency was more susceptible to the administration of malathion than that of well-fed animals (Bulusu and Chakravarty 1984). The compound also caused a decrease in liver protein and lipids in these animals. The significance of such findings to public health has yet to be evaluated.

Causes of pesticide intoxication

Many surveys of pesticide poisonings identify lack of knowledge and improper practices on the part of handlers as key factors in most cases of intoxication. This holds true for occupational accidents as well as for accidental poisonings in the home or elsewhere. Drs Condarco (Bolivia), Rubin de Celis (Peru), Mwanthi (Kenya), and Sansur (Jordan) describe their work on this subject in this volume. Dr Condarco's study was concerned with the practices of farmers using pesticides in the three distinct ecological zones of Bolivia where altitude, climate, and cultures differ. Dr Rubin de Celis has assembled a multidisciplinary team of investigators and is studying the knowledge and practices of apple farmers in Peru who are affected by the modernization of cultivation methods such as multiple crops and chemical inputs. Dr Sansur has studied the agricultural practices of Arab farmers on the West Bank, who have to contend with such problems as pesticide labels written in Hebrew.

Drs Kimani and Mwanthi have carried out an interesting study in Kenyan rural communities, specifically assessing people's knowledge, attitudes, and practices in spraying, storing, and disposing of pesticides. They have used the information to design an education program that may reduce the hazards of pesticide use by farmers and other members of rural communities.

Practices in packaging, labeling, and using pesticides safely were also discussed. Dr Boonlue discusses education and training with respect to communication. Dr Dollimore's paper on safer labeling and packaging raises issues that are pertinent to some of the problems that are continually reported in studies of pesticide poisoning — mainly users' lack of understanding of the precautions necessary for the safe use of any chemical input in agriculture.

Laws, regulations, and access to pesticides

The distribution of pesticides to end-users is not often addressed by health professionals. Researchers frequently carry out surveys based on official records of importation and sales collated by the government and its affiliated agencies. However, as is shown by Professor Mbiapo, official data often do not reflect the real situation. Rather, access to pesticides follows the vagaries of the black market, smuggled goods, and alternative marketing. Good regulatory control must be based on solid data. In his paper, Professor Maroni discusses the use of human data as a basis for pertinent regulation of pesticide use.

Several industrialized countries have recently undertaken to review their pesticide-registration procedures. Such experiences could form the basis of either similar reviews or the actual establishment of up-to-date regulations in developing countries. Dr Versteeg's paper expands on Canadian experience in this area. However, even in countries where the regulation is nonexistent or its enforcement too weak to make a difference, it is necessary to protect users. The FAO has published a voluntary code of conduct for pesticide manufacturers (FAO 1986). Dr Loevinsohn discusses a solution to making that code effective in countries where the implementation of regulations is difficult.

Information will remain the major weapon in the prevention and treatment of pesticide poisoning. Professor Mercier, manager of WHO's International Programme of Chemical Safety provides up-to-date information on this aspect of health promotion in relation to pesticide use.

How are we to promote prevention in developing countries? Many avenues have been looked at and others are being explored. Dr Sekimpi discusses the role of occupational-health services in the promotion of safe pesticide use and the infrastructural requirements of this option. Professor Xue and colleagues describe their experience near Shanghai where a primary health-care service is engaged in a program to reduce the intoxication of local farmers by pesticides.

Modern alternatives for pest management

New substances

As pests develop resistance to existing pesticides, the search continues for new compounds and methods to combat them. Dr Tordoir presents some interesting material on the development of safe pesticides. His paper concisely describes the process that should be followed to ascertain the safety and efficacy of new substances.

One new avenue of research is in the area of botanical substances that have pesticidal activity. Many such compounds have been identified and are being investigated further. For example, pyrethrin, which is extracted from chrysanthemums, is used in the preparation of natural insecticidal formulations (Gombe and Ogada 1988). However, a number of synthetic analogues, pyrethroids, are also being manufactured and are proving effective.

Some plant-derived substances do not appear to need sophisticated processing. A number of natural molluscicides that show promise in the control of snail-transmitted schistosomiasis have been identified in Africa and the Middle East (El Sawy et al. 1987; Lemma 1965). Dr Legesse presents information on one such molluscicide, endod. Dr Legesse was awarded the 1989 Swedish Right Livelihood Award for his work on this natural compound and is involved in a major project to conduct toxicity trials on endod. Some results obtained in a study of another natural molluscicide, damsissa, are presented by Dr Duncan, who has been involved as a consultant with the testing of this product.

Other products are not ready for use, but offer some promise for both agricultural and public health applications. Professors Philogène and Lambert describe the process involved in the development of such botanical compounds.

Biological pest control

Pest species have pathogens to contend with and it is possible to make use of some of them both for agricultural and public health purposes. For example, *Bacillus thuringiensis* and *B. sphaericus* are being used to control mosquito larvae in the fight against such arthropod-borne diseases as malaria, dengue, yellow-fever, filariasis, and encephalitis (Amonkar et al. 1988). New information on the use of this bacterium in Canada and Egypt is presented by Dr Morris in his paper on biological alternatives to chemical pesticides.

Bacillus thuringiensis has been considered innocuous to nontarget species. However, the microbe has been isolated in clinical cases where an effect could not be ruled out (Green et al. 1990). Although this microbial pesticide may not affect healthy individuals, patients who are already immunocompromised may be more susceptible to infection. Given the proportions of the acquired immune deficiency syndrome (AIDS) epidemic in some areas of Africa, the repercussions of using *B. thuringiensis* should be investigated carefully.

Another effective way to combat pests in both agriculture and public health is to promote their destruction by natural parasites and predators such as *Trichogramma chilonis* and a number of larvivore fish. Mr Hagerman discusses the use of this parasite and another, *Euborellia annulata*, in the protection of eggplants and describes a comparison of biological control and chemical control methods in the Philippines.

Integrated vector and pest management

Although pesticides are now a part of our lives, their use can be made safer by careful planning. One method is integrating them with sound ecological practices. Although this has already been explored for many years by agronomists and farmers, the public-health specialists have only recently joined the fray. Dr McKay provides an overview of integrated pest management as practiced in agriculture, specifically discussing examples of work being carried out in the Philippines and in China. Dr Wijeyaratne broaches the subject in relation to environmental management aimed at controlling vector-borne diseases. He elaborates on the history of vector control, discusses past successes and failures, and focuses on some promising strategies now being tested.

Old methods can be modernized through the use of pesticides. For example, bed netting and curtains are being impregnated with pyrethroids in Africa to prevent malaria transmission (Rozendaal 1989). In China, a number of trials of impregnated bednets have been successful.

Safe equipment, education, and training

For the Third World farmer, safe, affordable spraying and protective equipment are not always available. Equipment is often difficult to keep in good repair because parts for imported products are not easy to obtain in developing countries (Jusoh et al. 1987). Developing a national capacity to produce good quality equipment with easy access to parts would go a long way toward reducing occupational risk in these countries. Dr Jusoh et al. discuss such a program undertaken by the Malaysian Agricultural Research and Development Institute.

There is also a dire need for research on protective clothing that is appropriate for hot, humid tropical conditions. Many investigators have shown that pesticide poisoning can often be traced to the absence of body protection during spraying. Drs Chester and Sabapathy present data from studies of protective clothing developed and tested in tropical settings.

Drs Chase and Supapong have been involved in a project aimed to increase the awareness of medical students in Thailand of the problem of pesticide poisoning. The objectives of this project are to sensitize physicians and prompt them to think critically about the public-health aspects of this problem. Details of the project and the lessons to be gained from it are described in their paper.

Conclusions

The subject of this volume is the impact of pesticide use on health in developing countries. Many related and relevant topics are discussed. The symposium organizers hope that the many discussions carried out during the meeting will act to promote programs that ensure safer pesticide use. Issues relating to alternative strategies for vector and pest management will also encourage research into such practices, not as replacements to chemical management, but rather as adjuncts to it.

The research presented at the symposium in many cases points to the gaps in our knowledge, which may will lead to pertinent studies to allow safer pest and vector control. Dr White's paper provides an excellent perspective of current epidemiological approaches, discussing the potential of research as a basis for community action. Suggestions for further research addressing some missing knowledge are invaluable to those of us concerned with this problem.

Dr Durham was given the difficult task of preparing an overview. In his review of all the papers, he summarizes the findings and needs in this field.

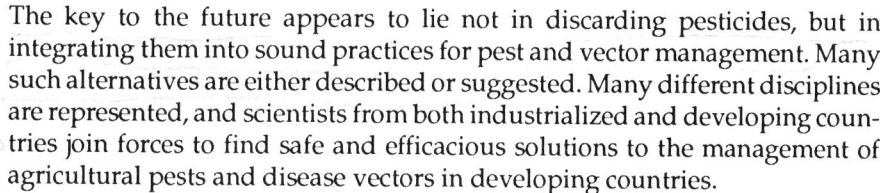

The key to the future appears to lie not in discarding pesticides, but in integrating them into sound practices for pest and vector management. Many such alternatives are either described or suggested. Many different disciplines are represented, and scientists from both industrialized and developing countries join forces to find safe and efficacious solutions to the management of agricultural pests and disease vectors in developing countries.

Amonkar, S.V.; Rao, A.S.; Narayaman, V. 1988. Application of *Bacillus sphaericus* in the control of *Culex fatigans. In* Pesticides: food and environmental implications. International Atomic Energy Agency, Vienna, Austria.

Atuma, S.S. 1985. Residues of organochlorine pesticides in some Nigerian food materials. Bulletin of Environmental Contamination and Toxicology, 35, 735–738.

Ault, S.K. 1989. Effects of malaria on demographic patterns, social structure and human behaviour. *In* Service, M.W., ed., Demography and vector-borne diseases. CRC Press, Boca Raton, FL, USA. Pp. 271–282.

Bulusu, S.; Chakravarty, I. 1984. Augmented hepatic susceptibility to malathion toxicity in rats on low protein diets. Environmental Research, 35, 53–65.

Carlson, G.A.; Castle, E.N. 1972. Economics of pest control. *In* Pest control: strategies for the future. National Academy of Sciences, Washington, DC, USA.

Chapin, G.; Wasserstrom, R. 1981. Agricultural production and malaria resurgence in Central America and India. Nature, 292, 181–185.

Cramer, H.H. 1967. Plant protection and world crop production. Pflanzenschutz Nachrichten Bayer, 20, 1–524.

Decker, G.C. 1974. Costs and benefits of pesticides: an overview. *In* Khan, M.A.Q.; Bederka, J.P., ed., Survival in toxic environments. Academic Press, New York, NY, USA. Pp. 447–471.

Deuse, J. 1988. La protection des cultures dans les regions chaudes: succès et incertitudes pour l'avenir. Phytoma, 400.

El Lakwah, F.; Meuser, F.; Suckow, P.; Abdel Gawaad, A.A.; Abdel Karim, K. 1989. Organochlorine insecticide residues in potatoes used for consumption in Egypt. *In* Abstracts of the 3rd World Conference on Environmental and Health Hazards of Pesticides, Cairo, Egypt. P. III-1-0.

FAO (Food and Agriculture Organization of the United Nations). 1986. International code of conduct on the distribution and use of pesticides. FAO, Rome, Italy. 31 pp.

FAO/WHO (Food and Agricultural Organization of the United Nations and World Health Organization). 1988. Guidelines for predicting the dietary intake of pesticide residues. Bulletin of the World Health Organization, 66, 429–434.

Fernando, R. 1990. National poisons information centre in a developing Asian country: the first year's experience. Human and Experimental Toxicology, 9 (3), 161–164.

Foo, G.S. 1985. The pesticide poisoning report: a survey of some Asian countries. International Organization of Consumers' Unions, Kuala Lumpur, Malaysia.

Gombe, S.; Ogada, T.A. 1988. Health of men on long term exposure to pyrethrins. East African Medical Journal, 65, 734–742.

Green, M.; Heuman, M.; Sokolow, R.; Foster, L.R.; Bryant, R.; Skeels, M. 1990. Public health implications of the microbial pesticide *Bacillus thuringiensis*: an epidemiological study, Oregon, 1985–86. American Journal of Public Health, 80, 848–852.

He F.; Sun J.; Han K.; Wu Y.; Yao P.; Wang S.; Liu L. 1988. Effects of pyrethroid insecticides on subjects engaged in packaging pyrethroids. British Journal of Industrial Medicine, 45, 548–551.

Headly, J.C. 1968. Estimating the productivity of agricultural pesticides. American Journal of Agricultural Economics, 50, 13.

Jensen, A.A. 1983. Chemical contaminants of human milk. Residue Reviews, 89, 2–128.

Jeyaratnam, J.; de Alwis Survitnatne, R.S.; Coppleston, J.F. 1982. A survey of pesticide poisoning in Sri Lanka. Bulletin of the World Health Organization, 60, 615–619.

Jeyaratnam, J.; Lun, K.C.; Phoon, W.O. 1986. Blood cholinesterase levels among agricultural workers in four Asian countries. Toxicological Letters, 33, 195–201.

_____ 1987. Survey of acute pesticide poisoning among agricultural workers in four Asian countries. Bulletin of the World Health Organization, 65, 521–527.

Jusoh Mamat, M.; Anas, A.N.; Heong, K.L.; Chan, C.W.; Nik Mohd Nor, N.S.; Ho, N.K.; Zaiton, A.S.; Fauzi, A. 1987. Features of lever operated knapsack sprayer considered important by the Muda rice farmer in deciding which sprayer to buy. *In* Proceedings of the international conference on pesticides in tropical agriculture, 23–25 September 1987, Kuala Lumpur, Malaysia. Malaysian Plant Protection Society, Kuala Lumpur, Malaysia.

LeBaerre, R.; Walsh, F.; Philippon, B.; Pangalet, P.; Henderickx, J.; Guillet, P.; Seketeli, A.; Quillevere, D. 1989. The WHO onchocerciasis control program: retrospectives and prospects. World Health Organization, Geneva, Switzerland. 7 pp.

Lemma, A. 1965. Preliminary report on the molluscicide properties of endod (*Phytolacca dodecandra*). Ethopian Medical Journal, 3, 187–190.

Loevinsohn, M. 1987. Insecticide use and increased mortality in rural Luzon, Philippines. Lancet, 1987 (June 13), 1 359–1 362.

Mahatevan, R. 1987. Pesticide hazards in the tropics. East Africa Newsletter of Occupational Health and Safety, 1987 (April), 4–7.

McNews, G.L. 1963. Pest control in relation to human society. *In* New developments and problems in the use of pesticides. National Academy of Sciences, Washington, DC, USA. NRC publication 1082.

Metcalf, R.L. 1970. Role of pesticides in the integrated control of disease vectors. American Zoologist, 10, 583–593.

_____ 1980. Changing role of insecticides in crop protection. American Review of Entomology, 25, 219–256.

Mowbray, D.L. 1986. Pesticide poisoning in Papua New Guinea and the South Pacific. Papua New Guinea Medical Journal, 29, 131–141.

Mugambi, J.; Kanja, L.; Maitho, T.E.; Skaare, J.U.; Lokken, P. 1989. Organochlorine pesticide residues in domestic fowl (*Gallus domesticus*) eggs from central Kenya. Journal of the Science of Food and Agriculture, 48, 165–176.

Odhiambo, T.R. 1984. International aspects of crop protection: the needs of tropical developing countries. Insect Sciences Applications, 5, 59–67.

PAHO (Pan American Health Organization). 1986. Pesticides and health. Human Ecology and Health, 5, 2–3.

PAHO/WHO (Pan American Health Organization and World Health Organization). 1973. The control of lice and louse-borne diseases. PAHO/WHO, Washington, DC, USA. 311 pp.

Pimentel, D. 1976. World food crisis: energy and pests. Bulletin of Entomological Society of America, 22, 20–26.

Plucknett, D.L.; Smith, N.J.H. 1986. Sustaining agricultural yields. Bioscience, 36, 40–45.

Remme, J.; Zongo, J.B. 1989. Demographic aspects of the epidemiology and control of onchocerciasis in West Africa. *In* Service, M.W., ed., Demography and vector-born diseases. CRC Press, Boca Raton, FL, USA. Pp. 368–386.

Reuben, R. 1989. Obstacles to malaria control in India: the human factor. *In* Service, M.W., ed., Demography and vector-born diseases. CRC Press, Boca Raton, FL, USA. Pp. 143–154.

Rozendaal, J.A. 1989. Self-protection and vector control with insecticide-treated mosquito nets. World Health Organization, Geneva, Switzerland. WHO/VBC/89.965.

El Sawy, M.F.; Duncan, J.; Amer, S.; El Ruweini, H.; Brown, N.; Hills, M. 1987. The molluscicidal properties of *Ambrosia maritima* L. (Compositae). 3 — A comparative field trial using dry and freshly-harvested plant material. Tropical Medicine and Parasitology, 38, 101–105.

Senanayake, N.; Johnson, M.K. 1982. Acute polyneuropathy after poisoning by a new organophosphate insecticide. New England Journal of Medicine, 306, 155–157.

WHO (World Health Organization). 1982. Biological control of vectors of disease: 6th report of the WHO Expert Committee on Vector Biology and Control. WHO, Geneva, Switzerland. Technical Report Series 679.

_____ 1984. Chemical method for the control of arthropod vectors and pests of public health importance. WHO, Geneva, Switzerland. 108 pp.

_____ 1986. Informal consultation on planning strategy for the prevention of pesticide poisoning. WHO, Geneva, Switzerland. WHO/VBC/86.926.

_____ 1988. Urban vectors and pest control: 11th report of the WHO Expert Committee on Vector Biology and Control. WHO, Geneva, Switzerland. Technical Report Series 767.

World Resources Institute and the International Institute for Environment and Development. 1986. World resources, 1986. Basic Books, New York, NY, USA.

Zweig, G.; Aspelin, A.L. 1983. The role of pesticides in developing countries. *In* Formulation of pesticides in developing countries. United Nations Industrial Development Organization, New York, NY, USA.

Health-risk assessment of pesticides: development of epidemiologic approaches

F. White

Caribbean Epidemiology Centre,
Port-of-Spain, Trinidad

Decisions regarding the manufacture and use of pesticides require input from two developing sciences: toxicology and epidemiology. This paper reviews the complementary nature of these disciplines and their underlying assumptions, strengths, and limitations. However, fully adequate solutions to the complex problem of preventing harmful exposures also require critical contributions from the social and behavioural sciences. A major focus of discussion is an overview, from the epidemiological perspective, of the other papers in this publication. The importance of the development of information and the potential of epidemiological research as a basis for community action are reviewed.

Fundamental ingredients for a healthy society include economic prosperity, combined with fair and equitable distribution of the benefits and risks of production. Aspirations, working conditions, living standards, and distribution of wealth are all related to individual and community health. Pesticide use today is widely recognized as indispensable to commercial agriculture, food supply, forestry, and public health in virtually all parts of the world. Defined as any substance or group substances for preventing or controlling unwanted species of plants or animals, pesticides thus contribute to this economic prosperity, so important to the health of a society.

The ideal pesticide should be highly species specific, cost effective, and have no deleterious effects on the environment. However, all pesticides are inherently toxic and, therefore, constitute a potential risk to the producer, the user, the bystander, the consumer, and the environment. This risk may result in immediate harm or the effects may take longer to be felt or may not occur at all.

In practice, most pesticides fall far short of the ideal and exhibit a wide range of toxicity for many species, including humans. The contemporary approach, therefore, should emphasize safe management in their production and use, with particular regard to the health of nontarget species and the environment.

The decisions to manufacture and use pesticides are economic and socio-political in nature and require input from two developing sciences: toxicology and epidemiology. In addition, adequate solutions to the complex problem of preventing harmful exposure require critical contributions from the social and behavioural sciences.

Toxicology and epidemiology as complementary sciences

In the context of occupational and environmental risk assessment, toxicology and epidemiology address two rather different questions: could it? and did it? Toxicology is based primarily on animal and cell studies, normally before human exposure has taken place, whereas epidemiological studies necessarily follow human exposure. A more complete series of questions relevant to health-risk analysis would also include: what price? (economics) and should it? (sociopolitics).

The complementary nature of these scientific disciplines can be appreciated historically. Epidemiology first revealed such relations as renal cancer to coke-oven emissions, mesothelioma to asbestos exposure, and phocomelia to thalidomide (Taussig 1962; Redmond et al. 1972; McDonald 1990). Toxicology has taken the lead in establishing the risk potential of a wide range of chemicals and their byproducts, including formaldehyde and nasal septal cancer in rats, 2,3,7,8-tetrachlorodibenzo-p-dioxin (TCDD) and renal agenesis in rabbits, and benzo[a]pyrene and mutagenesis in cell culture (Giavini et al. 1982; Clary et al. 1983; O'Donovan 1990). Working along parallel paths, the two disciplines have elucidated particular relations such as in the classic case of vinyl chloride, shown by toxicology to be carcinogenic in the mouse (liver and lung), rat (brain), and hamster (lymphatic tissue), whereas virtually simultaneous epidemiologic studies revealed human cancers at all these sites (Lilis 1986). However, both disciplines have underlying assumptions, strengths, and limitations that are important to acknowledge.

Two basic assumptions lie at the foundation of toxicology: that the effects produced by a chemical in laboratory animals when properly qualified are applicable to humans, and that it is possible to estimate risks associated with low-level exposure by observing effects at high-level exposure. The first assumption is confounded by considerations such as selective toxicity, differing portals of entry, variations in metabolism and toxicokinetics, questions of methodology relating to interspecies extrapolation, and complex problems of multiple exposures in the human setting. Contentious also are the results of studies in cell cultures, in the absence of host-defence or DNA-repair mechanisms, and it is difficult to know how to interpret this type of information, especially in the absence of complementary evidence from animal or human studies.

The assumptions and limitations of epidemiology are no less complex: human experimentation is not acceptable and the mobility and participation of individuals cannot be neatly controlled; exposures must have taken place and may be diluted, intermittent, combined, or multiple; some potential outcomes may take many years (perhaps generations) to develop and may not even be foreseeable, e.g., diethylstilbestrol and transitional cell carcinoma of the vagina in first-generation offspring of exposed women (Bornstein et al. 1988); and loss of information can occur at any level (identification of population, adequacy of records, tracing persons, nonresponse, survivor bias, and sample-size limitations).

The findings of both disciplines are frequently at variance. For example, caffeine can produce chromosomal aberrations in cell culture, but there is no evidence of a relation to human or animal cancers, stillbirths, or birth defects (Thilly and Call 1986). Such a finding may be of no pathologic importance, but should at least stimulate further study by both disciplines. The use of saccharine was banned in many countries based on rat studies indicating a link to bladder cancer (Gaylor et al. 1988). This was met by initial scepticism in some quarters due to a lack of supporting data.

The role of epidemiology

Epidemiology is viewed by many as the core science of public-health practice and is of equal relevance to the more specialized fields of occupational and environmental health. In assessing human exposures, including pesticide use in developing countries, epidemiologic techniques can be used to address six aims:

- To characterize distributions of risk factors, disorders, or disease;
- To reveal associations between factors and disease;
- To assess evidence of causality;
- To develop strategies for prevention, intervention, and control;
- To evaluate impact of potentially harmful exposures and corrective measures; and
- To contribute to regulation.

The application of epidemiological-study designs to risk assessment, prevention, and control of potentially harmful occupational and environmental exposures is conceptually straightforward, but can be very difficult in practice because of problems of technique as well as logistics. Furthermore, the results of such studies in themselves constitute only one input into a complex

19

analysis. Paradoxically, where pesticides are used to obtain a health benefit (such as vector control), some risks to human health may be judged acceptable. The continued use of dichlorodiphenyltrichloroethane (DDT) to control malaria in some countries is a case in point.

Overview of international projects

The focus of this proceedings is the impact of pesticide use on health in developing countries. The term "use" should be interpreted in its broader sense to extend from actual production, through distribution, marketing, storage, application, and disposal.

The International Development Research Centre (IDRC) has supported several pesticide-evaluation projects in developing countries, and these serve to illustrate the development of epidemiological approaches to the assessment of associated health risks and benefits.

Reconnaissance

In carrying out an initial review of reports from these projects, attention was given to the following questions:

- What was the context of the study (e.g., occupational or environmental)?
- Were the aims, objectives, and hypotheses clearly stated?
- What study designs were used?
- Were details of methods sufficiently clear and appropriate? and
- Did the study meet its aims and objectives?

Certain general characteristics of the studies were elicited by this initial approach. As in more developed countries, the greatest emphasis at this stage is on occupational as distinct from environmental exposures. Most studies had clear statements of aims and objectives, directed toward the immediate context of worker health and safety, with very few addressing more classically stated hypotheses. The dominant study design appears to be the cross-sectional survey, most frequently applied to biological measurements, e.g., cholinesterase levels, or the assessment of knowledge, attitudes, and practices. The level of detail provided on the methods used was not always sufficient (e.g., population denominators, sample- and control-selection characteristics) to assess the design or its implementation fully. Occasionally, terminology appears to have been used in an nonstandardized fashion (e.g., incidence vs prevalence). Nonetheless, most aims and objectives appear to have been met.

One might also observe that all studies had the potential to contribute to improved health, at least through the provision of relevant information on particular populations. However, whether such potential will be realized depends largely on external factors, such as the link between research and policy development in the countries concerned, and related issues, such as the adequacy of the occupational health and educational infrastructure.

Originality and innovation

In the field of pesticides research, as in other scientific endeavours, intellectual contributions can be made across a spectrum ranging from "originality" through "innovation" to "perseverance," and a judicious mix of these qualities is most likely to lead to human progress.

At the Caribbean Epidemiology Centre, a good example of innovation in vector control is the use of small-diameter styrofoam beads to obliterate the surface of selected peridomestic mosquito breeding sites such as latrine pits. This technique is surely as close as possible to the earlier stated concept of an "ideal pesticide" as one can currently get. Also at the Centre, work is ongoing in the biological control of *Aedes aegypti* mosquitoes through the selective use of a predator species, *Toxorhynchites montezuma*, which consumes *Aedes* larvae. Both these examples show evidence of originality and innovation, whereas their eventual evaluation in various settings, including their application within integrated pest-management strategies, will require perseverance.

On the international scene, it is often assumed that most of the truly original work will come from well-funded laboratories in developed countries, and that the type of science that will take place in developing countries will be applied in nature. Yet the potential for originality in the developing countries is no less real.

Among the IDRC-funded studies, two are focused on the use of botanical species to control schistosomiasis. The most critical step in this development was the first: the astute observation that one could not find snails in areas where particular plants were growing. The identification of such natural molluscicides, and their subsequent evaluation as potentially cost-effective alternatives to synthetically derived chemicals, well illustrates the opportunities for valuable research in developing countries. When combined with studies of community participation in cultivating the plant and of related economics, hyperendemic areas such as Egypt and the Sudan may indeed have a remedy for a long-standing problem literally growing in their own backyard. Whether this is really the case, however, will require continued attention to a systematic plan for further research, evaluation, and development.

Information development

A significant obstacle to the ongoing assessment of human health and safety in relation to pesticides is the problem of obtaining reliable information on the characteristics of their use. This is an area of difficulty shared by all countries, because the production and use of these products for economic benefit has generally preceded recognition of risk or harmful outcomes, and the need for developing appropriate surveillance systems has not yet been fully accepted.

Even in Canada, where the lead role for pesticide regulation has been accorded to ministries of agriculture, the industry itself has been found deficient in its ability to provide a level of information on aspects such as distribution, storage, application, and disposal sufficient for the needs of community health and safety. Thus, it is of interest to observe the emphasis on data-base development in Cameroon, highlighting the fact that international projects in this field are likely to yield experience and insights that should be of mutual benefit to donor and recipient country alike.

Comprehensive surveillance

Knowing the essential features of pesticide use is only the first step in the development of a system of epidemiological surveillance. An adequate system must include information on such issues as the patterns of poisoning and bioaccumulation and the underlying human factors obtained through surveys of knowledge, attitudes, beliefs, and practices.

The value of an integrated pesticide-information system, linking these various elements, will be seen in improved policies and practices protecting the health and well-being of communities, extending from the worker to the eventual consumer. Clear progress toward integrated surveillance and research is illustrated by studies such as that of the Regional Collaborative Pesticide Research Initiative of Malaysia, the Philippines, and Thailand. These include estimates of pesticide use by type of crop, animal and cell studies, hospital and laboratory reports of acute poisoning (including food poisoning), investigations into possible associations, studies of dermal exposure in the occupational setting, safety assessments, monitoring programs, determination of pesticide residues in foods, plasma cholinesterase levels, knowledge–attitudes–practices studies, economic and policy research, as well as the establishment of safety campaigns and poison-control programs.

Focused studies

Most of the IDRC-funded studies were designed to assess the dimensions of pesticide impact among farm workers. They include reports from Bolivia, China, Colombia, the Jordan Valley, Kenya, and the Philippines. The other occupational group studied consisted of workers in two formulating plants in Egypt. Several of these studies gave rise to new observations or hypotheses that deserve to be investigated further. For example, the association of paraquat use, smoking habits, and obstructive lung disease determined in one study may prove to be an important advance in our knowledge.

Specific situations may require specific remedies, especially when social, educational, and cultural factors are intertwined. It is important, therefore, that baseline descriptive studies of pesticide impact be carried out in all countries, as it is not sufficient simply to extrapolate from elsewhere. However, no amount of study will be useful unless the findings are eventually used in policies and programs. A logical link must connect these studies to some form of feasible intervention, such as worker education or improvements in regulation. Although workers themselves may be the current focus of study, it is important not to overlook other groups, as the potential for exposure also extends to the workers' families, consumers, and bystanders.

A good example of this chain of logic is provided by a Kenyan study of small-scale farmers (Mwanthi and Kimani, this volume). The specific objectives were:

- To identify the types of pesticides found in the community;

- To observe where and how these chemicals are handled and stored;

- To note the disposal methods used for empty containers and leftover chemicals;

- To find out if the farmers understand and follow the instructions on container labels;

- To establish protective clothing used during handling as well as application time;

- To measure awareness and assess practices;

- To develop baseline data about the type of pesticides and determine the extent of use in the selected rural, agricultural community of Kenya; and

- To develop a health-education package based on the knowledge, attitudes, and practices and to disseminate it within the community.

Research as a basis for community action

It is important to view research in a managerial context that relates to decision-making. This principle is well expressed in the Peruvian project (Rubin de Celis et al., this volume) whose authors state:

> Three characteristics of the project reflect the nature of the problem of pesticide intoxication and the conditions necessary for its solution: the multicausal character of the problem and the resulting need for multidisciplinary cooperation; the successive and simultaneous relations of the three principal components of the project (research, education, and epidemiologic monitoring); and the key importance of three results of the project (making self-diagnosis accurate, changing the conduct of individuals and of the group, and consolidating community organizations to ensure behavioural changes and the effective operation of the monitoring program).

It is clear that the scientific assessment of the impact of pesticide use on human health does not stop with toxicology and epidemiology, but requires equal attention from the behavioural and social sciences.

Conclusion

This volume reports an impressive array of initiatives addressing various aspects of the impact of pesticide use on health in developing countries. Perhaps even more pertinent for a world perspective is the evidence presented here of a common recognition by all countries, regardless of their state of development, of the usefulness of epidemiologic techniques in assessing the impact of pesticide use. Progress in the application of toxicology, epidemiology, and social and behavioural sciences has been advanced and will be of ultimate benefit to both donor and recipient countries, as the safe development and effective use of pesticides is to everyone's benefit.

Nonetheless, the present studies (with some exceptions) also reveal a relative lack of attention to contexts other than occupational exposures, a tendency that is shared with Canada and other developed countries. This phenomenon is perhaps justifiable as occupational exposures are likely to be more intense and harmful to persons directly exposed. However, a challenge for all awaits in the study of environmental exposures of groups such as families, bystanders, and consumers. Such studies are often fraught with more difficult design, measurement, and logistic problems.

Designs other than those demonstrated in this series of studies are available to epidemiologists. In addition to cross-sectional surveys, there is a full spectrum from case reports to cluster analyses, the investigation of outbreaks and of accidental exposures, correlational studies, cohort analyses, case–control studies, prospective studies, and randomized trials of intervention procedures. These are noted here for the sake of completeness, although their appropriateness depends on the nature of the question or hypothesis posed.

All epidemiological approaches have inherent strengths and limitations, and an adequate discussion of this reality can hardly be achieved in this brief overview. Most of the problems encountered in the assessment of pesticide impact on human populations fall into one or more of the following areas: formulation of questions or hypotheses, appropriate choice of study design, identification of plausible pathobiological endpoints (short-, medium-, and long-term), the problem of exposure measurement, identification and control of confounding variables, recognition and avoidance of bias (systematic error), sufficient sample size to be able to answer the question, appropriate choice of statistical analysis, and rigorous interpretation of resulting data.

In other words, the full potential of epidemiology is available to those who address the impact of pesticide use on health in both developing and developed countries, and in making use of such potential we must strive to ensure that the best possible work is done. I congratulate the participants in this conference for responding to this challenge.

Bornstein, J.; Adam, E.; Adler-Storthz, K.; Kaufman, R.H. 1988. Development of cervical and vaginal squamous cell neoplasia as a late consequence of in utero exposure to diethylstilbestrol. Obstetrical and Gynaecology Survey, 43(1), 15–21.

Clary, J.J.; Gibson, J.E.; Waritz, R.S., ed. 1983. Formaldehyde: toxicology, epidemiology, mechanisms. Marcel Dekker, New York, NY, USA.

Gaylor, D.W.; Kadhebar, F.F.; West, R.W. 1988. Estimates of the risk of bladder tumour promotion by saccharine in rats. Regulatory Toxicology and Pharmacology, 8, 467–470.

Giavini, E.; Prati, M.; Vismara, C. 1982. Rabbit teratology study with 2,3,7,8-tetrachlorodibenzo-p-dioxin. Environmental Research, 27, 74–78.

Lilis, R. 1986. Organic compounds. In Last, J.M., ed., Public health and preventative medicine (12th ed.). Appleton Century Crofts, Norwalk, CT, USA.

McDonald, J.C. 1990. Cancer risks due to asbestos and man-made fibres. In Recent results in cancer research (vol. 120). Springer-Verlag, Berlin, Germany.

O'Donovan, M.R. 1990. Mutation assays of ethyl benzidine and benzo[a]pyrene using Chinese hamster V79 cells. Mutagenesis, 5 (supplement), 9–13.

Redmond, C.K.; Ciocco, A.; Lloyd, J.W. 1972. Long term mortality study of steelworkers. Journal of Occupational Medicine, 14, 621–629.

Taussig, H.B. 1962. A study of the German outbreak of phocomelia: the thalidomide syndrome. Journal of American Medical Association, 180, 1 106.

Thilly, W.G.; Call, K.M. 1986. Genetic toxicology. In Klaassen, C.D.; Amdur, M.O.; Doull, J., ed., Toxicology: the basic science of poisons (3rd ed.). Macmillan, New York, NY, USA.

Acute pesticide poisoning in Asia: the problem and its prevention

J. Jeyaratnam

Department of Community, Occupational and Family Medicine,
National University Hospital of Singapore,
Singapore

Acute pesticide poisoning, a major health problem in the countries of the developing world, is reviewed in light of information available at the global level and the results of a four-country study in Southeast Asia. Adequate resources for the control of pesticide use and information about pesticides in use are important if the number of cases of poisoning is to be reduced. Factors contributing to pesticide poisoning are identified and discussed in relation to the importance of worker education and the development of preventive programs. The role of the agrochemical industry and the primary health-care system in preventing acute pesticide poisoning is discussed.

A pesticide is any substance or mixture of substances that can deter, destroy, or control a pest or that can be administered to animals for the control of pests in or on their bodies. Pests include vectors of human and animal diseases, plant pathogens, and unwanted species of plants and animals that cause harm to or interfere with production, processing, storage, transport, or marketing of food, agricultural commodities, wood products, and animal feed. The two important groups of pesticides that affect human health are insecticides and herbicides.

Pesticide exposure may result in acute or chronic poisoning. Acute poisoning implies an incident where overt reactions follow closely upon exposure to an agent. In contrast, chronic poisoning refers to the situation where the toxic reactions appear gradually after prolonged exposure to the agent. In acute poisoning, the incriminating agent is usually readily identifiable, but this is not always the case in chronic poisoning. It is much more difficult to assess the significance of the small doses that contaminate workers daily over long periods, because they usually do not cause clearly defined clinical symptoms. Furthermore, little is known about long-term health effects of repeated small doses (chronic poisoning) or delayed effects of clinically defined episodes of acute poisoning (such as cancers).

Estimates of the extent of acute pesticide poisoning have shown this to be a significant problem and one that is virtually confined to the countries of the developing world. Levine (1986) estimated that globally about 1.1 million cases of acute pesticide poisoning from accidental or unintentional factors occur each year resulting in about 20 000 deaths. An independent estimate cited 2.9 million cases of poisoning from all causes with 220 000 deaths; 25% of all poisonings are due to unintentional factors (Jeyaratnam 1985). The subject of this paper is mainly unintentional poisoning of occupational origin.

Pesticide types causing acute poisonings

The main classes of pesticides responsible for acute poisoning must be identified to focus attention on these compounds in prevention programs. In Sri Lanka, 76% of all acute poisonings were caused by organophosphorous compounds (Jeyaratnam et al. 1982). Similar findings were observed in other Asian countries (Jeyaratnam et al. 1987) indicating the important role of this class of compounds (Fig. 1). More recent results from Sri Lanka (Fernando, this volume) seem to confirm this conclusion.

However, in a significant percentage of poisoning episodes, the class of substance responsible was not identified, leading to problems in therapy. A

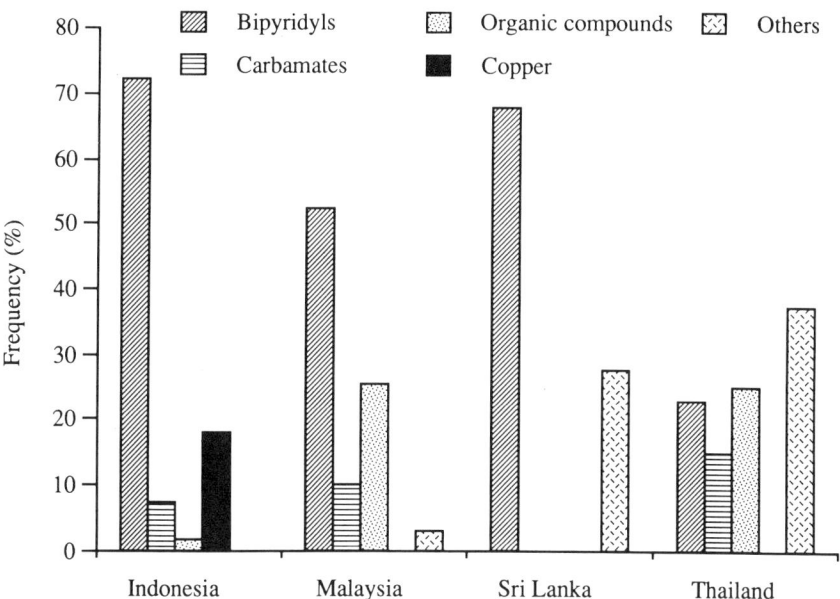

Fig. 1. Cause of poisoning, by class of chemical, in people admitted to hospital in four Asian countries.

case fatality rate of 28.5% was observed for patients in whom the class of pesticide responsible for poisoning was not identified at the time of treatment (Jeyaratnam et al. 1982). These findings lead to two conclusions:

- In view of the inadequacy of resources for controlling the use of pesticides in developing countries, available resources should be aimed specifically at control of organophosphorous pesticides.

- To reduce fatalities from poisoning, each country must develop and maintain a list of the trade names of pesticides and the corresponding chemical class. In parallel, physicians should be trained in the appropriate therapeutic interventions against organophosphate poisoning.

Identification of factors influencing poisoning

Workers' knowledge of hazards is an important factor for the prevention of acute poisoning. However, this knowledge must be factual and correct. Erroneous beliefs can seriously impair workers' capacity to protect themselves from health risks. For example, the most important route of absorption of phenoxy herbicides is dermal, whereas absorption through inhalation is relatively unimportant (Kolmodin-Hedman 1983). However, farm workers believe that the inhalation route is the most important and that their knowledge of the health risks associated with pesticide use is adequate. This misconception may have arisen because of the obvious odour of pesticides during spraying.

Clearly, it is urgent that worker education be established for the safe handling of pesticides in countries of the developing world. For these workers to be motivated to use safe practices, e.g., protective clothing, they must understand the mechanism of poisoning. In four Asian countries, agricultural workers identified other farmers and retail outlets as the most important sources of information on the safe use of pesticides (Jeyaratnam et al. 1987). These workers should be receiving instruction from health workers. Alternatively, health services could make use of this data to integrate participatory mechanisms in farmer education and use vendors as extrabudgetary resources for dissemination of information on safe practices.

Poisoning episodes occur largely during spraying, mixing, and diluting of pesticides (Table 1). The use of malfunctioning or defective equipment is also an important factor contributing to accidental acute pesticide poisoning among agricultural workers (Jeyaratnam 1982). Preventive measures should focus on these areas. Copplestone (1982) states, "We are more likely to be effective in preventing accidental poisoning by pesticides if we concentrate our activities on those areas where hazard is really high than by trying to give blanket coverage to all pesticide users."

Table 1. Poisoning incidents in four Asian countries by type of activity.

Country	Spraying		Mixing or diluting		Other[a]	
	Incidents	% poisoned	Incidents	% poisoned	Incidents	% poisoned
Indonesia	15	7.6	23	7.0	3	11.5
Malaysia	181	5.7	107	3.4	2	0.2
Sri Lanka	157	11.9	147	11.5	18	11.5
Thailand	361	19.4	352	19.8	32	17.6

[a] Includes equipment repair.

Some of the specific factors contributing to acute pesticide poisoning (Jeyaratnam 1985) are:

- Lack of protective clothing suitable for tropical climates;

- Poor knowledge and understanding of safe practices in pesticide use;

- Use of pesticides (by farmers) in concentrations in excess of requirements;

- Poor maintenance facilities for spray equipment, giving rise to hazard-ous contamination; and

- Use of pesticide mixtures.

Action for prevention and control

Having established that acute pesticide poisoning is an important health problem in the countries of the developing world and having identified some of the factors responsible, it is appropriate to consider prevention and control of this problem. Control of the factors listed above would contribute greatly to minimizing unnecessary misery for a large number of people. This is an area where the agrochemical industry could take direct responsibility. For example, industry could take the steps necessary to provide suitable backup services for spray equipment, develop protective clothing suitable for use in tropical climates, promote farmers' awareness of the need to use the correct concentration of pesticide, stop marketing mixtures unless they are absolutely necessary, and develop containers that are less likely to be misused.

The ultimate responsibility for the control of pesticides must, however, rest with national governments. Unfortunately, in many developing countries, pesticide controls either do not exist or are poorly implemented. In 81 member countries of the Food and Agriculture Organization of the United Nations (FAO), there were either no control measures or no information available, 6 countries were in the process of introducing some control measures, and 26 countries had control measures but implementation was poor (Bates 1981). Most of these were developing countries.

Bull (1982) has suggested that, ideally, Third World governments should enact strict legislative controls over the import, formulation, distribution, advertising, promotion, and use of pesticides. In addition, they should ensure adequate resources for effective enforcement of this legislation. Legislation should be complemented by an efficient and well-trained agricultural research, training, and extension service especially geared to the needs of small and marginal farmers and farm labourers. Finally, a primary health-care system should be closely allied with agricultural extension services and include occupational health care, especially in the prevention, recognition, and treatment of pesticide poisoning.

The concept, strategies, and activities of a primary health-care approach have been clearly set out by the World Health Organization. The characteristics of such a health-care system are based on accessibility, relevance, integration within the health system, people's participation and education, cost-effectiveness, and collaboration with other sectors. These characteristics make this approach the most appropriate form of health care for the nations of the Third World to embark on if they hope to contain the problem of acute pesticide poisoning.

Bates, J.A.R. 1981. Paper presented at an international seminar on the control of chemicals in importing countries, Dubrovnik, Yugoslavia.

Bull, D. 1982. Growing problem: pesticides and the Third World poor. Oxfam, Oxford, UK. 192 pp.

Copplestone, J.F. 1982. Problems in education on the safe handling of pesticides. In Van Heemstra, E.A.H.; Tordoir, W.F., ed., Education and safe handling in pesticide application. Elsevier, Amsterdam, Netherlands. Pp. 59–64.

Jeyaratnam, J. 1982. Health hazard awareness of pesticide applicators about pesticides. In Van Heemstra, E.A.H.; Tordoir, W.F., ed., Education and safe handling in pesticide application. Elsevier, Amsterdam, Netherlands. Pp. 23–30.

_____ 1985. Health problems of pesticide usage in the Third World. British Journal of Industrial Medicine, 42, 505–506.

Jeyaratnam, J.; De Alwis Seneniratne, R.S.; Copplestone, J.F. 1982. Survey of pesticide poisoning in Sri Lanka. Bulletin of the World Health Organization, 60(4), 615–619.

Jeyaratnam, J.; Lun, K.C.; Phoon, W.O. 1987. Survey of acute pesticide poisoning among agricultural workers in the four Asian countries. Bulletin of the World Health Organization, 65, 521–527.

Kolmodin-Hedman, B. 1983. Studies in phenoxy acid herbicides in agriculture. Archives of Toxicology, 54, 257–265.

Levine, R.S. 1986. Assessment of mortality and morbidity due to unintentional pesticide poisonings. World Health Organization, Geneva, Switzerland. WHO/VBC/86.929.

Pesticide research for public health and safety in Malaysia, the Philippines, and Thailand

K.Y. Lum,[1] Md. Jusoh Mamat,[1] U.B. Cheah,[1]
C.P. Castaneda,[2] A.C. Rola,[3]
and P. Sinhaseni[4]

[1]Fundamental Research Division, Malaysian Agricultural Research and Development Institute, Kuala Lumpur, Malaysia; [2]Department of Pharmacology, College of Medicine, University of Philippines, Manila, Philippines; [3]Center for Policy and Development Studies, University of Philippines, Los Baños, Laguna, Philippines; [4]Science and Technology Development Board and Department of Pharmacology, Chulalongkorn University, Bangkok, Thailand

In Malaysia, the Philippines, and Thailand, concern is growing about the impact of increased use of agropesticides on public health and safety, especially in farming communities. Based on data from public hospitals and clinics, the extent of pesticide poisoning is considered to be grossly underestimated, primarily due to underreporting. A review of public-health and safety research in these countries suggests that there is commonality in the research needs in this area. Three aspects have been identified as requiring attention: exposure assessment, application technology, and farmer education and training. Initial steps have been taken to establish a collaborative network comprising scientists from the three countries to address the problems identified. Research proposals are being drawn up to develop appropriate personal protective equipment, improved knapsack sprayers, and farmer education and training modules.

Because of their growing populations, agriculture remains an important component of the developing economies of Malaysia, the Philippines, and Thailand. In pursuit of higher agricultural productivity, the introduction of modern farming technologies in these countries has meant increased use of fertilizers and pesticides.

Growing global concern about the impact of increased use of agricultural chemicals, especially in developing countries, has raised questions about the extent and nature of the adverse effects of pesticides on farmers, consumers, and the environment. The need for more comprehensive information on the status of public-health and safety-related aspects of pesticide research has been reinforced by the recognition of the importance of regional collaboration to address pesticide problems common to the agricultural activities of these three countries.

Profile of pesticide use

In Malaysia, over 3 000 retail outlets are currently involved in the manufacture, formulation, and packaging of pesticides. In 1980, sales of pesticides amounted to 65 million USD. The continued demand for pesticides in Malaysian agriculture is evident from the increasing sales of agricultural chemicals. Sales of herbicides, insecticides, fungicides, and rodenticides increased from 95 million USD in 1984 to nearly 120 million USD in 1988 (MACA 1989). Herbicides account for nearly 80% of these sales in Malaysia; insecticides, about 15%; and rodenticides, nematicides, and others, 5% (MACA 1989).

Herbicides are used mainly on plantation crops, such as rubber, oil palm, and cocoa; the common ones are paraquat, glyphosate, 2,4-D (2,4-dichlorophenoxyacetic acid), diuron, MSMA (methylarsonic acid), picloram, and dalapon. Insecticides may also be required for disease control on rubber plantations, and fungicide and insecticide treatments are often used on cocoa plantations. Pesticides are also employed on a wide range of other agricultural crops, such as vegetables, rice, fruit, and tobacco.

Pesticide consumption in the Philippines, based on sales value of total imports, increased steadily from 1983 to 1987 (Table 1). Crops receiving the highest proportion of pesticides (based on value) in 1984 were rice (36.6%), bananas (25.5%), and vegetables (14.3%). Insecticides were used on most crops, except bananas and pineapples, where more fungicides were used.

Table 1. Pesticide sales values in the Philippines, based on total imports (USD '000 cost including freight).

Type	1983	1984	1985	1986	1987
Insecticides	16 196	14 077	12 151	16 379	23 270
Herbicides	3 637	4 562	4 016	6 042	7 139
Fungicides	9 058	11 528	7 953	9 691	10 589
Others	7 391	5 230	6 733	8 331	11 466
Total	36 282	35 397	30 853	40 443	52 464

Source: Fertilizer and Pesticide Authority, Philippines.

Current pest control practices in Thailand rely primarily on chemical methods. Cotton and vegetable crops, in particular, require large amounts of pesticides. The intensification of cotton production has led to an increase in the incidence of pests and diseases (FAO 1988), resistant species, and pest resurgence problems.

Thailand imported 65.2 million USD (cost including freight) of pesticides in 1987 (Fig. 1). Typically, 50% of the imports, by value, were insecticides, 30% herbicides, and the rest fungicides and other miscellaneous pesticides. Most of the insecticides are used on rice, cotton, and vegetables. The herbicide market is focused on sugarcane, pineapple, rubber, and rice (ADB 1987). Except for paraquat, which is produced locally from imported intermediates, pesticides are either imported as finished products or as technical grade ingredients and formulated in Thailand.

Prevalence of pesticide poisoning

In general, epidemiologic description of poisoning in Malaysia is hindered by lack of coordination in data collection. The best sources of information have been in-patient data and laboratory reports from government hospitals and records from the Chemistry Department of the Ministry of Science, Technology and the Environment.

Data from this Ministry from 1979 to 1986 suggest that most poisonings are due to pesticides: mainly the widely used herbicide, paraquat (Fig. 2). Ministry of Health information indicates that, in some cases, multiple agents are involved. Circumstances surrounding these poisonings show that 49.1% were intentional and 37.8% were accidental.

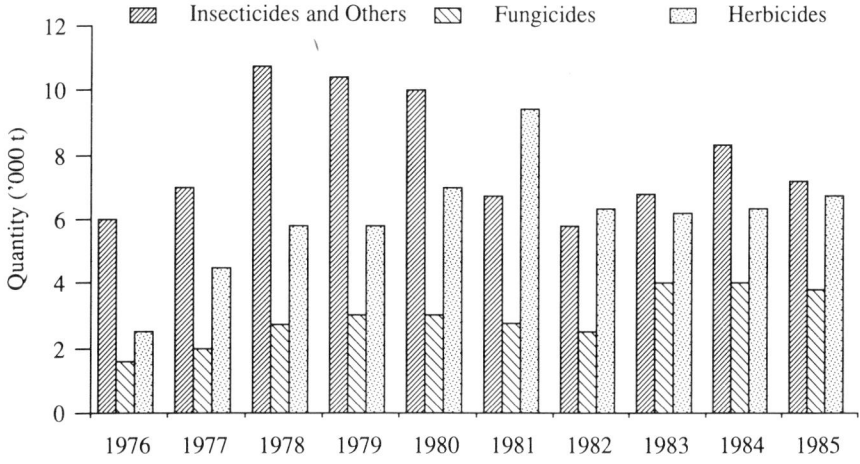

Fig. 1. Quantities of insecticides, fungicides, and herbicides imported by Thailand, 1976–1985.

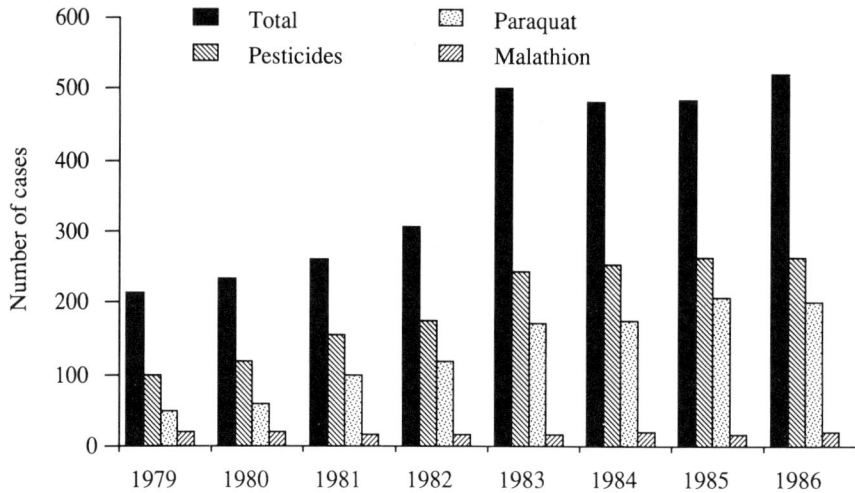

Fig. 2. Cases of poisoning from pesticides, malathion, paraquat, and all substances in Malaysia, 1978–1986 (Lum et al. 1990).

Regional differences in the incidence of poisonings within a country reflect the extent of toxic substance use in these areas (Sinnaia 1989). Areas that include large numbers of plantations and farms (where pesticides are likely to be used in large quantities) tend to record high levels of mortality due to pesticide poisoning. Also, mortality from pesticide poisoning is closely correlated with suicidal intent (72.6%), suggesting that some regulatory intervention would be useful. This is further supported by the fact that farm and plantation workers constitute 45% of the reported pesticide deaths.

More recent studies of Malaysian rice-farming communities indicate that a large percentage of farmers develop symptoms associated with pesticide poisoning. In the intensely cultivated area of the Cameron Highlands, where pesticides are widely used, 95% of all poisoning cases are attributed to pesticides. In 1980, there were 17 cases of pesticide poisoning resulting in 4 deaths; in 1981, the number of cases increased to 20 with 11 deaths; and between January 1982 and April 1983, there were 11 cases and 4 deaths (Asna et al. 1989).

Malaysia recently established a National Poison Centre in the Department of Pharmacology, Universiti Kebangsaan Malaysia. The main functions of this Centre are to identify the risk of poisoning in the local population, to establish preventive measures, and to ensure proper diagnoses and provide treatment for poison victims (Tariq 1989).

Department of Health hospitals in the Philippines reported 4 031 cases of acute pesticide poisoning and 603 deaths between 1980 and 1987 (Fig. 3). Most of these were suicidal (64%); accidental and occupational poisonings accounted for 18% and 14%, respectively. Death rates from acute pesticide poisoning ranged from 13% to 21%. The distribution by gender for all acute pesticide

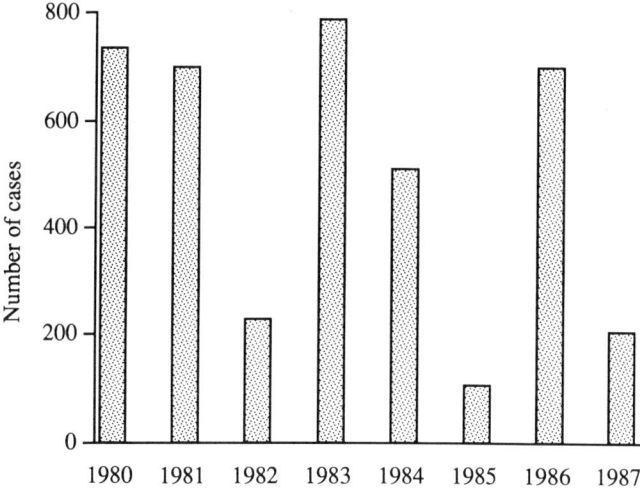

Fig. 3. Cases of pesticide poisoning in the Philippines, 1980–1987 (Castaneda and Rola 1990).

poisoning cases from 1980 to 1987 showed a slight predominance of males (54.0%) over females (46.0%) (Castaneda and Rola 1990).

Studies of civil registry information have also been carried out in the Philippines to determine the relation between causes of mortality and pesticide use. Strong evidence has been found linking a marked increase in mortality in Central Luzon with occupational exposure to insecticides (Bantilan and Rola 1989). Municipalities with high levels of pesticide use generally have higher mortality rates than those using smaller amounts.

In response to the findings of these studies, an occupational-health and monitoring program has been initiated for workers who apply nematicide on banana plantations. It is hoped that, in the future, the industry itself will sustain the program.

In Thailand, the Division of Epidemiology in the Ministry of Public Health monitors pesticide poisoning cases and deaths from unintentional causes. Data for 1975–1987 show an increase in the number of reported pesticide intoxication cases from 518 to 4 633, accompanied by a decrease in mortality rate from 3.47 per 100 000 in 1975 to 1.00 per 100 000 in 1987. In 1981–1987, the Ministry of Public Health reported 20 outbreaks of food poisoning attributed to pesticides, involving 722 cases (Swaddiwuthipong et al. 1989).

Although the incidence of pesticide poisoning in the Thai agricultural communities of Rayong province can be as high as 8 268 per 100 000, only 2.4% of this group spent time in hospital, emphasizing the possible extent of under-reporting in hospital statistics (Wongphanich et al. 1985). A 3-year study by the Office of the National Environment Board (Sinhaseni 1990) determined that organophosphate insecticides such as parathion were the leading pesticides implicated in cases of poisoning. Half the cases of exposure were due to attempted suicide; 7–26% were attributable to occupational exposure. Fewer than 10% of the cases resulted from accidental exposure. In contrast, animal poisoning is usually accidental and caused by direct ingestion of highly toxic pesticides. Cattle and goats are often affected by grazing on grasses treated with sodium arsenite, a toxic herbicide formerly used on rubber estates (now withdrawn from the market).

Summary of past research

Malaysia

Studies of occupational exposure of spray operators have been limited in Malaysia. Swan (1969) carried out two field trials on Malaysian rubber plantations to study the exposure of operators applying paraquat with hand-operated knapsack sprayers. Howard and colleagues (1981) studied the health of Malaysian plantation workers, especially those using paraquat sprayers. Chester and Woolen (1981) reported on the occupational exposure to paraquat of Malaysian plantation workers.

More recently, Lee and Chung (1985) and Lee (1987), using either Ciba-Geigy water-sensitive paper or fluorescent-tracer dye, carried out an extensive study on the potential contamination of various parts of a sprayer operator's body, using different types of applicators and under various crops or crop-spraying situations. They concluded that the extent of chemical contamination is affected by the type of applicators used, height and position of spray nozzle, and the speed and direction of the prevailing wind at the time of spraying. A conventional knapsack sprayer caused more extensive contamination than that resulting from a control droplet applicator (CDA).

Using the method of the Operator Protection Research Group (Ministry of Agriculture, Food, and Fisheries, UK) with lissamine green tracer dye, Tan et al. (1988) carried out a study to determine quantitatively the potential dermal exposure of spray operators. Operators using knapsack sprayers were exposed to 27.84 mL/ha, on average, compared with 31.78 mL/ha for those using CDA sprayers. Potential inhalation was 0.0010–0.0066 mL/ha for knapsack sprayers and 0.0001–0.0047 mL/ha for CDA sprayers. The front leg of the operator was most exposed, with exposure significantly higher for those using CDA sprayers (86%) than those operating knapsack sprayers (59%).

Pesticide application techniques used for annual crops (Lim et al. 1983) and perennial crops (Sidhu et al. 1987) have been discussed in detail. For many crops, pesticides are applied by spraying the foliage although the inefficiencies and the high risk of operator exposure with this technique have long been known (Matthews 1983; Hislop 1988; Zeren and Moser 1988). In Malaysia, spraying of foliage is mainly done with conventional knapsack sprayers designed 30–40 years ago (Jusoh Mamat et al. 1987). A high percentage of sprayers owned by farmers in the "rice bowl" of Malaysia had serious faults and exhibited a volume discharge efficiency of less than 75% (Anas et al. 1987). These factors contribute to the problems arising from underdosing, incorrect timing, and contamination of spray operators. Of the current application techniques, only a few, such as trunk injection, granular application to the soil, and pouring, dripping, and tea-bag techniques are relatively safe to use (Jusoh Mamat et al. 1985; Ooi 1988). However, these safer techniques are situation specific and not as versatile as conventional foliar spraying.

Protective clothing made of plastic or rubber material causes discomfort when worn for more than 3 h under the hot, humid field conditions of the tropics (Lee 1987; Jusoh Mamat and Anas 1988). Yet protective apparel is essential if minimal exposure to sprayed pesticides is to be achieved, especially when perennial crops taller than 150 cm are sprayed using ground applicators (Lee 1987). There is some interest in searching for a more suitable material for protective clothing. To date, only two lightweight disposable materials have been developed: Dupont Tyvek and Kimberley Clark Kleengard EP (Lee 1987).

The Malaysian Agricultural Research and Development Institute (MARDI), under the Ministry of Agriculture, conducts research on various pesticide-related problems in major agricultural ecosystems. In addition to studying application technology and exposure of operators, MARDI research includes investigation of environmental fate, effects, and bioefficacy. Recommendations on pesticide use for pests and diseases of specific crop systems are also made. Information and technology generated from the research are channeled to the Department of Agriculture for dissemination to farmers.

Farmers and the general public pay little attention to the proper use of pesticides, especially with regard to safety (Zain 1977; Zam 1980; Basri 1981; Heong 1982; Normiya 1982; Ooi et al. 1983; Heong et al. 1987; Anon. 1990). As a consequence of this apathy, reports of users suffering ill effects due to improper use of pesticides are common (Dawson 1985; Umakanthan 1985; Indrani 1988).

In August 1984, in response to these findings, the Malaysian Department of Agriculture launched a nation-wide campaign to promote the safe use of pesticides. The campaign had three main objectives (Esa and Ramasamy 1988):

- To create awareness in the general public of the dangers associated with the use of pesticides;

- To educate consumers and end users about safety procedures for handling pesticides; and

- To inform the public about the 1974 *Pesticide Act*.

The campaign included lectures and talks, exhibitions, distribution of pamphlets and posters, radio programs, and television screenings of documentary films on the safe use of pesticides (Esa and Ramasamy 1988). To educate farmers about the specific danger of using monocrotophos in rice and vegeta-bles crops a "Safe Use of Pesticides" campaign in Sabak Bernam district has also been initiated (Asna 1990).

Other semigovernmental agencies, such as MARDI, the Rubber Research Institute of Malaysia, Federal Land Development Authority, Muda Agricultural Development Authority, and Federal Land Consolidation and Rehabilitation Authority, also organize short training courses for their junior technicians on the safe application of pesticides. In the private sector, since 1982, societies such as the Malaysian Plant Protection Society and the Malaysian Agricultural Chemicals Association have also held courses on pesticide application technology. These have been mainly aimed at profession-als in crop protection, such as researchers, university lecturers, estate manag-ers, and pesticide-industry personnel (Lim and Ramasamy 1983; Teoh 1985).

The Philippines

Health and safety-related research activities in the Philippines include des-criptive studies on pesticide poisoning based on reviews of hospital records, occupational exposure studies in rice- and vegetable-farming communities, development of an occupational health and monitoring program, and health economics and policy research studies.

Two retrospective studies reviewing hospital records (Castaneda and Maramba 1980; Gonzales and Chua 1984) analyzed data on pesticide poison-ings by age and gender distribution, common presenting signs and symptoms, correlation between signs and symptoms and cholinesterase level in red blood cells, average dose of atropine administered, and toxicity ratings of the pesti-cides. Of a total of 1 074 cases of poisoning reported from 1982 to 1985, 42% involved organophosphates, 19% organochlorines, and 14% carbamates (Castaneda 1988). Peak incidence occurred among those under 40 years old and was slightly higher for males. Suicide accounted for 63% of the cases, occupational exposure 18%, and accidental exposure 16%. The highest case fatality rate was associated with organochlorine poisoning (49%) followed by organophosphates (16%).

Research efforts have also been directed toward characterizing occupational pesticide exposure among vegetable and rice farmers by examining demo-graphic data, agricultural practices, and general health conditions. Only one study has dealt with farmers' perceptions of the health impact of pesticide use. Generally, research on how farmers' attitudes can affect the way in which pesticides are used is lacking.

Studies by Castaneda (this volume) determined the potential exposure of rice farmers to an organophosphate compound and concluded that workers' hands are most exposed to dermal contamination and penetration during mixing and loading activities. Castaneda also examined the protection afforded by locally designed clothing. Such clothing was often uncomfortable and tore easily. One investigation found no difference in protection between people wearing "Gardsman" protective clothing and an unprotected group.

Several pesticides have been studied in the Philippines to determine their behaviour in model ecosystems (Tejada and Magallona 1985; Zukifli et al. 1985; Varca and Magallona 1987). These included deltamethrin, carbosulfan, chlorpyrifos (Brodan), and fenobucarb (BPMC, 2-sec-butyl-N-methyl carbamate). Carbosulfan does not have pollution potential, although its major metabolite, carbofuran, remains in soil and fish for up to 30 days at low levels. Brodan was found to be rapidly assimilated and concentrated in fish. Repeated applications of BPMC at recommended levels, on the other hand, were absorbed into the soil.

In general, research on pesticide residues in the Philippines has focused on insecticides. Both field residue trials and "market-basket" studies have generated information for a residue data base that includes insecticide residue levels in cabbages, pechay, string beans, green beans, tomatoes, bush sitao, cotton, tobacco, bananas, mangoes, fish, and the various components of the rice-paddy environment, such as soils, paddy water, rice plants, fish, and snails. Residue levels have also been measured in lactating goats and in human milk.

Other ongoing research projects in the Philippines include a project to provide training for rice and vegetables farmers in integrated pest management. Preliminary results indicate that the practices of these farmers affect the extent of their exposure to pesticides. Another study, funded by the Rockefeller Foundation, is to examine the health effects of pesticide exposure among Laguna farmers at the International Rice Research Institute (IRRI).

Research on the benefits and risks associated with pesticide use among rice and vegetable farmers has resulted in a recommendation for more government regulation and training of extension agents and farm workers (Rola 1989). Mechanisms for crop insurance and further research on integrated approaches to plant protection have also been advocated.

Thailand

Studies on the exposure of Thai people to pesticides revealed dangers from organochlorine insecticides and lipid-soluble herbicides; as much as 90 ppm heptachlor was found in farmers' blood (Department of Agriculture, Division of Toxic Substances). Between 1980 and 1986, at least 10 such studies were conducted.

Surveillance of pesticide poisoning by the Ministry of Public Health's Division of Epidemiology has generated more than 20 reports on incidents of pesticide poisoning and accidental deaths. Among the identifiable pesticides involved, organophosphate compounds caused 67.7% of the cases, carbamate 13.0%, herbicides 8.2%, pyrethroids 2.1%, rodenticides 1.3%, and chlorinated hydrocarbons 1.3% (Sinhaseni 1990).

In Thailand, research on the effects of pesticide exposure have generally concentrated on acetylcholinesterase inhibition. Efforts have been made by the Department of Medical Sciences (Sinhaseni 1990) to estimate average normal values for plasma cholinesterase levels in Thai people. Researchers have found a range of 1 500–4 000 milliunits (Ellman's method), with significant differences between males (2 760 ± 712 milliunits) and females (2 516 ± 665 milliunits). The Shell Company (Thailand) has supported a surveillance program measuring blood cholinesterase levels in farmers in various areas and at special functions such as agricultural fairs. The Occupational Health Division, Ministry of Public Health, under the Green Esarn Project, has developed and distributed paper-strip cholinesterase test kits to village health-care personnel. About 300 tests per village in six provinces were conducted in 1990.

An integrated knowledge, attitudes, and practices (KAP) survey incorporating clinical investigations as well as cholinesterase studies is currently being conducted by a multidisciplinary team from Chulalongkorn University. The project involves 150 vegetable farmers in Bang Bua Thong, near Bangkok. In a survey of 658 agriculturalists (150 families), the most common complaints were weakness (61%), dizziness (47%), headache (39%), shortness of breath (35%), poor memory (34%), and nausea or vomiting (33%) (Suwanabun et al. 1986).

Assessment of pesticide exposure has been carried out by local scientists as well as international teams (Working Group 1989; Shell Thailand 1989). Perhaps the most comprehensive field study was that conducted by Tuinman and Eadsforth (1987) on exposure and health effects after application of phosdrin formulations on vegetable farms in Thailand. No health effects were observed in any of the workers and it was recommended that attention be paid to the packaging of the concentrate to minimize the possibility of contamination of hands during preparation of formulations.

Tongsakul and Punepan (1988) conducted an exposure-assessment study using high-volume spray dye to compare exposure of the different parts of operators' bodies when spraying crops of different heights using a knapsack mistblower. Sinhaseni and Tesprateep (1988) similarly conducted an assessment of exposure to endosulfan by gardeners who sprayed large trees.

The pesticide application research team in the Entomology and Zoology Division, Department of Agriculture, is responsible for providing information related to pesticide-application technology in Thailand. Various types of lever-operated knapsack sprayers are currently being evaluated. In addition, the Division of Plant Protection Services operates pesticide clinics; they gather farmers' and government-owned out-of-service equipment for repair by six rotating units. These units also offer training in equipment repair.

There have been at least 16 KAP studies of Thai farmers. They have revealed that farmers usually do not use protective apparel according to guidelines (Working Group 1989). The studies have emphasized the importance of education, mass-communication tools, peer influence, and extension services to promote safe pesticide-handling practices in farming communities.

Ecological contamination has been monitored by the Division of Environmental Health of the Ministry of Health in 38 rivers in Thailand from 1978 to 1985. Organochlorine levels above World Health Organization (WHO) standards were detected in a number of instances. Pesticide levels in food have also been monitored by the Food and Drug Administration and the Department of Medical Sciences from 1982 to 1985. Multiresidue analyses revealed that 52.5% of the samples contained pesticides.

Various training activities have been carried out to promote the safe use of pesticides. With support from external agencies, such as the Agricultural Requisites Scheme for Asia and the Pacific, the Groupement international des associations nationales de fabricants de produits agrochimiques, the United Nations Environment Programme (UNEP), the Food and Agriculture Organization (FAO), and WHO, programs for training trainers and raising public awareness have been conducted.

Research needs

For Malaysia, there is an obvious need to establish a mechanism for collating all statistics on the health and safety-related aspects of pesticide use. While research to improve application technology, education, and communication continues, work on exposure and ecological contamination in relation to human health must be given additional emphasis and support. There is insufficient emphasis on investigations into the acute and chronic effects of pesticides on both the farmer-user and the consumer population at large. In this context, the involvement of medical researchers is considered vital.

Technological improvements should be aimed at the small-scale farmer, with regard to application systems and safety. Education and communication must be carried out in parallel with research into improved technology. Small-farmer communities can benefit from more frequent KAP studies and education programs. Research into remediation of the contaminated environment as an approach to improving health and safety for farmers and the general population should concentrate on gathering information on the extent and status of these problems and the adaptation and adoption of available technology relevant to regional needs.

In reviewing past research in the Philippines, it is notable that health-related pesticide and residue studies have attracted the interest of only a few people. Despite this, there is sufficient baseline information for investigators to proceed to intervention studies, although data on the chronic effects of pesticide exposure and residues are still lacking. Likewise, the effects of incidental pesticide exposure, for which the general public is at risk from contaminated food and water, air pollution, and occupational exposure, also warrant investigation.

Educating farmers in the safe and proper use and handling of pesticides must be a research priority. With education, concern for the chronic effects of pesticide exposure of farmers and consumers should also be addressed. Other areas where research is needed in the Philippines include:

- KAP studies that can form the basis for education modules and approaches;

- Policy research to provide a basis for pesticide-policy decisions;

- Technology development to address the issue of minimizing pesticide exposure;

- Protective clothing to safeguard the health of the farmer-user;

- Investigation of the relation between pesticide-residue levels in fruits and vegetables and farmers' crop-protection practices to determine areas of intervention; and

- Health economics of pesticide exposure to provide cost–benefit analyses for decision-makers.

In Thailand, the emphasis has been on exposure assessment in health- and safety-related pesticide activities. Nevertheless, Thai scientists believe that much remains to be done. Research needs in Thailand can been summarized as follows.

Mitigation of risk

- Exposure assessment, especially in crops where pesticide use is high (cotton, vegetables, fruit trees, and rice);

- Equipment maintenance, including appropriate design, repair services, and training.

Better identification and assessment of risk

- Integration of epidemiologic information on dose-response and appropriate control;

- Consideration of reproductive effects, such as spermatogenesis, and cardiac effects;

- Standardization and improvement of cholinesterase measurements, exposure-assessment protocols, etc.;

- Technical assistance in the identification of pesticide metabolites in blood and urine;

- Improve existing data-handling systems to accommodate community data and accurately define mortality and morbidity due to pesticides.

Increase and improve training activities

- Training the trainers to be more efficient using KAP strategy and communicative skills;

- Pest management practices;

- Information exchange;

- Sharing and standardizing information and materials required for training.

Better regulation of highly toxic pesticides

Regional collaborative pesticides research

A review of the past and current activities in Malaysia, the Philippines, and Thailand with respect to health- and safety-related aspects of pesticide use in agriculture raises several important points. Although agriculture continues to play an important role in the economies of these countries, the major crops differ. However, the health and safety problems in these countries are similar and the identified research needs are common. Three areas deserve attention: the need for comprehensive exposure assessment; the need to instil safety consciousness, especially in farming communities; and the need to upgrade pesticide-application methods, especially among small farmers.

Recognizing the commonality of research needs has been the driving force behind an initiative to develop a regional strategy for pesticides research through a network comprising scientists from Malaysia, Thailand, and the Philippines. At a meeting in July 1990 in Bangkok, Thailand, the research group reconfirmed the priorities for the region and reached consensus on the following specific research proposals:

- Exposure assessment — design and development of personal protective equipment for rice- and vegetable-farm labour, appropriate to the environmental and sociocultural conditions of the three countries;

- Pesticide-application technology — design and development of a safe, efficient, and cost-effective knapsack sprayer for small farmers in the Asian region; and

- Education and training — using a KAP survey approach, educate and train trainers and farmers in the safe and efficient use of agricultural pesticides.

The collaborative effort of scientists from the three countries is summarized in their mission statement:

> Development of appropriate application technology and generation of exposure-assessment information, reinforced through farmers'

education and training as a strategy for the minimization of health risks in agricultural pesticide usage.

Regional collaboration is to be facilitated by a coordinated network that will improve linkage through regular meetings of the scientists, a data base of relevant research, and facilities for information exchange. More specifically, the network will aid in the development of common approaches to identified problems. It has already agreed upon common core protocols for baseline KAP surveys for all three research proposals; basic specifications for personal protective equipment and knapsack sprayers; and joint development and use of training modules and communication materials.

Collaborating scientists in each country have been charged with the task of developing projects that will be combined into a regional proposal to solicit funding from international aid agencies. The within-country and between-country linkages resulting from this effort are expected to generate information and technology that directly address the need to upgrade public health and safety in the use of agropesticides in the ASEAN (Association of South East Asian Nations) region.

Acknowledgment — This review was made possible by a grant from the International Development Research Centre, Canada.

ADB (Asian Development Bank). 1987. Handbook on the use of pesticides in the Asia–Pacific region. Asian Development Bank, Manila, Philippines. 294 pp.

Anas, A.N.; Jusoh Mamat, M.; Heong, K.L.; Ho, N.K. 1987. Field observation of lever operated knapsack sprayers owned by the rice farmers in the Muda Irrigation Scheme. *In* Proceedings of the international conference on pesticides in tropical agriculture, 23–25 September 1987, Kuala Lumpur, Malaysia. Malaysian Plant Protection Society, Kuala Lumpur, Malaysia.

Anonymous. 1990. Laporan ringkas bancian kegunaan racun makhluk perosak di kalangan peserta dan pekerja FELCRA Seberang Perak. Jabatan Pertanian, Kuala Lumpur, Malaysia. Mimeo, 12 pp.

Asna, B.O. 1990. Cadangan aktiviti kempen keselamatan penggunaan racun makhluk perosak daerah sabak bernam. Jabatan Pertanian, Kuala Lumpur, Malaysia. Mimeo, 5 pp.

Asna, B.O.; Balasubramaniam, A.; Rabirah, A. 1989. Survey on the use of pesticides among non IPC farmers in Tanjung Karang. Unpublished.

Bantilan, M.C.S.; Rola, A.C. 1989. Farmers' mortality statistics in the Philippines: patterns in causes of death and possible linkages with chemical input use. International Rice Research Institute, Los Baños, Philippines.

Basri, M.W. 1981. Study on the use of and hazards posed by certain insecticides on tobacco and vegetables in peninsular Malaysia. Crop Protection Branch, Department of Agriculture, Kuala Lumpur, Malaysia. Mimeo, 33 pp.

Castaneda, C.P. 1988. Pesticide poisoning data collection in the Philippines. *In* Teng, P.S.; Heong, K.L., ed., Pesticide management and integrated pest management in Southeast Asia: proceedings of the Southeast Asia pesticide management and integrated pest management workshop, 23–27 February 1987, Pattaya, Thailand. Island Publishing House, Manila, Philippines. Pp. 311–314.

Castaneda, C.P.; Maramba, N.C.P. 1980. A three-year retrospective study of acute poisoning due to anti-cholinesterase insecticide. Filipino Family Physician, 18(2), 1–10.

Castaneda, C.P.; Rola, A.C. 1990. Regional pesticide review — Philippines. Paper presented at the Regional pesticide review meeting, 24 March 1990, Genting Highlands, Malaysia. International Development Research Centre, Ottawa, ON, Canada. 87 pp.

Chester, G.; Woollen, B.H. 1981. Studies on the occupational exposure of Malaysian plantation workers to paraquat. British Journal of Industrial Medicine, 38, 23–33.

Chuapanich, P. 1986. Pesticide supply and consumption in Thailand. Division of Agricultural Regulations, Department of Agriculture, Ministry of Agriculture and Cooperatives, Bangkok, Thailand.

Dawson, F. 1985. Estate workers suffer from using poisons: 13 women complain of falling sick often. New Straits Times, Singapore, 1 April.

Esa, Y.M.; Ramasamy, S. 1988. Pesticide safe use training program in Malaysia. *In* Teng, P.S.; Heong, K.L., ed., Pesticide management and integrated pest management in Southeast Asia: proceedings of the Southeast Asia pesticide management and integrated pest management workshop, 23–27 February 1987, Pattaya, Thailand. Island Publishing House, Manila, Philippines.

FAO (Food and Agriculture Organization of the United Nations). 1988. Report of the expert consultation on integrated pest management in major vegetable crops. FAO, Rome, Italy. 11 pp.

Gonzalez, E.Z.; Chua, R.H.C. 1984. Organophosphate poisoning: Cebu (Velez) General Hospital experience, 1974–1983. Philippine Journal of Internal Medicine, 22, 262–269.

Heong, K.L. 1982. Pest control practices of rice farmers in Tanjung Karang, Malaysia. Paper presented at the Conference on the perception and management of pests and pesticides, 21–25 June 1982, Nairobi, Kenya. Mimeo.

Heong, K.L.; Jusoh Mamat, M.; Ho, N.K.; Anas, A.N. 1987. Sprayer usage among rice farmers in the Muda area, Malaysia. *In* Proceedings of the international conference on pesticides in tropical agriculture, 23–25 September 1987, Kuala Lumpur, Malaysia. Malaysian Plant Protection Society, Kuala Lumpur, Malaysia.

Hislop, E.C. 1988. Electrostatis ground-rig spraying: an overview. Weed Technology, 2, 94–105.

Howard, J.K.; Sabapathy, S.S.; Whitehead, P.A. 1981. A study on the health of Malaysian plantation workers with particular reference to paraquat sprayers. British Journal of Industrial Medicine, 38, 110–116.

Indrani, A. 1988. Mangsa racun serangga kini terus bertambah. Berita Harian, Rabu 29 Jamadilakhir, 1408, 17 February 1988.

Jusoh Mamat, M.; Anas, A.N. 1988. Herbicide application technology for irrigated rice in Malaysia. *In* Lam, Y.M.; Cheong, A.W.; Azmi, M. ed., Proceedings of the national seminar and workshop on rice field weed management, Penang. Malaysian Agricultural Research and Development Institute, Kuala Lumpur, Malaysia. Pp. 221–229.

Jusoh Mamat, M.; Anas, A.N.; Heong, K.L.; Chan, C.W.; Nor, N.M.; Ho, N.K.; Zaiton, A.S.; Fauzi, A. 1987. Features of lever operated knapsack sprayers considered important by Muda rice farmers in deciding which sprayer to buy. *In* Proceedings of the international conference on pesticides in tropical agriculture, 23–25 September 1987, Kuala Lumpur, Malaysia. Malaysian Plant Protection Society, Kuala Lumpur, Malaysia.

Jusoh Mamat, M.; Heong, K.L.; Rahim, M. 1985. Principles and methodology of pesticide application techniques. *In* Proceedings of the workshop and course on pesticide application technology, 21–26 October 1985, Universiti Pertanian Malaysia, Serdang, Selangor, Malaysia. Malaysian Plant Protection Society, Kuala Lumpur, Malaysia. Pp. 9–26.

Lee, S.A. 1987. Protection of the spray operator from pesticide contamination. Malaysian Agriculture Digest, 5(2), 11–15.

Lee, S.A.; Chung, G.F. 1985. Safety aspects of LV, VLV, ULV and CDA application. *In* Teoh, C.H., ed., Recent developments in pesticide application technology: proceedings of the workshop, 21 January 1985, Serdang, Malaysia. Malaysian Plant Protection Society, Kuala Lumpur, Malaysia. Pp. 46–66.

Lim, G.S.; Hussein, M.Y.; Ooi, A.C.P.; Zain, M.B.A.R. 1983. Pesticide application technology in perennial crops in Malaysia. *In* Lim, G.S.; Ramasamy, S., ed., Pesticide application technology. Malaysian Plant Protection Society, Kuala Lumpur, Malaysia. Pp. 13–41.

Lim, G.S.; Ramasamy, S., ed. 1983. Pesticide application technology. Malaysian Plant Protection Society, Kuala Lumpur, Malaysia. 182 pp.

Lum, K.Y.; Jusoh Mamat, M.; Cheah, U.B. 1990. Health and safety-related aspects of pesticide usage in Malaysia. Paper presented at the Regional pesticide review meeting, 24 March 1990, Genting Highlands, Malaysia. International Development Research Centre, Ottawa, ON, Canada. 17 pp.

MACA (Malaysian Agricultural Chemicals Association). 1989. Annual report 1988/89. MACA, Kuala Lumpur, Malaysia.

Matthews, G.A. 1983. Problems and trends in pesticide application technology. *In* Lim, G.S.; Ramasamy, S., ed., Pesticide application technology. Malaysian Plant Protection Society, Kuala Lumpur, Malaysia. Pp. 163–170.

Normiya, R. 1982. Problems in the transfer, delivery and acceptance of rice technology. Malaysian Agricultural Research and Development Institute, Kuala Lumpur, Malaysia. Rural Sociology Bulletin, 12.

Ooi, A.C.P.; Heong, K.L.; Lim, B.K.; Mazlan, S. 1983. Adoption of pesticide application technology by small-scale farmers in peninsular Malaysia. *In* Lim, G.S.; Ramasamy, S., ed., Pesticide application technology. Malaysian Plant Protection Society, Kuala Lumpur, Malaysia. Pp. 148–158.

Ooi, G.H.C. 1988. NC-311 — a revolution in rice herbicide technology. *In* Lam, Y.M.; Cheong, A.W.; Azmi, M., ed., Proceedings of the national seminar and workshop on rice field weed management, Penang. Malaysian Agricultural Research and Development Institute, Kuala Lumpur, Malaysia. Pp. 131–139.

Rola, A.C. 1989. Pesticides, health risks and farm productivity: a Philippine experience. University of the Philippines, Los Baños, Philippines. APRB Monograph 89-01.

Shell Thailand. 1989. Blood test activity summary, 1988–1989. Chem-Lab, Bangkok, Thailand.

Sidhu, M.S.; Sim, C.S.; Johney, K.V. 1987. Practical aspects of chemical spraying for cocoa pod borer management in Sabah. *In* Ooi, A.C.P.; Luz, G.C.; Khoo, K.C.; Teoh, C.H.; Jusoh Mamat, M.; Ho, C.T.; Lim, G.S., ed., Management of the cocoa pod borer. Malaysian Plant Protection Society, Kuala Lumpur, Malaysia. Pp. 19–42.

Sinhaseni, P. 1990. Problem of pesticides from a public health and safety viewpoint — Thailand. Paper presented at the Regional pesticide review meeting, 24 March 1990, Genting Highlands, Malaysia. International Development Research Centre, Ottawa, ON, Canada. 69 pp.

Sinhaseni, P.; Tesprateep, T. 1988. Histopathological effects of paraquat and gill function of *Puntius gonionotus* Bleeker exposed to paraquat. Bulletin of Environmental Contamination and Toxicology, 38(2).

Sinnaia, S. 1989. Problem of poisoning in Malaysia. Paper presented at the First Malaysian–Asian international meeting on prevention and management of poisoning by toxic substances, 27–28 November 1989, Kuala Lumpur, Malaysia. Mimeo, 15 pp.

Suwanabun, N.; Ludersdorf, R.; Schache, G. 1986. Use of pesticides by Thai vegetative agriculturists and subjected complaints. Paper presented at the Regional workshop on environmental toxicity and carcinogenesis, 15–17 January 1986, Mahidol University, Thailand.

Swaddiwuthipong, W.; Kumasol, P.; Sangwanloy, O.; Srisomporn, D. 1989. Foodborne disease outbreaks of chemical etiology in Thailand 1981–1987. South-east Asian Journal of Tropical Medicine and Public Health, 20(1), 125–132.

Swan, A.A.B. 1969. Exposure of spray operators to paraquat. British Journal of Industrial Medicine, 26, 322–329.

Tan, S.H.; Ramasamy, S.; Yeoh, C.H.; Ahmad, K. 1988. A comparative study on the occupational exposure of spray workers using knapsack and spinning disc sprayers. *In* Teng, P.S.; Heong, K.L., ed., Pesticide management and integrated pest management in Southeast Asia: proceedings of the Southeast Asia management and integrated pest management workshop, 23–27 February 1987, Pattaya, Thailand. Island Publishing House, Manila, Philippines. Pp. 407–415.

Tariq, A.R. 1989. Pusat keracunan negara: konsep, fungsi, keperluan dan harapan. Penerbit Universiti Kebangsaan Malaysia, Bangi, Selangor, Malaysia.

Tejada, A.W.; Magallona, E.D. 1985. Fate of carbosulfan in a rice paddy environment. Philippines Entomology, 6(3), 255–273.

Teoh, C.H. 1985. Introduction to LV, VLV, ULV and CDA application. *In* Teoh, C.H., ed., Recent developments in pesticide application technology: proceedings of the workshop, 21 January 1985, Serdang, Malaysia. Malaysian Plant Protection Society, Kuala Lumpur, Malaysia. Pp. 7–14.

Thailand, Department of Agriculture. 1980. Cholinesterase levels in blood of farmers and other groups of people. Division of Toxic Substances, Ministry of Agriculture and Cooperatives, Bangkok, Thailand. Report 23-12-12-08-2329, 7 pp.

Tongsakul, S.; Punepun, V. 1988. Preliminary study on the amount of chemical deposit on various parts of the applicators' body from spraying soybean and cotton. Department Division, 28 pp.

Tuinman, C.P.; Eadsforth, C.V. 1987. Field study of exposure and health effects following application of phosdrin formulations to low crops in Thailand. Shell International, Health, Safety and Environment Division, Bangkok, Thailand. Report HSE 87.005, 28 pp.

Umakanthan, G. 1985. Major man-made threat to the environment. New Straits Times, Singapore, 30 June.

Varca, L.N.; Magallona, E.D. 1987. Fate of BPMC in paddy rice components. Philippines Entomology, 7(2): 177–189.

Wongphanich, M.; Prasertsud, P.; Samathiwat, A.; Kongprasart, S.; Kochavej, L.; Bupachanok, T.; Samarnsin, S. 1985. Pesticide poisoning among agricultural workers: research report to the International Development Research Centre, Ottawa, ON, Canada. 186 pp.

Working Group (Working Group on Protective Clothing for Hot Climates). 1989. Field evaluation of protective clothing materials in a tropical climate. Groupement international des associations nationales de fabricants de produits agrochimiques and Food and Agriculture Organization, Bangkok, Thailand. 4 pp.

Zain, M.B.A.R. 1977. Survey on the use of pesticides on tobacco in peninsular Malaysia. Crop Protection Branch, Department of Agriculture, Kuala Lumpur, Malaysia. Mimeo Report 1, 22 pp.

Zam, A.K. 1980. Bancian pengurusan dan kawalan serangga perosak padi di rancangan pengairan tanjung karang dan krian. Cawangan Pemeliharaan Tanaman, Jabatan Pertanian, Kuala Lumpur, Malaysia. Mimeo, 20 pp.

Zeren, Y.; Moser, E. 1988. Effects of electrostatic charging and vertical air current on deposition of pesticide on cotton plant canopy. Agriculture Mechanization in Asia and Latin America, 19, 55–60.

Zukifli, M.; Tejada, A.W.; Magallona, E.D. 1985. The fate of BPMC and chlorpyrifos in some components of the paddy rice ecosystem. Philippines Entomology, 6, 555–565.

Investigations into acute pyrethroid poisoning in cotton growers in China

Fengsheng He, Shuyang Chen, Zouwen Zhang, Jinxiu Sun, Peipei Yao, Yuqun Wu, Shaoguang Wang, Lihui Liu, and Hai Dang

Institute of Occupational Medicine,
Chinese Academy of Preventative Medicine,
Beijing, People's Republic of China

The use of pyrethroid insecticides for pest control in China has been increasing. Acute pyrethroid poisonings have occurred in cotton growers since 1982. In an epidemiological study of 3 113 cotton farmers of Gaochen County, the prevalence of acute pyrethroid poisoning was 0.31% in pyrethroid-exposed sprayers and 0.38% in those exposed to pyrethroid–organophosphate mixtures; blood cholinesterase levels were normal. Determination of exposure levels by measuring air concentrations and dermal exposure indicated that dermal exposure is the most important route of absorption. An increase in median-nerve excitability was found in cotton farmers spraying deltamethrin for 3 days, but there was no significant correlation with levels of urinary deltamethrin or its metabolite Br$_2$A. Diagnostic criteria for acute pyrethroid poisoning are proposed. Recommendations are made for preventive measures and safer use of pyrethroids by agricultural workers.

Since 1982, the application of pyrethroids in cotton-growing areas of China has been increasing. Compared with commonly used broad-spectrum insecticides, such as organophosphates and carbamates, synthetic pyrethroids are more powerful on a weight-per-activity basis, yet they metabolize rapidly and leave virtually no residue in the biosphere. As a result, they are used extensively to control a wide variety of agricultural insect pests. In China, the most frequently used pyrethroids are deltamethrin, fenvalerate, and cypermethrin.

Pyrethroids were originally classified as compounds unlikely to pose an acute hazard in normal use. However, evidence from experiments on animals indicates that deltamethrin and fenvalerate affect both the central and peripheral nervous systems; acute poisoning in mice and rats was manifested as choreoathetosis and convulsion (Vijverberg and Bercken 1982). Earlier studies suggested that industrial exposure to pyrethroids induced only local effects on human skin. After exposure, abnormal facial sensations described as burning, itching, or tightness and erythema or miliary red papules persist (Le Quesne and Maxwell 1980; Kolmodin-Hedman et al. 1982; Tucker and Flannigan 1983; He et al. 1988).

No clinical data about acute pyrethroid poisoning in occupationally exposed subjects had been reported until several incidents of acute intoxication occurred among farmers spraying deltamethrin and fenvalerate in China in 1982 (He et al. 1989). Having been told that these new insecticides were "nontoxic," the cotton farmers handled them without any safety precautions. More than 300 cases of acute pyrethroid poisoning occurred in 1982, when pyrethroids were first introduced and extensively used in cotton-growing areas of China, such as Hebei, Jiangsu, Hubei, Liaoning, Shandong, and Henan provinces. It is now evident that the compounds do have an inherent neurotoxicity.

Although the use of pyrethroids as pesticides is increasing, little research has been done on their acute and long-term effects on man. From 1987 to 1989, investigations were carried out on the diagnosis of acute pyrethroid poisoning, the prevalence of occupational pyrethroid intoxication in cotton growers, and the assessment of risks to people spraying under defined conditions (by measurement of exposure and median-nerve excitability). The aim of these studies was to reduce the risks associated with the use of pyrethroids by cotton farmers in China.

Methods

Diagnosis of acute pyrethroid poisoning

Between 1983 and 1988, 573 cases of acute pyrethroid poisoning were reported (He et al. 1989). We reviewed the case histories and medical records of these patients and interviewed some of them.

Epidemiology

To assess the extent of the problem of pyrethroid poisoning in cotton growers, an epidemiologic survey was conducted in Gaochen County, Hebei Province, from June to August in 1987 and 1988. The procedures established by the World Health Organization (WHO 1982) were followed. The criteria derived from the diagnosis investigation, above, were used to classify patients. Eight villages, containing 3 390 families (total population of 19 692) in Gaochen County were identified as clusters for sampling. A total of 3 113 adults, who were exposed to pyrethroids during spraying, were studied.

Level of exposure

To assess the level of exposure and to quantify the exposure–response relation, 50 cotton growers in three groups from clusters in Gaochen and Langfang counties were selected for study in 1987 and 1988. Subjects had no history of exposure to pyrethroids in the week before the investigation, and all were confirmed to be healthy by interview and physical examination. Each subject sprayed pyrethroids in the cotton field for 5 h/day using single-nozzle pressure sprayers for 1 day in group 1, then for 3 days in groups 2 and 3, consecutively (Table 1). All subjects were monitored by project personnel throughout the spraying.

Air samples were collected from subjects in groups 1 and 2 using DuPont personal samplers with an air flow of 2 L/min throughout the 5 h of spraying; respiratory exposure (g/h) was calculated as air concentration at breathing zone (g/m^3) times ventilation volume (0.01 m^3/min) times 60 minutes. Deltamethrin and fenvalerate in the air were collected on polyester filters. Meteorological conditions were monitored with respect to temperature, relative humidity, wind speed, wind direction, and atmospheric pressure.

Five-layer gauze pads (5 × 5 cm), pretreated with n-hexane and Soxhlet's extractor, were used to measure dermal exposure of the head, chest, back, forearms, thighs, legs, hands, and feet of each subject in groups 1 and 2. Dermal exposure was estimated as the concentration in one gauze pad (g/cm^2) times total body surface area (cm^2) times percentage of body area sprayed divided by spraying time (h). Total exposure was also calculated, as respiratory exposure level plus dermal exposure level per kilogram of body weight.

Table 1. Controlled exposure of subjects to pyrethroids during spraying.

Group	No. of subjects	Amount of pyrethroid (mL ± SD)	Dilution	Spraying time (h/day × days)
1				
a	6	100 ± 14	1:1 250 (D)	5 × 1
b	6	117 ± 103	1:2 000 (D) + 1:600 (M)	5 × 1
c	6	96 ± 77	1:1 250 (F)	5 × 1
2				
a	5	376 ± 30	1:1 000 (D)	5 × 3
b	5	420 ± 45	1:1 000 (D) + 1:600 (M)	5 × 3
c	5	376 ± 50	1:1 000 (F)	5 × 3
d	5	364 ± 64	1:1 000 (F) + 1:600 (M)	5 × 3
3	12	428 ± 90	1:1 000 (D)	5 × 3

Note: D, deltamethrin; F, fenvalerate; M, methamidophos; SD, standard deviation.

Urine samples were collected at 3, 6, 9, 12, 24, 48, and 72 h after the start of spraying for each subject of group 1; at 12, 24, 36, 48, 60, 72, 96, and 120 h for people in group 2; and at 24, 48, 72, 96, and 120 h for group 3.

Air, gauze, and urine samples were analyzed for deltamethrin and fenvalerate by gas chromatography (Varian 3700) with a 63 Ni electron-capture detector. The detection limit was 2 ng/mL hexane for air and gauze samples and 0.2 g/L in a 10-mL urine sample for both urinary deltamethrin and fenvalerate (Han et al. 1988; Wu et al. 1992). The analysis of deltamethrin metabolite, dibromovinyl-dimethyl-cyclopropane carboxylic acid (Br_2A) in urine was determined by high-performance liquid chromatography (HPLC; Beckmann 344) (Yao et al. 1992).

To determine the toxic effects of pyrethroids, all participants completed a questionnaire and were interviewed. A physical examination was also conducted after pyrethroid exposure. Blood cholinesterase levels were measured in subgroups 1b, 2b, and 2d, before and after spraying.

Nerve excitability in workers spraying deltamethrin

Twenty-four adult male cotton growers from Langfang County, aged 17–57 years, sprayed 2.5% deltamethrin emulsifiable concentrate (EC) diluted 1:1 000 in water for 5 h/day for 3 days in the cotton fields. To assess exposure, urine samples were collected from 19 of the subjects at 0, 24, 48, 72, and 120 h after the beginning of spraying. Deltamethrin was determined by gas chromatography and its metabolite, Br_2A, by HPLC. Median-nerve excitability was detected with a mobile electromyograph (Medelec MS 92a), with pairs of stimuli using a percutaneous technique described by Gilliatt and Willison (1963) in each member of the exposed group before and after the 3-day spraying period and 2 days after cessation of exposure. Twenty-nine unexposed people in Langfang County, aged 17–62 years, were also measured for median-nerve excitability at 2- and 3-day intervals. The difference in nerve threshold between two examinations was kept at less than 1.5 mA for each subject in both groups.

Results

The 573 cases of acute pyrethroid poisoning were classified by specific substance and by cause (Table 2). The clinical manifestations due to acute deltamethrin and fenvalerate poisoning, whether occupational or accidental, were similar. Abnormal facial sensations were obvious first symptoms in occupationally exposed subjects, whereas accidentally intoxicated patients displayed more epigastric disorders due to ingestion of pyrethroids.

Symptoms in mild cases included dizziness, headache, nausea, anorexia, fatigue, listlessness, and muscular fasciculation. Frequent attacks of

Table 2. Cases of acute pyrethroid poisoning, 1983–88.

Cause	Deltamethrin	Fenvalerate	Cypermethrin	Others	Total
Occupational	158	63	6	2	229
Accidental	167	133	39	5	344
Total	325	196	45	7	573

convulsion, pulmonary edema, and coma were seen in severe cases. To identify acute poisoning cases, the following criteria, graded by severity, were proposed (they were adopted by the Chinese Ministry of Public Health in 1989 as the national diagnostic criteria for acute pyrethroid poisoning):

- Verified exposure to pyrethroids within 3 days before onset of symptoms.

- Suspected acute poisoning — abnormal facial sensations (burning, itching, or tingling, often exacerbated by sweating or washing with hot water, lasting no longer than 24 h) or miliary papules and contact dermatitis, without systemic symptoms or signs.

- Mild acute poisoning — in addition to the above symptoms: dizziness, headache, nausea, anorexia, and fatigue, as well as signs of listlessness or muscular fasciculation in limbs requiring leave from work for more than 1 day.

- Severe acute poisoning — convulsive attack, pulmonary edema, or coma in addition to the above systemic symptoms and signs.

- Reasonable exclusion of other diseases (e.g., common cold, heat stroke, food poisoning, or other pesticide poisonings).

The age of the 3 113 cotton growers studied ranged from 15 to 72 years, with the majority between 25 and 44 years. There were twice as many males (2 230) as females (883). Of the participants, 77.8% had primary and secondary school education.

The major kinds of pyrethroids used by the farmers were: 2.5% deltamethrin EC (Roussel-Uclaf Company), 20% fenvalerate EC (Suimoto Company), and 10% cypermethrin EC (Shell Oil Company), which were used by 2 588 farmers (83.1%); and pyrethroids mixed with organophosphates, i.e., DDVP (2,2-dichlorovinyl dimethyl phosphate, 9%), methamidophos (4.4%), o-methoate (2.4%), or others, used by 525 farmers (16.9%) (Table 3). Four types of spraying equipment were used to apply the pesticides: single-nozzle pressure type (83.2%); bearing and mobile type (16.2%); mechanical type (0.4%); and ultra-low-volume type (ULV, 0.2%).

Pyrethroids of a concentration higher than 1:2 000 were used by 37.7% of the farmers; the highest concentration being 1:50 for the ULV-spraying apparatus. The application rates were 10–30 L/h, and spraying time was no more than

Table 3. Number of farmers using various types of pyrethroids.

	1987	1988	Total no.	%
Deltamethrin				
Pure EC[a]	980	379	1 359	43.7
Mixed with —				
organophosphates	318	—	318	10.2
DDVP	157	—	157	
Methamidophos	95	—	95	
o-methoate	45	—	45	
Others	21	—	21	
Fenvalerate				
Pure EC	492	729	1 221	39.2
Mixed with —				
organophosphates	203	—	203	6.5
DDVP	123	—	123	
Methamidophos	43	—	30	
o-methoate	30	—	7	
Others	7	—	7	
Cypermethrin				
Pure EC	7	1	8	0.3
Mixed with				
organophosphates	4	—	4	0.1

[a]EC, emulsifiable concentrate.

4 h/day for 69.1% of the farmers. The mean rate of exposure to pure pyrethroids was less than 40 mL/h for two-thirds of the farmers.

The equipment of 778 cotton farmers leaked or became blocked during spraying; 65.2% of them cleared stoppages using their mouths or hands, leading to potential topical contamination. Moreover, 91.6% of the cotton farmers prepared the pyrethroids by hand, using the lid of the pyrethroid container instead of a measuring glass, presenting another opportunity for skin contact.

The survey of knowledge, attitudes, and practices (KAP) of pyrethroid users showed that 2 137 cotton farmers (69.8%) were unaware of the toxicity of pyrethroids. In addition, their personal protection was not satisfactory. None wore a mask or gloves. Approximately half the farmers kept their upper extremities or feet uncovered during spraying. The majority of spraying operations did not comply with safe-handling practices: 67% did not spray every other row of cotton; 84% sprayed while walking backward; 45% sprayed into the wind; and 9% ate or smoked in the field without washing their hands first. Because of sloppy handling, 67% of the farmers had their shoes and trousers soaked with pyrethroids.

Of the subjects, 98% (3 051) were in good health before the study. Symptoms, such as abnormal facial sensations (appearing in 742 subjects) and dizziness (in 121 subjects), usually developed within 4 h after exposure during spraying. A total of 834 farmers (26.8%), including 696 spraying only pyrethroids and 138 spraying pyrethroid–organophosphate mixtures, developed burning and tingling sensations, mainly on the face, but also on the neck or upper extremities. These skin symptoms often disappeared within 24 h. Systemic symptoms were also observed, but with minimal signs (Table 4).

After excluding other possible diseases, only 10 subjects who developed all five systemic symptoms (dizziness, headache, fatigue, nausea, and anorexia) and signs (listlessness and muscular fasciculation) in addition to abnormal facial sensations were diagnosed as having mild acute pyrethroid poisoning (Table 5). Based on the diagnostic criteria, four were cases of mild acute deltamethrin poisoning; four were diagnosed as mild acute fenvalerate poisoning; and two were deltamethrin–DDVP poisoning with normal blood cholinesterase activity. All patients recovered within 2–4 days.

In this study, the prevalence rate of occupational acute pyrethroid poisoning was 0.31% among subjects exposed to pyrethroids only and 0.38% among those exposed to pyrethroid–organophosphate mixtures. The risk of acute pyrethroid poisoning did not differ significantly ($p > 0.05$) between the two groups.

Investigation revealed that 2 131 cotton growers in the study had been previously exposed to pyrethroids since 1982. Of them, 25.7% (547 subjects) were exposed to pyrethroids every year between 1982 and 1987 and experienced abnormal facial sensations, dizziness, headache, or nausea after each spraying, but all recovered within 2 days. No sequelae were found. This suggests that pyrethroids are unlikely to have chronic toxicity for humans.

Table 4. Number of farmers showing symptoms and signs of pyrethroid poisoning ($n = 834$).

	No.	%
Symptom		
Abnormal sensations	768	92.1
Dizziness	121	14.5
Headache	99	11.9
Fatigue	91	10.9
Nausea	68	8.2
Anorexia	44	5.3
General malaise	33	4.0
Blurred vision	23	2.8
Chest tightness	18	2.2
Sign		
Listlessness	10	1.2
Muscle fasciculation	1	0.2

Table 5. Effects of pyrethroids on 3 113 subjects.

Treatment	Sample size	Adverse effects No.	%	Acute poisoning No.	%
Exposure to pyrethroids only					
1987	1 840	347	23.4	4[a]	0.27
1988	1 109	349	31.5	4	0.36
Total	2 589	696	26.9	8	0.31
Exposure to pyrethroid– organophosphate mixture					
1987	524	138	26.3[b]	2[c]	0.38[d]

[a] Three cases of deltamethrin poisoning and one case of fenvalerate poisoning.
[b] $p > 0.05$ in chi-square test.
[c] Deltamethrin–DDVP mixture.
[d] $p > 0.05$ in Poisson probability distribution test.

Table 6. Exposure levels (SD) of 38 subjects spraying pyrethroids in Gaocheng County.

Group	Respiratory exposure No.	Level (μg/h)	Dermal exposure No.	Level (mg/h)	Total exposure No.	Level (μg/kg per h)
1						
a (D)	6	0.01 ± 0.00	6	0.46 ± 0.39	6	4.00 ± 4.00
b (D+M)	6	0.19 ± 0.22	5[a]	0.72 ± 0.49	5[a]	8.00 ± 6.00
c (F)	6	0.41 ± 0.44	6	4.24 ± 1.96	6	44.00 ± 22.00
2						
a (D)	5	0.07 ± 0.06	4[a]	0.59 ± 0.30	4[a]	7.33 ± 3.33
b (D+M)	5	0.06 ± 0.05	5	0.27 ± 0.13	5	4.67 ± 2.67
c (F)	5	0.65 ± 0.32	5	3.97 ± 1.88	5	68.00 ± 32.67
d (F+MS)	5	0.41 ± 0.20	5	2.83 ± 1.20	5	48.00 ± 18.00

Note 1: Group 1 = 1 day; group 2 = 3 days.
Note 2: D, deltamethrin; F, fenvalerate; M, methamidophos.
[a] One subject excluded because gauze pad was lost.

Dermal exposure was more frequent than respiratory exposure and represented the most important route for pyrethroid absorption (Table 6). Skin contamination by pyrethroids was heaviest on legs, feet, and hands. Because air temperature was about 30 °C, some of the farmers wiped sweat from their faces with their hands and sleeves, increasing secondary dermal exposure.

Deltamethrin was detectable in the urine of subjects exposed in 1 day of spraying for up to 12 h after the beginning of exposure; urinary fenvalerate was still detectable more than 24 h after the beginning of exposure in subjects spraying for 1 day. This suggests that deltamethrin is more rapidly metabolized than fenvalerate in humans. However, urinary levels of both pyrethroids were higher in the groups spraying for 3 days, even 2 days after cessation of exposure.

The deltamethrin metabolite Br_2A could be detected in the urine of the 12 spray operators in groups 1a and 1b by HPLC 3–12 h after exposure. Levels peaked 3–9 h after spraying, with concentrations in the range of 8–110 g/L, i.e., 150 times higher than the level of deltamethrin in urine. This indicates that urinary Br_2A is a better indicator for biological monitoring than urinary deltamethrin. Among groups spraying deltamethrin for 3 days, deltamethrin and its metabolite Br_2A appeared in urine samples early on the 1st day of spraying and persisted for up to 2 days after cessation of the 3-day exposure.

In the groups spraying for 3 days, measurement of changes in nerve excitability showed that the supernormal period in the median nerve was prolonged compared with that before spraying. This difference became more significant 2 days after cessation of exposure (Table 7). No corresponding change in supernormal period was found in the control group at 3-day intervals. After spraying, nearly half of the operators displayed a prolongation of the supernormal period in the median nerve of more than 4 ms, whereas almost none of the control group showed similar changes. However, there was no correlation between the nerve excitability changes and urinary deltamethrin or Br_2A excretion, nor diagnosis of acute deltamethrin poisoning.

Discussion

Pyrethroids have been classified as pesticides of high insecticidal activity and low toxicity to mammals. However, pyrethroids can produce symptoms, mainly involving the central and peripheral nervous systems, in humans exposed to heavy concentrations in their work. The clinical data on the 573 reported cases of acute pyrethroid poisoning in China, including 344 accidental poisoning cases and 229 occupational cases, provided a scientific basis for developing the diagnostic criteria of acute pyrethroid poisoning (He et al. 1989). Exposed subjects having abnormal facial sensations without any systemic symptoms can only be diagnosed as suspicious cases of acute poisoning. Those who have systemic symptoms (i.e., headache, dizziness, nausea, fatigue, and anorexia) and signs (i.e., listlessness and muscular fasciculation) in addition to the abnormal facial sensations, causing them to miss work for more than 1 day, can be diagnosed as mild acute pyrethroid poisoning after exclusion of other diseases. Subjects with severe acute poisoning have disturbance of consciousness, convulsive attacks, or pulmonary edema. However, so far there have been no specific diagnostic indicators available for acute

Table 7. Nerve excitability in median nerve among farmers spraying deltamethrin for 3 days and the control group.

Time of measurement	Absolute refractory period (ms)	Relative refractory period (ms)	Supernormal period (ms)	No. with supernormal period >4 ms longer than previous detection	
				No.	%
Spray operators ($n = 24$)					
Before spraying	1.11 ± 0.61	4.51 ± 1.64	6.50 ± 3.77	—	—
After 3 days spraying	0.93 ± 0.53	5.04 ± 1.84	8.93 ± 4.91	10	41.7[a]
2 days after cessation	0.90 ± 0.12	5.69 ± 2.07	10.41 ± 6.02[b]	11	45.8[a]
Control group ($n = 29$)[c]					
1st detection	0.87 ± 0.24	6.62 ± 3.42	8.16 ± 4.67	—	—
2nd detection	0.93 ± 0.50	6.04 ± 2.79	7.64 ± 3.43	0	0
3rd detection	0.84 ± 0.13	7.03 ± 2.69	9.12 ± 3.53	1	5.3

[a] $p < 0.01$, compared with control group in χ^2 test.
[b] $p < 0.001$, compared with value before spraying in paired test by SYSTAT.
[c] Timing of measurements corresponded with those of spray operators, i.e., 3 and 5 days after first detection. For third detection, $n = 19$.

pyrethroid poisoning. Therefore, the importance of differential diagnosis should be emphasized. Overdiagnosis, misdiagnosis, and maldiagnosis should be avoided.

The diagnostic criteria were used in the epidemiologic study to assess the extent of the acute pyrethroid poisoning problem in cotton farmers. The results showed that 834 out of 3 113 sprayers surveyed (27%) were clinically affected by exposure to pyrethroids, complaining of abnormal facial sensations (a symptom specific to pyrethroid exposure) as well as dizziness, headache, or fatigue, but with minimal signs. Only 10 sprayers were diagnosed with mild acute pyrethroid poisoning according to the diagnostic criteria. In this study, the prevalence of acute poisoning was 0.31% for subjects exposed to pyrethroids only and 0.38% in those exposed to pyrethroid and organophosphate mixtures. The risk of acute pyrethroid poisoning did not differ significantly between the two groups. The prevalence of acute pyrethroid poisoning in this study is lower than that of acute organophosphate poisoning in cotton farmers, reported to be 0.16-11.6% in China. Cases of acute pyrethroid poisoning in this survey were mild and the prognosis was good.

In 38 of the subjects selected from the studied clusters, the measurement of air concentration, dermal exposure, and urinary excretion of pyrethroids or deltamethrin metabolites showed that dermal exposure was the most significant route of absorption. This indicates that minimizing dermal exposure is of paramount importance in the prevention of acute pyrethroid poisoning. Urinary deltamethrin and fenvalerate were both detectable, confirming the absorption of pyrethroids by spray operators, although these two pyrethroids metabolize rapidly. After 1 day of spraying, deltamethrin was present in urine up to 12 h after the beginning of spraying, but the deltamethrin metabolite Br_2A was detectable at a higher level and for a longer time. Urinary fenvalerate was also present longer — up to 24 h after exposure began. Hence, the deltamethrin metabolite Br_2A and fenvalerate in urine may be possible indicators for biological monitoring in pyrethroid-exposed subjects.

Of 2 131 spray operators who had used pyrethroids every year from 1982 to 1987, 25.7% experienced adverse effects due to pyrethroid exposure similar to the symptoms in this survey, but they recovered rapidly and entirely. No sequelae or long-lasting symptoms were found. This indicates that pyrethroids are unlikely to have chronic toxic effects in humans. However, the biological monitoring of 32 cotton growers spraying pyrethroids for 3 days showed that urinary pyrethroids were not detectable more than 24 h after exposure.

In addition, 24 spray operators exposed to deltamethrin showed a prolonged supernormal period of median-nerve excitability after 3 days of spraying and 2 days after spraying had stopped; this increase of nerve excitability was reported by Parkin and Le Quesne (1982) and Takahashi and Le Quesne (1982) in animal experiments. Prolongation of the supernormal period and urinary pyrethroid excretion were not significantly correlated. However, these findings may imply an increased risk of intoxication as a result of consecutive exposures to pyrethroids, and might well explain why the convulsive attacks in severe acute pyrethroid-poisoning cases can persist for 1 week (He et al. 1989).

Higher risk of acute pyrethroid poisoning is related to the unsatisfactory knowledge, attitudes, and practices among farmers. These include ignorance of pyrethroid toxicity; spraying pyrethroids in high concentrations; cleaning equipment with mouth and hands; leakage of sprayers; lack of personal protection; and unsafe spraying habits. This study confirmed these to be causal factors in identified cases of intoxication.

Recommendations

An educational strategy for the safe use of pesticides in rural areas is advocated. Spray operators, local agricultural technicians, village-health personnel, and those involved in transportation, storage, and distribution of pesticides should be taught the risks involved in the handling of toxic pesticides, the toxicity and adverse effects of pyrethroids, preventive measures, and the importance of personal hygiene. Pamphlets, videocasts, films, and training courses are suggested because these are feasible and useful approaches to implement in the rural areas.

To avoid pyrethroid contamination of sprayer's hands, spraying equipment should be improved to prevent leakage. Packaging pyrethroids in disposable 10-mL plastic eye-drop bottles is also suggested.

Because the heaviest dermal exposure occurs on legs and polyester trousers offer 66–88% protection against pyrethroids, it is advisable for spray operators to wear a long plastic apron in addition to their polyester trousers during spraying in the fields.

Regulations concerning safe production, transportation, storage, distribution, and use of pyrethroids and other pesticides should be established. Moreover, multisectoral efforts should be devoted to the implementation of these regulations.

Occupational health in rural areas should be integrated into primary health care to improve occupational health services for agricultural workers by prevention, early diagnosis, and treatment of acute pyrethroid poisoning.

Acknowledgment — This work was carried out with the aid of a grant from the International Development Research Centre, Ottawa, Canada.

Gilliatt, R.W.; Willison, R.G. 1963. The refractory and supernormal periods of the human median nerve. Journal of Neurology, Neurosurgery and Psychiatry, 26, 136–147.

Han K.; Xiang B.; Xu B.; Wang C.; Wu Y.; Gao X. 1988. Measurements of delta-methrin and fenvalerate in air by gas chromatography. Journal of Hygiene Research, 17(2), 13–17. [In Chinese].

He F.; Sun J.; Han K.; Wu Y.; Yao P.; Wang S.; Liu L. 1988. Effects of pyrethroid insecticides on subjects engaged in packaging pyrethroids. British Journal of Industrial Medicine, 45, 548–551.

He F.; Wang S.; Liu L.; Chen S.; Zhang Z.; Sun J. 1989. Clinical manifestations and diagnosis of acute pyrethroid poisoning. Archives of Toxicology, 63, 54–58.

Kolmodin-Hedman, B.; Swensson, A.; Akerblom, M. 1982. Occupational exposure to some synthetic pyrethroids (permethrin and fenvalerate). Archives of Toxicology, 50, 27–33.

Le Quesne, P.M.; Maxwell, I.C. 1980. Transient facial sensory symptoms following exposure to synthetic pyrethroids: a clinical and electrophysiological assessment. Neurotoxicology, 2, 1–11.

Parkin, P.J.; Le Quesne, P.M. 1982. Effect of a synthetic pyrethroid deltamethrin on excitability changes following a nerve impulse. Journal of Neurology, Neurosurgery and Psychiatry, 45, 337–342.

Takahashi, M.; Le Quesne, P.M. 1982. The effects of the pyrethroids deltamethrin and cismethrin on nerve excitability in rats. Journal of Neurology, Neurosurgery, and Psychiatry, 45, 1 005–1 011.

Tucker, S.B.; Flannigan, S.A. 1983. Cutaneous effects from occupational exposure to fenvalerate. Archives of Toxicology, 54, 195–202.

Vijverberg, H.P.M.; Bercken, J.V.D. 1982. Action of pyrethroid insecticides on the vertebrate nervous system. Neuropathology and Applied Neurobiology, 8, 421–440.

WHO (World Health Organization). 1982. Field surveys of exposure to pesticides. WHO, Geneva, Switzerland. Standard Protocol VBC/82.1.

Wu Y.; Gao X.; Li C.; Wang Y. 1992. Determination of urinary deltamethrin and fenvalerate in man by gas chromatography. (In press).

Yao P.; Li Y.; Zhang X.; He F. 1992. Biological monitoring of deltamethrin metabolites in spraymen and one suicide by HPLC. (In press).

Field exposure during application of organophosphates using knapsack sprayers in the Philippines

Carmen P. Castaneda

Department of Pharmacology, College of Medicine,
University of the Philippines,
Manila, Philippines

Three field studies were undertaken in the Philippines to measure exposure to pesticides using World Health Organization (WHO) protocols. Farmers used methyl parathion (50% emulsifiable concentrate) and monocrotophos (28.5% water-soluble concentrate) under field conditions. Potential exposure and dermal exposure were determined by measuring the concentration of pesticide on pieces of clothing and gauze patches worn under the clothing. The protection afforded by locally designed protective clothing was assessed, the urinary metabolites, p-nitrophenol and dimethylphosphate were monitored, and cholinesterase activity in red blood cells was measured before and after exposure to the pesticides.

For the Philippines, a developing country with a population of about 60 million people, agriculture is the backbone of the national economy. To ensure crop protection, pesticides are an important agricultural input. Anticholinesterase insecticides are the most widely available and frequently used pesticides. With increased use, however, there is growing concern about the potential adverse effects of pesticides on man and the environment. Based on sales value of imports, pesticide consumption in the Philippines has increased steadily from 1983 to 1987 (Table 1). Estimates for 1984 show that 36.6% of total pesticide spending was for rice crops, 25.5% for bananas, and 14.3% for vegetables (Fig. 1).

In the Philippines, there are two growing seasons for rice: February to April and July to October. Sprayable insecticides are applied up to three times during a season. Applications normally take place about 2 weeks after planting and up to 4 weeks before harvest. The height of the crop during spraying varies from 20 to 100 cm.

Table 1. Value of imported pesticides (USD thousands).

Substance	1983	1984	1985	1986	1987
Insecticides	16 196	14 077	12 151	16 379	23 270
Herbicides	3 637	4 562	4 016	6 042	7 139
Fungicides	9 058	11 528	7 953	9 691	10 589
Others	7 391	5 230	6 733	8 331	11 466
Total	36 282	35 397	30 853	40 443	52 464

Source: Fertilizer and Pesticide Authority, Department of Agriculture, Philippines.
Note: 23 Philippine pesos (PHP) = 1 US dollar (USD).

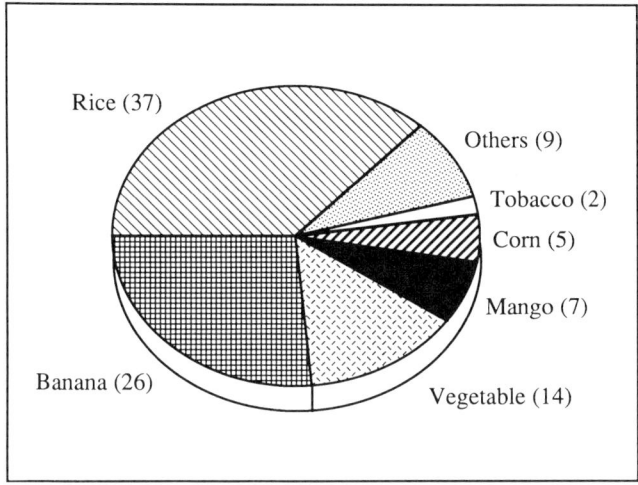

Fig. 1. Pesticide use by crop, based on 1984 spending levels (Rola n.d.).

Typically, a farmer sprays early in the morning, for 4–5 h, using a stainless steel, hand-pumped knapsack sprayer with a pressure of 15–20 pounds per square inch (1 psi = 6.89 kPa). It is common practice to use the lid of the sprayer to measure the concentrate, transfer it to a 16-L knapsack, and make up the volume with water. An average of 1 ha is sprayed per day. After application, the farmer cleans the equipment with water and stores it, together with the chemicals, in his home.

Occasionally, however, spraying is carried out by contract sprayers. They may apply a variety of insecticides over a maximum area of 1 ha/day for up to 3 consecutive days/week (van Sitter et al. 1989). They usually wear ordinary work clothes, consisting of long pants and long-sleeved cotton shirts. Gloves, masks, or boots are not normally worn during the application of insecticides (Castaneda et al. 1989).

The objective of this paper is to present the findings of three field studies on pesticide exposure undertaken in the Philippines between 1987 and 1989. In the first, the potential dermal exposure of rice farmers applying methyl parathion (50% emulsifiable concentrate, EC) was measured at various stages under field conditions. The applicability of the World Health Organization's protocol (WHO 1982) in conducting the study was assessed, and the necessity for protective clothing for rice farmers was investigated. The second exposure study assessed the protection afforded by locally designed protective clothing during the application of methyl parathion (50% EC) under field conditions. The third study assessed the exposure and resulting health risk arising from the application of monocrotophos (28.5% water-soluble concentrate, WC) to rice under field conditions. An attempt was made to study the benefits of "Gardman" protective clothing and high-density polyethylene (HDPE) bottles with dispensers.

Methods

The methods used in the three studies were based on the WHO protocol for field surveys of exposure to pesticides (WHO 1982). The three exposure studies were conducted in rice fields in the Philippine regions of Bulacan, Pampanga, and Laguna in the rice-growing area of the central plain of Luzon.

Study one

Forty-five farmers participated in the first methyl parathion study. They were divided into eight groups (about six farmers per group), one for each pesticide application activity:

- Mixing and loading;
- Spraying at the crop's vegetative stage (first of three replicates, R1);
- Spraying at the crop's maximum tillering stage (R2);
- Spraying at the crop's booting stage (R3);
- Cleaning and disposal;
- Day 1 field reentry;
- Day 7 field reentry; and
- All activities.

The last group represented the normal activities undertaken by rice farmers. All groups were exposed for a minimum of 1 h.

The farmers wore cotton overalls, hoods, and gloves with inner patches. The patches were analyzed for pesticide residue using gas–liquid chromatography. Samples of early-morning urine and urine collected 6 h after exposure were analyzed for the urinary metabolite, p-nitrophenol, using spectrophotometric methods (Kaye 1980).

Study two

Twelve farmers participated in the second methyl parathion study, comprising two groups of six subjects each. All farmers wore cotton T-shirts, pants, and gloves. The first group wore the locally designed protective clothing in the form of plastic vests over their T-shirts, plastic bags to cover hands, and plastic wrappings around their legs. Each group was exposed for at least 1 h.

Sampling gauze pads, backed with aluminum foil and another cotton patch, were placed in the inner clothing of all farmers. The purpose of the cotton patch was to secure the aluminum foil and gauze pads inside the clothing. Patch locations were based on the WHO protocol with the following modifications:

- Head — on the back part of the head or hood;

- Arm — upper surface of right and left forearms, as they are held with elbow bent at a right angle across body, midway between elbow and wrist;

- Forearm — back surface;

- Thigh — front of mid-thigh;

- Leg — front of mid-leg;

- Back — right and left shoulder blades; and

- Chest — right and left chest.

Patches were analyzed for pesticide residues using gas–liquid chromatography. Cholinesterase levels in red blood cells were determined before and after exposure using the Michel method (modified by Aldridge and Davies 1952). Urine samples were collected and analyzed as described for Study one above.

To determine the recovery of methyl parathion from the sampling media, field and laboratory recovery tests were conducted for both studies.

Study three

Twenty-eight farmers participated in the third study focusing on monocrotophos. They were divided into four groups of seven subjects. Group A wore normal work clothes, consisting of long pants and long-sleeved cotton shirts. Group B wore "Gardman" protective clothing over their normal working clothes. Group C wore normal work clothes, but were supplied monocrotophos concentrate in 1-L HDPE bottles with built-in dispensers. Group D was the control group. All groups were exposed for 3 days for about 4–5 h/day.

Cholinesterase levels in red blood cells were determined before and after monocrotophos exposure using the Michel method. Twenty-four hour urine was collected before spraying, during the 3 days of application, and 1 day after completion of the spraying. The urinary metabolite, dimethylphosphate, was determined by gas–liquid chromatography.

For all three studies, operational details (crop height, row spacing, area covered, and rate of application) and meteorological data (wind speed, wind direction, air temperature, and relative humidity) were recorded as recommended by the WHO protocol.

Results and discussion

Postexposure levels of p-nitrophenol in urine were not significantly different from baseline levels. Two factors explain this: over the previous 9 years, subjects have been exposed to a number of different pesticides, thus the relatively short exposure to methyl parathion during this study may not have been sufficient to have an additional effect; the method used to measure p-nitrophenol may not have been sensitive enough to detect changes.

In keeping with the WHO protocol, contamination of the cotton overalls and potential dermal contamination represented by the gauze patches were assessed. Mixing and loading activities, which were carried out three to five times during the 1-h study periods, accounted for 55% of the contamination of clothing; that of the "all activities" group represented 28% of total contamination. Spraying at any of the three stages did not account for a large proportion of the contamination of clothing. The risks were equal for the R1 and R2 stages (spraying at the vegetative and maximum tillering stages, respectively) and accounted for 8% each. The R3 group (spraying at the booting stage) represented only 1% of all contamination of clothing.

Potential dermal contamination, i.e., penetration of clothing, was assessed by measuring the amount of pesticide in the gauze patches. Of total contamination, 47% was found in the "all activities" groups. Farmers in these groups performed all activities in 1 h. They were exposed for a longer period and at greater pesticide concentrations than those undertaking specific activities. Mixing and loading accounted for 12.0%; R1 for 16.0%; R2 for 21.0%; R3 for 3.0%; and cleaning and disposal 1.0%. There was no dermal contamination during the reentry periods on days 1 and 7.

The distribution of methyl parathion contamination over different regions of the body was assessed by comparing the various cloth sections and patches for the "all activities" group and the R2 group (Fig. 2). Hands had the highest potential for dermal contamination and penetration. The high degree of contamination of the left hand of those in the "all activities" group was due to the farmers' practice of applying the pesticide with their right hand, while the left held the cover of the tank sprayers used for measuring the pesticide.

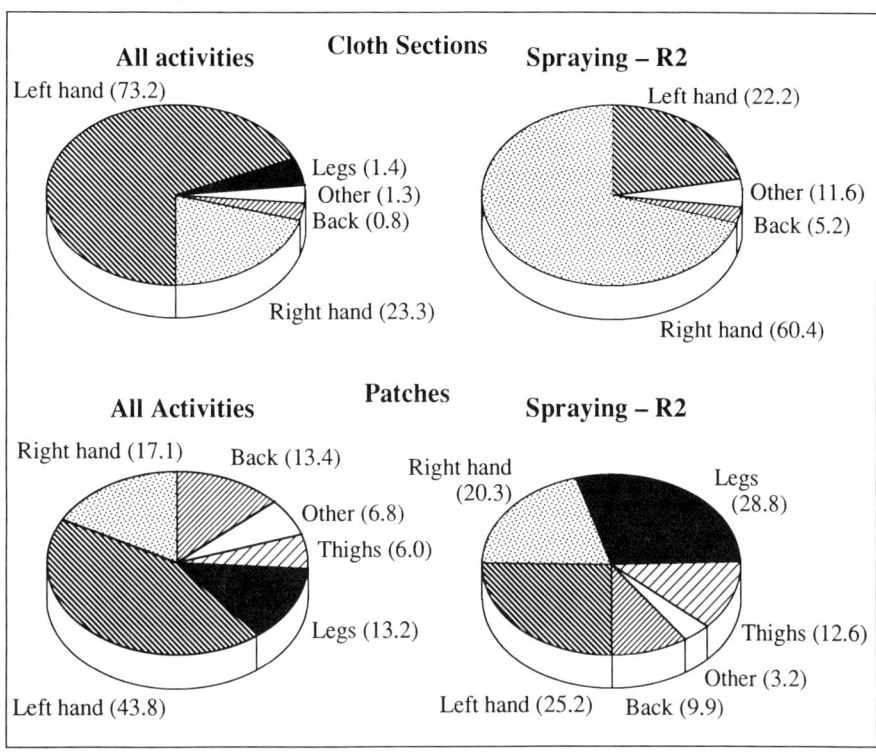

Fig. 2. Distribution of potential (cloth sections) and actual (patches) dermal contamination in "all activities" and R2 spraying groups.

During spraying, the level of contamination of the hands varied. The farmers practiced a swinging technique of spraying and shifted the sprayer extension pipes from the left to the right hand and vice versa. For those who cleaned the sprayer tanks, the right hand was the most contaminated, because it was used most to accomplish the task. There was no dermal contamination in the reentry groups.

Other areas of contamination included the back and lower leg. Exposure of the back may have resulted from leaking knapsack sprayers, which were poorly maintained, and from spillage during mounting. Contamination in the lower leg during spraying and mixing and loading activities may be due to wading in the treated areas during spraying operations.

Study two showed higher mean methyl parathion residues in the patches of farmers who wore no protective clothing than for those who wore protective clothing. Statistically significant differences were found in the level of residues in patches on the left back, right and left lower legs, and left hand between those who wore protective clothing and those who were unprotected.

Problems encountered with the locally designed protective clothing included a sensation of warmth and the plastic material at the sides of the vests and the bags covering the hands was easily torn.

There was no statistically significant difference in cholinesterase levels in red blood cells before and after exposure to methyl parathion among farmers with and without protective clothing. Mean urinary p-nitrophenol levels were not significantly different between exposure groups after spraying.

In the two methyl parathion studies, no variability in meteorological parameters and no clinical symptoms of organophosphate poisoning were observed in any of the participants.

Study three, using monocrotophos, indicated that hands were exposed during mixing and loading operations, particularly those of farmers not using protective measures. High exposure of uncovered hands and face to spray mists was observed during spraying activities. However, no acute adverse health effects were observed during application of monocrotophos. Cholinesterase activity in whole blood or red blood cells was not significantly affected, but in all subjects plasma cholinesterase level decreased by 40%.

Urinalysis for dimethyl phosphate confirmed absorption in all subjects, regardless of the use of protective measures. In this study, protective measures were mainly used during mixing and loading activities. However, because mixing–loading and spraying activities were carried out by the same person, no definite conclusions can be drawn regarding the effects of the protective measures.

Conclusions

These studies demonstrated that, in the course of a normal spraying operations, Filipino farmers are exposed to contamination of their clothing and potential dermal absorption by organophosphate pesticides. However, no adverse health effects were noted.

The area of the body with highest potential for dermal exposure and penetration is the hands. Mixing and loading activities had the highest potential for dermal contamination; undertaking all the activities was associated with the highest potential for dermal penetration.

The protective clothing studies had different results. In the methyl parathion study, pesticide-residue levels in the group who used protective clothing were lower than those who were not protected. In the monocrotophos study, no conclusion could be drawn. For both pesticides, there was no change in cholinesterase levels in red blood cells after exposure.

Acknowledgment — The three studies reported here were funded by the International Development Research Centre, the Philippines Council for Health, Research and Development, and the Shell Company.

Aldridge, W.N.; Davies, D.R. 1952. Determination of cholinesterase activity in human blood. British Medical Journal, 1952 (May–June), 945–947.

Castaneda, C.P.; Maramba, N.C.; Ordas, A.V. 1989. A field exposure study to methyl parathion among selected rice farmers. Acta Medica Philippina, 25(3), 71–77.

Kaye, S. 1980. Handbook of emergency toxicology. Charles Thomas, Springfield, IL, USA. Pp. 433–442.

Rola, A.C. n.d. Pesticides, health risks and farming productivity: a Philippine experience. Final report (Assessing the benefits and risks of pesticide use in Philippine agriculture) submitted to the Agricultural Policy Research Program, University of the Philippines at Los Baños. Philippines. 122 pp.

van Sitter, N.J.; Castaneda, C.P.; Dumas, E.P. 1989. Field study of exposure and health effects following knapsack application of an azodrin formulation to rice in the Philippines. Report HSE 89.008.

WHO (World Health Organization). 1982. Field surveys of exposure to pesticides. WHO, Geneva, Switzerland. Standard Protocol VBC/82.1.

Use and control of pesticides in Pakistan

J. Liesivuori

Regional Institute of Occupational Health,
Kuopio, Finland

With the development of the textile industry, consumption of pesticides has increased markedly in Pakistan: 90% of all pesticides are applied to cotton fields. Many compounds banned in other countries are still in use in Pakistan. Several epidemic pesticide poisonings have occurred and increased organochlorine pesticide levels have been found among Pakistani mothers and infants. Although regulations and provisions for the safety of workers handling pesticides have been established, there are no arrangements for the implementation of these rules. Despite the economic disincentive, it is recommended that the pesticide load in the general population and the environment be evaluated and the present regulations for the registration of pesticides be implemented. The safe use of pesticides and monitoring workers' exposure should be encouraged.

Pesticides constitute a potent health risk, especially in developing countries where working conditions are often poor and the educational level of the population low. Pakistan, in this respect, is typical. Although it is taking the first steps toward promoting health and safety in the workplace, only 26% of the adult population is literate, and no institutional system for evaluating the toxicity of pesticides exists. Many companies in Pakistan promote the use of pesticides through television and newspaper advertisements without effective instructions on safe handling and application.

Pesticide regulations

In Pakistan, the most important regulations governing the use of pesticides are the 1971 *Agricultural Pesticides Ordinance* and the 1973 *Agricultural Pesticides Rules*. Regulations for pesticides were also included as a part of the environmental protection programs of the *Pakistan Environmental Protection Ordinance* in 1983.

Since the creation of Pakistan in 1947, procurement and distribution of pesticides has been the responsibility of the federal government. Initially, pesticides were supplied to farmers free of charge as most were received as aid from abroad. In 1980, the management of pesticides was transferred to the private sector. The Ministry of Agriculture, Food and Cooperatives, through the Agricultural Pesticides Technical Advisory Committee (consisting of members drawn from various relevant federal and provincial agencies), however, is responsible for the implementation of rules. There are also detailed regulations on storage and labeling of pesticides (Baloch 1985).

Although provisions for the safety of workers handling pesticides were passed, no arrangements have been made to monitor the implementation of these provisions. A special Pest Warning and Quality Control of Pesticides Directorate has been established to control the market and offer guidance to farmers.

Pesticides in use

Pakistan began using pesticides in agriculture in 1954, with an initial consumption of 254 t. The quantity rose to 16 226 t of formulated pesticides in 1976 and 25 853 t in 1981. During the last decade, production plants for various formulations have been set up to use imported active ingredients. These facilities are capable of preparing up to 26 000 t of granules, 33 500 t of dusts and wettable powders (WP), and 25 million L of liquid emulsifiable concentrates (EC) annually (Baloch 1985).

Statistics collected by the Department of Agriculture in the province of Punjab, the most important area in agricultural production, indicate a huge increase in the use of pesticides: 641 t of active ingredient was used in 1980 and 3 543 t in 1988. About 90% of this is applied to cotton fields and 5% to rice fields. A total of 112 pesticide products are standardized and imported by 42 registered distributors in Punjab, including:

- Chlorinated hydrocarbons, e.g., dichlorodiphenyltrichloroethane (DDT), lindane, dieldrin, heptachlor, endosulfan;

- Organophosphorous compounds, e.g., parathion, methyl parathion, malathion, diazinon;

- Carbamates, e.g., carbaryl;

- Dithiocarbamates, e.g., mankotseb; and

- Pyrethroids, e.g., permethrin, fenvalerate, cypermethrin.

Although the consumption of pyrethroids has increased during the last few years, older, more persistent and toxic compounds, such as DDT and methyl bromide, are still in use.

The health-care system

The provision of health services in Pakistan has progressed faster than education, but there is still considerable room for improvement. Indicators available for recent years on health infrastructure and expenditure provide evidence of this. At present, less than 1% of the gross national product is spent on health. In 1985–1986, only 4.5% of total public spending was allocated to the health sector. The poor health situation in Pakistan is reflected in its high rates of infant mortality (95 per 1 000) and low life expectancy (61 years) compared with other countries (UNICEF Country Classification — Pakistan Basic Data from 1988 and earlier years).

Today, there is one public physician for every 3 400 people, one public nurse for 7.4 hospital beds or for 13 100 people, one public hospital for 150 000 people, and one hospital bed for 1 760 people. Primary health-care facilities other than hospitals are offered by 2 468 basic-health units, 455 rural-health centres, 869 maternal- and child-health centres, and 3 994 dispensaries (Government of Pakistan 1987). Along with public-health facilities, some health facilities are also made available by the private sector. However, these are expensive and beyond the reach of most of the population. Despite the official statistics, many physician posts remain vacant in the countryside. For these reasons, physicians seldom have the opportunity to treat farm workers and their experience with pesticide poisonings is often limited.

Outbreaks of pesticide poisoning in Pakistan

Several outbreaks of pesticide poisoning have been reported (Table 1). Peasants driven by starvation to eat treated grain have been the victims in most of these poisonings. In one instance, 34 people were affected and four died after eating mercury-treated grain (Hag 1963). In 1976, an outbreak of organophosphate insecticide poisoning from malathion occurred among 7 500 fieldworkers in the malaria control program. It is estimated to have affected 2 800 workers and resulted in at least 5 deaths.

Subsequent field studies have associated low red-blood cell cholinesterase activity with the signs and symptoms of organophosphate intoxication. The greatest toxicity was found with products containing high amounts of isomalathion, a toxic product of malathion degradation. When malathion and fenitrothion poisoning was diagnosed in exposed Pakistani workers, cholinesterase activity was found to have decreased by more than 50% (Miller and Shah 1982). This exposure was attributed to poor work practices, which had developed when DDT was the primary insecticide used for malaria control. These practices resulted in excessive skin contact and percutaneous absorption of the organophosphate pesticides (Baker et al. 1978). Evidence of the widespread effect of this has been found in the increased organochlorine

Table 1. Incidents of pesticide poisoning in Pakistan.

Year	Pesticide involved	Work or material contaminated	Number of cases	Number of deaths	Reference
1961	Organic mercury	Seed-grain	100	4	Haq 1963
1969	Methylmercury	Seed-grain	100		Elhassani 1982
1972	Phosphate	Unloading a truck		7	Baloch 1985
1976	Malathion Isomalathion	Application, mixing	2 800	5	Baker et al. 1978
1980	Organophosphate	Loading, mixing	200[a]		Baloch 1982
1982	Malathion Fenitrothion	Application	68[b]		Miller and Shah 1982

[a] Most of the workers had low cholinesterase activity.
[b] One-third of workers had cholinesterase activity below 62.5% of normal.

pesticide residues in the tissue of Pakistani mothers and their infants (Mughal and Rahman 1973; Skaare et al. 1988).

Because funds are lacking, few laboratories in Pakistan are equipped to conduct pesticide studies. However, hepatotoxicity of dieldrin has been studied in rats (Shakoori et al. 1982) and the genotoxicity of certain pesticides used in Pakistan has also been assessed (Sandhu et al. 1985). No laboratories are able to conduct studies on occupational exposure to pesticides.

Recommendations

Although some institutes in Pakistan are conducting research on pesticides, the need to evaluate pesticide load in the general population and the environment is urgent. Occupational exposure to pesticides must be determined and training in toxicologic evaluation must be provided. The registration system should be improved and the sale and import of pesticides should be controlled. The safe use of pesticides can be promoted by more effective training of those who handle and apply them. Doctors and other health personnel should also be trained in the treatment of poisonings. Monitoring systems for worker exposure should be established.

Acknowledgments — I thank Dr M.H. Akhtar, retired from the Malaria Control Programme, Lahore, and Tarig Sultan Pasha, industrial hygienist from the Centre for the Improvement of Working Conditions and Environment, Lahore, Pakistan.

Baker, E.L.; Zack, M.; Miles, J.W.; Alderman, L.; Warren, M.C.; Dobbin, R.D.; Miller, S.; Teeters, W.R. 1978. An epidemic of malathion poisoning in Pakistan malaria workers. Lancet, 1(8 054), 31–34.

Baloch, U.K. 1985. Problems associated with the use of chemicals by agricultural workers. *In* Muhammed, A.; von Borstel, R.C., ed., Basic and applied mutagenesis. Plenum Press, New York, NY, USA. Pp. 63–78.

Elhassani, S.B. 1982. The many faces of methylmercury poisoning. Journal of Clinical Toxicology, 19(8), 875–906.

Hag, I.U. 1963. Agrosan poisoning in man. British Medical Journal, 1, 1 579–1 582.

Miller, S.; Shah, M.A. 1982. Cholinesterase activities of workers exposed to organophosphorus insecticides in Pakistan and Haiti and evaluation of the tintometric method. Journal of Environmental Science and Health, B17, 125–142.

Mughal, H.A.; Rahman, R.A. 1973. Organochlorine pesticide content of human adipose tissue in Karachi. Archives of Environmental Health, 27, 396–398.

Pakistan, Government of. 1987. Economic review in agriculture: economic survey 1985–86. Government of Pakistan, Islamabad, Pakistan.

Sandu, S.S.; Waters, M.D.; Simmon, V.F.; Mortelmans, K.E.; Mitchell, A.D.; Jorgenson, T.; Jones, D.C.L.; Valencia, R.; Stack, F. 1985. Evaluation of the genotoxic potential of certain pesticides used in Pakistan. *In* Muhammed, A.; von Borstel, R.C., ed., Basic and applied mutagenesis. Plenum Press, New York, NY, USA. Pp. 185–219.

Shakoori, A.R.; Rasul, Y.G.; Ali, S.S. 1982. Effect of dieldrin feeding for six month old albino rats — biochemical and histological changes in liver. Pakistan Journal of Zoology, 14, 191–204.

Skaare, J.U.; Tuveng, J.M.; Sande, H.A. 1988. Organochlorine pesticides and polychlorinated biphenyls in maternal adipose tissue, blood, milk, and cord blood from mothers and their infants living in Norway. Archives of Environmental and Contamination Toxicology, 17, 55–63.

Pesticide poisoning among agricultural workers in Bolivia

G. Condarco Aguilar, H. Medina, J. Chinchilla,
N. Veneros, M. Aguilar, and F. Carranza

Instituto Nacional de Salud Ocupacional,
Ministry of Public Health and Social Security,
La Paz, Bolivia

The division of Bolivia into altiplano, valleys, and lowlands, with many intermediate subregions, favours the development of different agricultural practices and the use of different agricultural chemicals. Both traditional methods, using hand tools for personal food production, and modern agroindustrial methods, primarily in the lowlands and parts of the valley regions, are used in agriculture. This report presents the results of a study of the knowledge, attitudes, and practices (KAP) of workers in the altiplano, valley, and lowland regions of Bolivia. An assessment of the health status of workers exposed to organophosphate insecticides was also carried out. The lowland areas and the mesothermal valleys were found to be the regions with the greatest problems. The factors contributing to this situation are reviewed.

The well-known division of Bolivia into altiplano, valleys, and lowlands, with many intermediate subregions, has resulted in the development of various agricultural products for domestic or commercial use. Two main types of production are carried out: traditional, rudimentary, cultivation using manual tools to produce food for personal consumption (in the altiplano and parts of the valleys); and modern agroindustry (in the lowlands and parts of the valleys). With the economic crisis, which is not limited to Bolivia, agriculture is becoming the hope for improving social and economic conditions. This suggests that crop production should be increased, which is likely to lead to greater use of pesticides.

From the earliest times, farmers have fought pests that, at times, destroy 30% to 50% of the crops. Between 1930 and 1940, new stronger insecticides were introduced, including dichlorodiphenyltrichloroethane (DDT), toxaphene, and chlordane. Pest resistance to these chemicals has resulted in a search for more toxic products having less environmental permanence, such as the organophosphates and the pyrethroids.

Although pesticides are designed to eliminate undesirable and harmful organisms that damage crops, other living beings, including humans, are susceptible to them to a varying degree, so that their use can constitute a public-health problem. In 1973, the World Health Organization (WHO) estimated that there were over 500 000 pesticide poisonings per year in the world, with a death rate of 1%. In developing countries, it is estimated that 375 000 severe cases occur annually, with 10 000 deaths; this corresponds to about one case per minute and one death per hour (Henao and Carey 1986). Some examples are:

- In Chinguira (Colombia), 600 cases of poisoning by parathion and 88 deaths were reported (Vargara 1984);

- In Brazil, out of 5 000 workers examined, 900 showed reduced cholinesterase levels (Zanaga et al. 1984); and

- Also in Brazil, of 1 107 workers examined, 173 suffered from poisoning, 41.6% due to organophosphates (Zanaga et al. 1984).

The Instituto Nacional de Salud Ocupacional (INSO) began taking action in this field in 1976. In 1978, 11 farms were visited; of the 237 workers seen, 62% showed some level of poisoning, 6.7% of the cases were children. In the valleys of Tarija in 1977, 43% of those examined showed signs and symptoms of poisoning. In the Rio Abajo valleys, 14 deaths in 1977 were due to pesticide poisoning. During the period 1975 to 1985, the hospital serving the mesothermal valleys of Comarapa treated 47 pesticide poisonings; of these 15% died and, of the total, 45% were children (INSO 1987).

Other studies clearly demonstrate the extent of this problem in Bolivia (Condarco and Medina 1986a,b; INSO 1986; Malgarejo 1987). In the mesothermal valleys where tomatoes are cultivated, 25% of people examined had reduced levels of cholinesterase activity, 16% of them were under 18 years of age. Most (94%) agricultural workers had no knowledge about the safe use of pesticides.

Our concern was to determine the real incidence of pesticide poisoning in the three ecological areas and what factors affect this problem, and to find suitable alternatives for pest control in the Bolivian environment. Specifically, this study examined the extent of use of organophosphates and their effect on the health of the agricultural worker and on the ecology. The results of this project will provide a basis for future studies in this field, through the health-care network of the Ministry of Health.

Methods

The study was carried out in the four major ecological regions of Bolivia (altiplano, valleys, mesothermal valleys, and eastern lowlands). In total, 731 workers were studied: 235 from the altiplano, 247 from the valleys, 57 from the mesothermal valleys, and 197 from the lowlands.

The method used for choosing a sample of workers was identical in each ecological region. Two towns or centres were chosen from which workers would start out each day to work in the communities selected for the study. On the day before a community was to be studied, the commission met with its members and political leaders, parish priests, teachers, property owners, etc., to explain the reasons for the work and to request their collaboration.

The evaluation was initiated with an educational talk and distribution of pamphlets on the use and handling of pesticides and on how to prevent their harmful effects for all interested people, including women, children, and the elderly. This was followed by a medical evaluation and blood tests. The Edson method (Lovibond equipment) recommended for field studies by WHO was used to assess the level of exposure to organophosphates. We considered cholinesterase level in the blood to be normal when it was 75–100% of the mean value in the general population; values of 50–62.5% indicated mild poisoning, 25–50% severe poisoning, and less than 25% extremely severe poisoning.

In all regions, numerous attempts were made to take blood samples for the assessment of cholinesterase level before pesticide application as recommended by the WHO protocols. However, this proved impossible in the altiplano, the valleys, and the mesothermal valleys, because of the workers' resistance to having biological samples taken more than once and the relative difficulty of reaching the workers before spraying started. This paper, thus, only reports the cholinesterase study carried out in the tropical lowlands. The knowledge, attitudes, and practices (KAP) study, however, was carried out in all regions.

Results

In the four ecological regions, 66.5% of the people in the samples were agricultural workers, who handled pesticides or were directly exposed (DE); 19.3% did not handle pesticides, but were indirectly exposed (IE); and 14.2% did not use pesticides and were not exposed to them (NE), the latter group being the most difficult to find, because our rural workers tend to use pesticides for domestic purposes as well.

Of all workers in the samples, 11.2% were under 18 years of age, which explains the high known incidence of poisoning in the 12- to 18-year-old age group. In the mesothermal valleys, altiplano, and valleys the proportion of young workers was 15.8%, 6.8%, and 4.3%, respectively. These figures corroborate previous studies, where children doing farm work were exposed to risks and working stress unsuitable for their age. Although only 2.5% of workers were over 55 years of age in the lowlands, in the mesothermal valleys, valleys, and altiplano, this group accounted for 8.8%, 16.5%, and 12.8%, respectively. These differences are because work in the lowlands is more difficult, because of the extent of land cultivated, and because of the adverse climatic conditions.

The level of eduction encountered was low: only 48% of the agricultural workers in the tropical lowlands had completed primary school. There were 12 university professionals who were not involved in community work in the field of pesticides.

Migration from the countryside to the city, which occurs in all Latin American countries, has been increasing in recent years because of drought. At the same time, seasonal migration of agricultural workers from dry arid zones toward more productive areas, such as the lowlands and mesothermal valleys, during harvesting is becoming a regular occurrence. Of the agricultural workers studied in the lowlands, 51.8% originally came from other areas. In the mesothermal valleys, valleys, and altiplano, 38.6%, 17.4%, and 7.6%, respectively, had come from other parts. This high percentage of migrants in the eastern portion of Bolivia will be subject to other environmental risks, different types of dwellings, food, customs, climatic factors, etc., which may reduce their resistance and make them more susceptible to pesticide poisoning.

Previous poisonings

In the lowlands, mesothermal valleys, valleys, and altiplano, 9.6%, 10.0%, 4.5%, and 2.1% of the participants mentioned having suffered previous occupational poisoning, confirming that the highest incidence occurs in the mesothermal valleys and lowlands. The low figure obtained in the altiplano and valleys suggest that the information gathered in previous years (INSO 1987; Malgarejo 1987) referred to nonoccupational poisoning.

Knowledge and methods of prevention

In the lowlands, 81.2% of those studied were not aware of the dangers of poisoning associated with pesticide handling (Table 1). Looking at the differential distribution, 79.4% of the directly exposed group were not aware of this risk, 78.9% of those indirectly exposed and 92.9% of those not exposed, which demonstrates that agrochemicals are considered harmless by most agricultural workers. Even more serious, they are unaware of the risk they represent to their family, plants, domestic animals, and their immediate environment. This lack of knowledge is greater in the mesothermal valleys, valleys, and altiplano.

Table 1. Factors likely to affect the occurrence of pesticide poisoning in the three exposure groups from the lowlands (number of positive responses with percentage in parentheses).

Factor	DE ($n = 131$)	IE ($n = 38$)	NE ($n = 28$)
Aware of danger of poisoning	27 (20.6)	8 (21.1)	2 (7.1)
Took a course	22 (16.8)	7 (18.4)	2 (7.1)
Changes clothing	34 (26.0)	5 (13.2)	—
Eats at work	118 (90.1)	20 (52.6)	4 (14.3)
Washes hands before eating	120 (91.6)	21 (55.3)	5 (17.9)

Note: DE, directly exposed group; IE, indirectly exposed; NE, not exposed.

Table 2. Preventive measures of workers (number and %) directly exposed to pesticides in the four study regions.

		Preventive measure	
	n	Use personal protection during spraying	Bathe after spraying
Altiplano	165	11 (6.7)	20 (12.1)
Valleys	186	16 (8.6)	55 (29.6)
Mesothermal valleys	52	1 (1.9)	30 (57.7)
Lowlands	131	41 (31.3)	121 (92.4)

This ignorance is due to a lack of guidance and training; only 15.7% of all workers reported having taken a course or had training in pesticide use. The figures are alarming, particularly in the lowlands and mesothermal valleys, and show how susceptible our agricultural population is to risk. In the lowlands, 74% of the workers did not change their clothing after spraying, even though this is a necessity in that climate. In the mesothermal valleys, valleys, and altiplano the figures were 38.5%, 64.5%, and 77.0%. Likewise, the habit of serving food in the place of work was common: 90.0% for those directly exposed in the lowlands, and 65.4%, 10.8%, and 15.6%, respectively, in the mesothermal valleys, valleys, and altiplano.

This lack of knowledge encountered in the group of directly exposed workers has negative repercussions (Table 2). Of those studied, 68.7% in the lowlands did not use any method of personal protection when using agrochemicals, but 92.4% did bathe after spraying, probably because of the climate. In the mesothermal valleys, valleys, and altiplanos, 98.1%, 91.4%, and 93.3% did not take any measures for personal protection, and 57.7%, 29.6%, and 12.1% bathed after spraying. The latter information is not considered very reliable for the altiplano and may represent overreporting, given the water shortages and infrastructure of the workplaces, as well as the cold temperature (close to freezing or below).

Storage of pesticides

Although the storage of pesticides is subject to international standards as in other countries, our study shows that in Bolivia this is not adhered to. In the lowlands, 80.9% of directly exposed workers said that they stored pesticides in special storage areas for this purpose; however, in the mesothermal valleys, valleys, and altiplano, 80.8%, 79.0%, and 63.6% stated that they store them at home. Those storing pesticides at home greatly increase the risk of poisoning, not only for the agricultural workers but also for their families and animals. This may explain the high incidence of poisoning found in the altiplano and valleys (INSO 1987; Malgarejo 1987).

Mixing pesticides

Pesticides must be mixed in precise quantities and applied a specified number of times to eliminate pests. Otherwise, not only will the pest be only partially eliminated, but strains will begin to develop that are resistant to the pesticide.

In the lowlands, the preparation work (where *pongueaje*, the system of exploitation dating from the colonial times still exists) is generally carried out by supervisors, landowners, or experts; only 29.0% of the workers said that they mix their own pesticides. However, in the mesothermal valleys, valleys, and altiplanos, 90.4%, 78.0%, and 60.6% of the directly exposed farmers mix their own. It was further discovered that agricultural workers were mixing up to five or six products in the same container and testing the concentration by tasting it! Spraying is also often done just before harvest, so products taken to the markets for sale to consumers are contaminated by pesticides.

Cholinesterase levels in the exposed population

The level of poisoning in a study population is usually obtained after three tests of cholinesterase level before exposure to organophosphates establish the normal level; an examination after exposure will give the level of absorbtion of the toxin. This method is not always possible in countries such as ours, where there are no established pesticide-spraying programs, and where the health-care system does not reach the agricultural population. Great efforts were taken in this study to obtain samples, sometimes by covering over 200 km/day over almost impassable routes and roads.

Of the DE agricultural workers in the lowlands, 10.7% showed levels of blood cholinesterase activity below the set threshold of 75%; 0.8% were at the lower limit of severe poisoning. Of the IE workers, only 5.3% had reduced levels of cholinesterase activity and, in the NE group, 100% of the workers had normal levels of activity.

In the valleys, in a sample of 24 NE workers who work in roofed nurseries in the commercial cultivation of roses, 16.7% showed reduced cholinesterase activity (< 62.5%), which is of concern, as these women are of reproductive age and could suffer chronic damage that could be passed on to their offspring.

Of the agricultural workers with reduced cholinesterase activity, 15.4% were between 12 and 18 years of age.

Cholinesterase and spraying time

Of the DE workers in the lowlands, 24.4% have only been using pesticides for 1 year, 49.6% for 2–5 years, 15.3% for 6–9 years, and 10.7% have been using pesticides for over 10 years. In this same group, 46.2% of those with reduced cholinesterase levels have been using pesticides for 1 year, 38.5% for 2–5 years, and 7.7% for 6–9 years. This information confirms that the agricultural workers most affected are those who have been using pesticides for 1–5 years, who have little work experience, and who, for the most part, have come from other areas.

DE workers who had sprayed pesticides within the last 14 days were tested for cholinesterase activity. Of those exposed on the day of the test, 13.3% showed reduced levels of activity (62.5% of normal level); 7.4% of those who had used pesticides within the last 2–7 days were similarly affected. This high proportion of workers showing the biological effects of organophosphate poisoning dramatically illustrates the severity of this problem to public health in the eastern lowlands of Bolivia.

Symptoms and signs

The main symptoms found by order of importance are: severe headaches, abdominal colic, and palpitations; no significant differences were noted among the three groups studied, and the symptoms were not significantly correlated to reduced level of cholinesterase activity. In general, the signs were negative, demonstrating that these variables in this type of study are less important and may be the result of the exhausting work typical in agriculture and of the harshness of the climate.

Basic health and hygiene

In general, the dwellings of agricultural workers in the lowlands are improvised (pawichis), and do not provide any of the basic conditions for well-being.

During our travels, we encountered 12 health-care posts (two in the lowlands) without professional staff or a medical kit to treat cases of poisoning. Sanitation and hygiene are lacking. Pit latrines were used on 46% of the properties; however, most labourers do not use them, but prefer the open field. Water for domestic use is obtained from wells or nearby rivers. Leftover pesticides are disposed of, by washing the containers or tanks, in the same water.

Discussion

The lowland regions and mesothermal valleys are the areas most susceptible to risk of pesticide poisoning. In more-developed countries, such as Brazil, where the use of pesticides is for the most part controlled, the incidence of poisoning is 7.9% (Zanaga et al. 1984). The values we encountered in the tropical lowlands (10.7% of DE workers, but, more significantly, 5.3% of IE workers) are significant and indicate to us that measures must be taken to prevent higher levels of poisoning. In the four regions studied, there is a high degree of ignorance of risks and of appropriate handling of pesticides because training directed at rural workers is almost completely lacking. These factors must be considered in any study related to agriculture, so that hygiene, use of personal protection measures, preparation of pesticides without technical assistance, storage of pesticides, prevention of environmental pollution, and other factors related to poisoning, become known. Only in this way can the harmful effects be avoided in the future for agricultural workers, their families, and the surrounding environment.

Likewise, we should not neglect health care and the need to equip health-care posts to allow them to assist in emergency cases of poisoning. Further epidemiological studies must also be carried out under the Ministry of Health in the most affected areas, including greenhouses and nurseries in the altiplano and valleys.

There should be stricter controls on the sale of prohibited products. Studies must be planned for detecting cases of human poisoning and of environmental pollution due to organochlorine pesticides, which are being extensively used primarily in the northern altiplano.

As a result of our study, this project is being expanded to include: a seminar to review the existing laws on pesticides in Bolivia; and a course for educators and trainers, to be selected from employees in the Ministry of Health, teachers, companies selling pesticides, and other organizations involved in this issue. Courses will be implemented in the university, and epidemiologic follow-up will be done on cases of poisoning encountered through the health-care network of the Ministry of Health.

Acknowledgment — This research was carried out with the assistance of a grant from the International Development Research Centre, Ottawa, Canada.

Condarco, G.; Medina, H. 1986a. Control integrada de plagas — Toxicological report on workers exposed to pesticides in the region of Comarapa. Instituto Nacional de Salud Ocupacional, La Paz, Bolivia.

Condarco, G.; Medina, H. 1986b. Control de plagas integrado — Toxicological report on pesticide sales companies in the cities of Cochabamba and Santa Cruz. Instituto Nacional de Salud Ocupacional, La Paz, Bolivia.

Henao, S.; Carey, G. 1986. Plaguicidas organo-fosforados y carbamatos. Pan American Health Organization and World Health Organization, Mexico City, Mexico. Serie Vigilancia Z.

INSO (Instituto Nacional de Salud Ocupacional). 1986. History of the use of pesticides in Bolivia. Report submitted to the United States Agency for International Development, La Paz, Bolivia.

_____ 1987. Memorias bodas de plata (1962–1987). INSO, La Paz, Bolivia.

Malgarejo, V.H. 1987. Informe sobre la problematica regional de Cochabamba sobre uso de plaguicidas. Instituto Nacional de Salud Ocupacional, La Paz, Bolivia.

Vargara R., R. 1984. Use and abuse of pesticides in Bocaya: summary of the conference held for the 8th national meeting of RAPAL, June 1984, Palmeira, Colombia.

Zanaga, A.; et al. 1984. Proyecto de vigilancia epidemiológica en toxicología de los pesticidas (FUNDACENTRO). Revista Brasileira de Salud Ocupacional, 2 (47).

Pulmonary obstructive disease in a population using paraquat in Colombia

M.E. Arroyave

Instituto Nacional de Salud, Bogotá, Colombia

Paraquat, a herbicide of the bipyridil group, has been used in Colombia with increasing frequency since its registration in 1969. This study aimed to discern the effects of paraquat on agricultural workers. A questionnaire and a medical examination (including lung-function tests) were used to determine exposure. The results indicate that there is a positive association between the use of paraquat (at high levels of exposure), smoking habits, and chronic obstructive pulmonary disease.

Paraquat (1,1'-dimethyl-4,4'-bipyridylium dichloride) has been used as a broad-spectrum herbicide for nearly 30 years. Since its registration in Colombia in 1969, paraquat has been used principally for the control of weeds in cotton, sugarcane, potato, corn, and tomato cultivation. It is also used as a desiccator to facilitate the harvesting of rice, sugarcane, soybeans, corn, and sorghum crops and as a defoliant in the cultivation of bananas, cacao, oil palms, citrus, and other fruits. In nonagricultural areas, paraquat is used as a nonspecific contact herbicide.

Paraquat has been classified by the World Health Organization (WHO) as a class II (moderately hazardous) compound. In experimental animals, the systemic administration of this substance results in progressively degenerative and potentially lethal lesions in the lungs. In addition to degenerative lung fibrosis, observed effects in humans range from dermal toxicity and irritation of the skin and mucous membranes (Hearn and Keir 1971) to corrosive gastrointestinal effects, renal tubular damage, and liver dysfunction (Grant et al. 1980). At present, there is no antidote for paraquat poisoning and treatment is perforce symptomatic.

Paraquat is not destroyed by the biological processes of plants or mammals. In mammals, absorbed paraquat is largely eliminated unchanged by the kidneys because of its low molecular weight and nonvolatility. However,

because of its cationic nature, most of the compound filtered by the kidneys is reabsorbed by the body before it can be eliminated in the urine. This high rate of retention results in toxic effects for unusually low doses of paraquat. The renal tubular damage caused by the compound's redox cycling properties rapidly renders the kidney incapable of removing it from the blood, thereby increasing the harmful effects on bodily functions.

Paraquat is selectively toxic to the lungs, where it is found to accumulate (Levin et al. 1979). Humans are the most susceptible species to paraquat-induced lung damage, which is manifested as a loss of lung capacity and function. The accompanying liver impairment adds to a general loss of vitality. Brain damage has also been observed after paraquat intoxication.

The marked increase in the use of paraquat in Colombia (560 kL in 1980, 698 kL in 1981, 1 052 kL in 1982, and 2 483 kL in 1987) prompted the government to investigate its effects on the health of people working with the substance or indirectly exposed to it in the environment. It was expected that those most exposed would be found to suffer a pulmonary fibrosis type of pneumopathy (a condition in which the elastic tissue of the lungs is damaged and replaced by fibrotic, nonelastic tissue), as well as restricted pulmonary function leading to impaired respiration. After a review of regional data on paraquat consumption, target crops, and dosages, the municipality of Carmen de Viboral in the district of Antioquia was selected for study.

Antioquia is an agricultural region where paraquat use is atypically high, representing 10% of the national total. The working population of Carmen de Viboral had allegedly been trained in the use of paraquat by local manufacturers and the Secretariat of Agriculture of Antioquia. However, investigation revealed that only the heads of farms had received training even though an average of nine other people per farm were involved in plantation work using paraquat. Information obtained from the regional health service indicated a high prevalence of paraquat-related health problems in Antioquia, and these figures could be extrapolated to other agricultural regions of the country with a fair degree of accuracy.

Carmen de Viboral is located in the eastern part of Antioquia and covers an area of 448 km^2. Of its 24 657 inhabitants, 43% live in the municipality while 57% form the rural population. The entire municipal agricultural area is devoted to potato, bean, and corn cultivation. The crop area constitutes 32% of the township land; the rest is used for grazing or forestry. In contrast to other areas, where itinerant workers spray the crops, the agricultural workers in Carmen de Viboral are a stationary population.

The purpose of this study was to assess the frequency and severity of the pathologies associated with unrestricted use of the herbicide paraquat. The specific objectives were:

- To determine the prevalence of respiratory pathologies in the population of Carmen de Viboral;

86

- To investigate the presence and measure the concentration of paraquat in the blood and urine of the population;

- To assess the possible and actual levels of exposure in the population by measuring the concentration of paraquat in the air during spraying; and

- To establish an association between different exposure levels and the physiopathological manifestations observed in the population.

Materials and methods

The study population consisted of 21 932 inhabitants of the rural township of Carmen de Viboral. Due to geographical and socioeconomic differences, 2 725 inhabitants of this area were not included in the sampling.

Census and other information were obtained from the Ministry of Health, together with maps and plans from the Department of National Statistics Administration. All data were verified by the investigators. One in every four houses was chosen for the survey. The sample of 941 homes consisted of 5 483 people, all of whom were interviewed. For medical examination, 896 people (a random subsample of about 20% of the original population) aged 7 years and older (except women more than 3 months pregnant) were selected. For the pulmonary function study, this sample was further reduced to 485 by including only men between the ages of 15 and 70 years.

Between April and July 1986, a questionnaire was administered to obtain information on sociodemographic status, occupational or other exposure to paraquat, and general health (particularly regarding the respiratory system). Variables likely to influence the effect of paraquat, such as smoking habits and previous exposure to other lung risk factors, were also determined. In consultation with medical personnel at the San Juan de Dios University Hospital in Bogotá and on the basis of current knowledge of the effects of paraquat on humans, the American Thoracic Society's 1978 revision of the Medical Research Council's questionnaire for respiratory function was adopted for this study. The questionnaire was pretested in a pilot study and refined accordingly.

All 896 subjects in the subsample underwent a general medical examination during this same period. Lung function was evaluated by spirometry. Thoracic x-ray examination was not performed for ethical reasons. According to the values for the components of the test, lung function was classified as follows:

- Restrictive = forced vital capacity (FVC) $\leq 80\%$ of the predicted value; forced expiratory volume per second (FEV$_{1.0}$) divided by FVC $\geq 75\%$; and residual volume (RV) or total lung capacity (TLC) or both $\leq 120\%$ of the predicted value;

- Obstructive = FVC > 80% of the predicted value; $FEV_{1.0}/FVC < 75\%$; and RV > 80% of the predicted value; and

- Normal = FVC ≥ 80% of the predicted value; $FEV_{1.0}/FVC ≥ 75\%$; RV 80–120% of the predicted value; and TLC 80–120% of the predicted value.

Based on an exposure-to-paraquat score obtained from the questionnaire, blood and urine samples were collected during December 1986 and March 1987 from 603 paraquat users. The concentration of paraquat in these samples was then determined.

During the paraquat spraying period, December 1986 to March 1987, air in the breathing zone of 14 workers was collected and the concentration of paraquat measured. Sampling was done with the aid of personal sampling pumps that passed the air over collection filters. Paraquat was extracted from the filters and concentrations determined by the National Institute of Occupational Safety and Health analytical method S294: paraquat (NIOSH 1979). Samples were analyzed for paraquat content at the National Institute of Health, Bogotá, with a 10% external control carried out in the laboratory of toxicology of the University of Laval, Quebec, Canada.

Because data were generated from a random sample of the population, it is possible to compare the group of people that displays the outcome of interest (disease) with the group that does not, according to the presence or absence of exposure to paraquat in a 2 × 2 table. The estimate of the risk of exposed people relative to unexposed people is the cross product of the entries in this table. It represents a relative risk or odds ratio, for which variances, standard errors, confidence limits, and significance tests (chi squared) were calculated.

Results

Of the 5 483 people interviewed, 61.2% (3 357 people) lived in the rural area of Carmen de Viboral and 38.8% (2 126 people) in the municipality. By age distribution the largest group consisted of people 15–44 years of age, followed by those under 15 years. In general, 11.3% of the population was found to be illiterate; 18.7% of this group consisted of men who lived in the rural area. Of the rural working population, 86.3% worked in agriculture or with cattle and 2.4% performed activities related to the production of ceramics. In the township, the corresponding proportions for these two groups of activities were 26.0% and 13.6%, respectively.

Of the population interviewed, 11.0% (603 people) used paraquat: 4.4% (94 people) of the urban population and 15.2% (509 people) of the population in the rural area. The distribution of paraquat users by age and level of exposure (high and low) was determined by the questionnaire (Table 1).

Table 1. Distribution of paraquat users by age and level of exposure.

| Age (years) | Level of exposure | | | |
| | Low | | High | |
	No.	%	No.	%
5–9	4	1.2	1	0.4
10–14	16	4.9	6	2.2
15–24	120	36.8	87	31.4
25–34	82	25.2	76	27.4
35–44	50	15.3	50	18.1
45–54	29	8.9	35	12.6
55–59	13	4.0	13	4.7
60+	12	3.7	9	3.2
Total	326	100	277	100

Of the 603 paraquat users surveyed, six were women; four in the 25–34 years age group and two 35–44 years of age. One woman was in the high-exposure group.

Commonly, 94.5% of the paraquat users applied the herbicide with sprayers; 98.4% used knapsack-type sprayers. Of these, 2.8% considered their spraying equipment to be in poor condition, 21.4% in regular condition, and 75.8% in good condition. However, 71.2% of the users reported that paraquat wet their bodies during spraying. Generally, from the answers reported in the question-naire, the safety instructions provided by the producers or recommended by international health agencies were not properly followed in Carmen de Viboral.

Questions about present and previous exposure to dust, fumes, mists, gases, or vapours revealed that 6.7% of the full-time workers were exposed to dust in the workplace in the year preceding the study. One-third of these workers were from the rural area.

Among those 10 years old and above, 21% were regular smokers, consuming at least one cigarette every day for the past 5 years or more. Of all smokers, 80% had smoked for 5 years or more. Of the total population surveyed, 33.9% of the men and 11.4% of the women were regular smokers.

When asked whether they had experienced illness during the 2 weeks before the study, 17% of those interviewed gave positive replies. Of the problems cited, 7.2% were related to the respiratory tract in both the rural and urban areas. With regard to duration of illness, 62.5% had been ill for less than 15 days; 22.7% between 15 days and 3 months; and 10.1% for at least 1 year. The major complaints associated with respiratory problems were coughing, runny nose, fever, chills, phlegm, and dyspnea.

Respiratory diseases were also diagnosed by a physician (Table 2). Chronic bronchitis accounted for 12.8% of the responses, asthma 2.7%, and tuberculosis 0.2%. Official data for the region confirmed the prevalence of tuberculosis.

Compared with nonusers, a higher prevalence of colds that became localized in the chest was observed in users exposed to high levels of paraquat. This difference was not significant when the values for high- and low-exposure users were combined and compared with nonusers. This phenomenon can be explained by differences in the handling of paraquat, i.e., some users apply paraquat correctly, whereas others do not.

In the subsample examined, chronic bronchitis, one of the diseases of the chronic obstructive pulmonary disease (COPD) group and the pathology of greatest concern, was more prevalent in paraquat users. Chronic bronchitis was also more prevalent in both smoking and nonsmoking paraquat users compared with nonusers in both categories (Table 3). The highest prevalence

Table 2. Physical condition of paraquat users compared with nonusers.

Diagnosis	Nonusers		Users	
	No.	%	No.	%
Hypertension	11	1.4	1	0.8
Cold	38	4.9	2	1.6
Chronic bronchitis	9	1.2	14	10.9
Asthma	5	0.7	0	0
Allergic rhinitis	4	0.5	0	0
Other	27	3.5	5	3.9
Subtotal	94	12.2	22	17.2
Healthy heart and lungs	674	87.8	106	82.8
Total examined	768	100	128	100

Table 3. Effect of smoking on the medical condition of paraquat users and nonusers (number and %).

Diagnosis	Smokers		Nonsmokers	
	Nonusers	Users	Nonusers	Users
Hypertension	2 (1.1)	1 (1.8)	9 (1.8)	0
Cold	9 (5.0)	1 (1.8)	21 (4.2)	1 (1.3)
Chronic bronchitis	7 (3.9)	13 (23.2)	1 (0.2)	1 (1.3)
Asthma	1 (0.5)	0	4 (0.8)	0
Allergic rhinitis	0	0	4 (0.8)	0
Other	8 (4.4)	2 (3.6)	19 (3.8)	3 (4.0)
Subtotal	27 (15.0)	17 (30.4)	58 (11.7)	5 (5.7)
Healthy heart and lungs	153 (85.0)	39 (69.6)	439 (88.3)	70 (93.3)
Total examined	180 (100)	56 (100)	497 (100)	75 (100)

was among paraquat users who smoked. The relative risk or odds ratio for paraquat users who smoked was 7.47 (95% CI = 2.9–19). From this, an additive effect of paraquat use and smoking can be postulated.

The attributable risk or maximum proportion of chronic bronchitis that can be attributed to paraquat use in the study sample corresponds to 84 cases per 1 000 people. Consequently, for the population (21 932) from which the study sample was taken, it can be extrapolated that there are 1 116 cases of chronic bronchitis in the community resulting from paraquat use.

Because the reference values used in Colombia for lung function tests are for people 15–70 years of age, these tests were done only for the 486 men in this age group (Table 4). Women were excluded from this analysis because only six used paraquat.

Paraquat users in Carmen de Viboral suffered from pulmonary obstructive lung disease in greater proportion than nonusers. The figures displayed give users of paraquat an odds ratio for COPD of 3.11 (95% CI = 1.41–6.78).

Controlling for exposure to dust and other irritants, 15.6% of paraquat users who were smokers were found to have COPD compared with no smoking nonusers (Table 5). The odds ratio of developing COPD for paraquat users who smoked was 16.6 (95% CI 0.9–298). Statistically significant differences

Table 4. Lung condition of men, 15–70 years old, who use paraquat compared with nonusers.

Lung function	Users		Nonusers	
	No.	%	No.[a]	%
Restrictive	6	5.4	16	4.3
Obstructive	12	10.8	14	3.7
Both	2	1.8	2	0.5
Normal	91	82.0	342	91.4
Total	111	100	374	99.9

[a] No diagnosis for one person.

Table 5. Number (and %) of male smokers and nonsmokers, 15–70 years old, with impaired lung function among those using paraquat compared with nonusers.

Lung function	Smokers		Nonsmokers	
	Users	Nonusers	Users	Nonusers
Restrictive	1 (2.2)	4 (9.5)	5 (9.6)	2 (4.7)
Obstructive	7 (16.5)	0	3 (5.8)	3 (7.0)
Both	1 (2.2)	1 (2.4)	1 (1.9)	0
Normal	36 (80.0)	37 (88.1)	43 (82.7)	38 (88.4)
Total	45[a] (100)	42 (100)	52 (100)	43[b] (100)

[a] One person not diagnosed.
[b] Three people not diagnosed.

were not observed for COPD or restrictive pulmonary disease among paraquat users and nonusers who were nonsmokers (Table 5).

Air sampling tests, in which filters from 14 people sampled at the breathing level under normal working conditions were analyzed, showed the concentration of paraquat to be consistently below 0.5 mg/m^3, the acceptable level of exposure. All samples of blood and urine taken from users of paraquat in this study gave negative results.

Discussion

Those exposed to a high level of paraquat consistently displayed a higher prevalence of symptoms related to colds that affected the lower respiratory tract. Subsequent medical examination showed that this symptom complex was a good indicator of the presence of COPD. Clinically diagnosed chronic bronchitis was the predominant disease category with a higher prevalence among paraquat users than nonusers.

The pulmonary function tests clearly demonstrated the statistically significant occurrence of COPD among paraquat users. The prevalence of COPD was higher among paraquat users who smoked compared with nonsmoking users, but the statistical significance was weak, probably because of the small sample size.

There were no significant differences with regard to restrictive lung disease between users and nonusers of paraquat. The largest proportion of those with restrictive disease were among those not using paraquat; exposure to dust and other irritants may have contributed to the condition. However, only two paraquat users reported previous dust exposure and their pulmonary function tests were within normal limits.

The occurrence of COPD in this occupational group was unexpected. This is the first time such an association has been described; poisoning by paraquat more commonly produces restrictive lung disease. It is not known to what extent duration and levels of exposure may have influenced the study findings. Less than 5 years has elapsed since paraquat was introduced in Carmen de Viboral and the levels of observed exposure were within suggested limits.

Acknowledgment — This research was carried out with the assistance of a grant from the International Development Research Centre, Canada.

Grant, H.C.; Lantos, P.L.; Parkinson, C. 1980. Cerebral damage in paraquat poisoning. Histopathology, 4, 185–195.

Hearn, C.E.D.; Keir, W. 1971. Nail damage in spray operators exposed to paraquat. British Journal of Industrial Medicine, 28, 399–403.

Levin, P.J.; Klaff, L.J.; Rose, A.G.; Ferguson, A.D. 1979. Pulmonary effects of contact exposure to paraquat: a clinical and experimental study. Thorax, 34, 150–160.

NIOSH (National Institute for Occupational Safety and Health). 1979. Manual of analytical methods (2nd ed.). US Department of Health, Education and Welfare, Cincinnati, OH, USA. Vol. 5, S294, pp. 79–141.

Control of pesticide intoxication through research and rural education in a Peruvian valley

E.L. Rubin de Celis, J.L.B. Robles, J.Q. Paredes, J.B. Garcia, and C.D. Chavez

Huayuna Institute for Development, Lima, Peru

The unique climatic conditions of the coastal valley of Mala, Peru, combined with the use of agricultural chemicals such as dinitro-o-cresol (DNOC), which stimulates blossoming and promotes uniform flowering, allow farmers to produce several apple crops per year. However, the increased use of pesticides results in insect resistance and adverse impact on health and the environment. A research project is under way to determine the prevalence of pesticide effects on health and to assess the knowledge and practices of fruit growers with regard to intoxication and the use of pesticides. An educational program will be designed and an epidemiologic monitoring system, to be operated by peripheral health-care workers, will be developed. In this paper, we discuss the basic elements of the project and report progress and difficulties to date.

This paper outlines the preliminary results of the occupational health and rural-community education project conducted by the Huayuna Institute of Peru. After 15 months of research, the project team is in the second phase of the 30-month project. Although it is impossible to talk of definite successes at this early stage, a discussion of the difficulties that have been encountered is appropriate and of value. These difficulties are common to most research and health-education projects and, for this reason, it is important to foster the interchange of experiences and ideas.

Study area and main problems

The project site is in the coastal valley of Mala, 87 km south of Lima, Peru. The study area includes the middle and part of the lower sections of the valley and extends from the village of Viscas at the head of the river to the city of Mala at the delta. The total population of this area is about 7 000, most of whom live in small dispersed villages sometimes up to 90 km from the main road connecting Mala and Viscas.

Since 1983, the Huayuna Institute (a nonprofit organization) has carried out health promotion and development projects in the Mala valley. To improve the living conditions of the people of this area, a program of primary health care (PHC) is being developed. The occupational health project is an offshoot of the larger PHC project and is being implemented with the joint collaboration of the Ministry of Health and the local communities.

The main economic activity in the area is the cultivation of apple trees. This generally takes place in small family units (the average unit is 1 ha), although for certain tasks casual labourers are also hired. The climate of this area is ideal for cultivating 'Delicious Viscas', a variety of tropical apple that can be grown year round. This allows the farmers to divide their fields and stagger production of 'Delicious Viscas' as desired.

After many years of experience, most growers have adopted a level of cultivation that includes the use of manure and chemical pesticides. In addition, chemical defoliants are widely used, especially dinitro-o-cresol (DNOC). This product also stimulates blossoming, permitting the trees to flower uniformly, and it is a nonsystemic stomach poison and contact insecticide, killing the eggs of certain insects and spider mites. DNOC is poisonous to man and animals, especially by ingestion. It acts as a cumulative poison in humans, although there is little evidence of accumulation in laboratory animals (Worthing and Walker 1987). Acute exposure to large amounts of DNOC affects the body's heat regulatory system causing overheating (IOCU 1986).

The climate, agricultural inputs, and regulation of flowering allow three harvests in 2 years, instead of one harvest each year. The immediate advantage of this intense cultivation is increased annual profits for the growers. However, the economic gain is accompanied not only by serious problems of environmental contamination but also by risks of moderate to severe pesticide intoxication. (Cases of fatal intoxication are well documented, while the outcome of moderate intoxication is less well known.)

To reduce the effect of general hyperinflation in the Peruvian economy, there is a trend toward subdividing more land and distributing harvests over the entire year. Depending on the overall production strategy, e.g., maintenance or reduction of the number of hired workers and immediate or delayed use of various pesticide applications in other orchards, the risk of pesticide intoxication can vary.

Moreover, because of the lack of extreme fluctuations in seasonal temperatures in the Mala valley, there are more generations of insects and pests per season than normally found in apple-producing countries. As a result, the Mala apple growers depend on the use of more and more pesticides to keep pest populations in check. The environmental consequence of this practice is a major reduction in the numbers of organisms that provide biological control naturally.

In addition, pesticides are contaminating the water channels and rivers of the Mala valley. Villagers have observed the effects and are worried about the decrease in prawn production and the death of domestic animals drinking from irrigation ditches.

The project

The general objective of the occupational health and rural community education project is to investigate human intoxication caused by the use of pesticides in agricultural work and to control the effects through community education and the creation of a system of epidemiologic monitoring. The specific objectives are threefold: research, education, and epidemiologic monitoring.

Research

- To study the prevalence of human intoxication by pesticides;

- To study the risk factors, practices, and knowledge of the farmers using pesticides; and

- To study and analyze the potential for self-diagnosis of intoxication and improve methods to obtain an accurate tool for self-diagnosis.

Education

- To design, conduct, and evaluate an educational health program for farmers that will introduce appropriate pesticide practices and spread information about the use of the method for self-diagnosis of pesticide intoxication.

Epidemiologic monitoring

- To develop a program to monitor the toxic effects of pesticides and to provide a model monitoring system based on the participation of health promoters and communal committees.

Three characteristics of the project reflect the nature of the problem of pesticide intoxication and the conditions necessary for its solution:

- The multicausal nature of the problem and the resulting need for a multidisciplinary approach;

- The successive and simultaneous relations of the three principal components of the project (research, education, and epidemiologic monitoring); and

- The key importance of three results of the project (making self-diagnosis accurate, changing the conduct of individuals and of the group, and consolidating community organizations to ensure behavioural changes and the effective operation of the monitoring program).

The multidisciplinary approach of the project is clearly perceived in its research objectives. The medical, agricultural, and social sciences are jointly employed in investigating the prevalence of pesticide intoxication and the validity of self-diagnosis, the agricultural practices involving the use of pesticides, and the social, economic, and cultural factors that affect these practices. A multidisciplinary approach is essential for the educational process because the content, interchange, and transfer of knowledge must address the diverse dimensions of the pesticide problem.

The consecutive implementation of the project's phases serves to link its objectives. The research component fuels the education process and this enriches and feeds back to the research activities. By design, the monitoring program is not formally implemented, but rather is initiated from, and conditioned by, the first phase of research. The education component also provides feedback and ensures that the population understands and is motivated to participate in the project.

The three expected results of the project (accurate self-diagnosis, a change in individual and collective behaviour, and the strengthening of existing organizations) correspond to its design. The common basis of these results is the education process, which allows the researchers:

- To discover what the population knows about aspects of the problem; to organize this knowledge and make people conscious of the magnitude of the problem; and to detect adequate and inadequate practices;

- To exchange information with the population and transfer knowledge about the problem to increase knowledge and change individual behaviour in the handling of pesticides;

- To motivate collective changes in behaviour to prevent contamination of the village and the environment; and

- To initiate dialogue about self-diagnosis to improve it and create a scientific instrument that can form the basis of an epidemiological monitoring program.

The monitoring program, in particular, depends on the education process. A successful monitoring program requires that the population participate actively by taking precautions and notifying the local health promoter about personal or rumoured cases of intoxication. Such involvement is possible only after an improvement in self-diagnosis has been achieved. Health promoters also require specialized knowledge and technical skills in first aid and in the registration and referral of pesticide-intoxicated patients.

At a higher level, state institutions and private doctors involved in the control and treatment of pesticide intoxication must receive training and be sensitized to the purpose and goals of a monitoring program. Technicians and professionals in the agricultural sciences must also participate to promote the safe and rational use of toxic pesticides.

Methodological difficulties and the multidisciplinary approach

The research process

A multidisciplinary approach was one of the principal characteristics of the project. However, this can define several different ways of working. For example, combining research results from two overlapping research designs can correctly be called multidisciplinary. Alternatively, a single research design can use a multidisciplinary approach in the process of obtaining data. This second strategy is perhaps more properly called interdisciplinary and it is the approach employed by this study.

Of the seven localities on the left side of the Mala valley, three were selected to constitute the study area. This selection ensured the manageability of the study without compromising its statistical rigour. The selection took into account differences in climate and the history of land tenure. The unit of analysis was defined as agricultural workers who apply pesticides (about 750 workers). A random sample of 150 agricultural workers from the three localities was chosen, representing 20% of the study population.

Surveys were conducted to gather information about the level of intoxication and risk factors; the preferred practices of the workers with regard to protective and application equipment were observed directly. A subsample of those who had applied DNOC were subsequently questioned about the level of intoxication experienced.

From a social sciences perspective, two variables of special relevance for determining the risk level of intoxication by pesticides were the size of plots and the status of workers (owners, labourers, and mixed cases). Other variables, such as length of exposure, dosage, etc., depended on these two factors. To study the conditional factors of risk, the project's social scientists wished to stratify the study population before sampling, taking into account these two variables.

The medical scientists and physicians, on the other hand, were interested in guaranteeing statistical results by determining the level of intoxication based on clinical studies and laboratory tests. They argued against stratifying the sample fearing it would compromise the representativeness of their results.

Ultimately, the position of the physicians dominated, and the sample was not stratified. As a result, it was not possible to establish the influence of plot size or occupation on risk levels and application or prevention practices. (This was corrected in the second phase of research in which the population was stratified according to orchard size.) The clinical analysis and laboratory tests were conducted on all those who were fumigating at the time of observation.

Difficulty in harmonizing the approaches of the professionals in the social and medical sciences was encountered on another point as well. The social scientists proposed that risk studies should be concerned not only with application practices, protection, and prevention, but also should consider the point of view of the growers. This information would serve to orient the educational process and probe the importance of conditional elements such as economic factors. Accordingly, the survey included a series of open-ended questions that, with difficulty, would permit a quantitative analysis. In contrast, the medical scientists felt that the growers' opinions could be useful only if they were categorized quantitatively as "adequate/inadequate" or "relation/no relation with DNOC."

The open-ended questions of the survey yielded general comments, e.g., "It bothers me to use gloves, masks, or boots." What, precisely, bothered the person (the inconvenience, the expense, etc.) was not included. Therefore, it was not possible to see the logic of these answers or compare them with others. Because of this deficiency, the debate was arbitrarily resolved in favour of the physicians.

On the issue of self-diagnosis, the same problem arose. The medical scientists were interested in tabulating, as faithfully as possible, the perceived annoyances, the type of annoyances, and their perceived cause to categorize them as "right/wrong" and "related/not related to DNOC." They proposed direct questioning, clinical examination, and corroboration of results with laboratory analysis. The social and educational scientists, on the other hand, were interested in studying this aspect through open-ended questions that would permit the discovery of "the logic of folk wisdom" and enable future comparison with the medical sciences. In their opinion, the classification of "right/wrong" or "related/not related to DNOC" would not give precise grounds for educational intervention. It would only identify areas of intervention.

For example, if a grower has complained about "suffocation" and he or she is asked the reason for that complaint, the response "it is because of DNOC" or "because of the day's work" is sufficient for the physician. Those answers reveal the percentage of the population that is able to identify the root cause of the complaint. However, the social scientist is interested in obtaining answers that indicate, for example, that the "suffocation" was produced because the DNOC was "hot." It is reported as "hot" because it "burned" the tree buds. This characteristic of DNOC may change perception of "normal" body temperature. Knowing this, cases of workers reporting "no annoyance" or "resistance to bathing immediately" can be explained more fully. For the social scientist, understanding the rationale behind responses is essential because the education process attempts to modify ways of thinking, not just a limited number of practices.

The results of the study of self-diagnosis in the population that used DNOC were poor. Only 2 of 15 cases reported any annoyance at the end of the work day and they were not correlated with the level of DNOC in the blood. More instances of annoyance were reported among the 135 people who did not usually use pesticides than among those who had used DNOC in the past (20% said they were bothered after using DNOC). For the second phase of the project, the time of clinical observation and blood sampling was extended to 24 h after DNOC application.

This raises a point related to the process of self-diagnosis, i.e., the concept of "normal." Self-diagnosis is the identification of an illness by individuals who recognize in themselves certain symptoms or signs of that illness. However, what individuals take to be a normal state of health directly affects their self-diagnosis ability.

For example, if a grower thinks that it is normal (i.e., logical and not problematic) to find at the end of a day that he is "suffocated" and "hot" because DNOC is also "hot," he may not report any symptoms when asked "did you experience any health effects the day you applied DNOC?" For the grower, the so-called "abnormal" state is considered normal. It is expected and does not cause alarm, even if body temperature is elevated more than after prolonged physical exercise.

Only when the logic by which growers interpret all aspects of their contact with DNOC is understood can researchers begin to piece together an accurate assessment of self-diagnosis. Competing concepts of normality must be recognized. In the case of the normalities employed by the physicians (the statistical normality and the deontological normality), there is a consciousness of the difference. In the case of the growers, there is only the consciousness of a normality, which may deviate from medical concepts.

The project team encountered several other difficulties. The agricultural season in Mala traditionally begins in March–May and ends in October–November. The research team began gathering data in mid-April. However, irregular temperatures forced the growers to move the time of DNOC application ahead to March and the beginning of April when the research team was still engaged in the design stage of the project.

As a consequence, it was often not possible to establish the exact date of DNOC application. The growers knew only approximate dates. The decision to apply DNOC was often taken only 1 or 2 days in advance, depending upon the state of blossoming, weather conditions, and the availability of water and labourers.

This made the collection of blood samples problematic because the growers could not always be contacted when necessary.

In addition, the researchers were initially unable to locate a laboratory that could perform gas chromatography, the method of analysis selected to determine levels of DNOC in the collected blood samples. Chromatography had to be replaced by the less specific and sensitive procedure of photocolorimetry. This technique, however, can be carried out in rural health centres. For the second phase of research, a laboratory able to do gas chromatography was located.

Because of the laboratory problems and the small number of workers applying DNOC, the epidemiologic component was reduced to a pilot study to test the method of laboratory analysis and the survey. Only 15 growers who had applied DNOC were interviewed. These interviews were conducted at the end of the 1st day of work. The risk survey, however, included all 150 workers in the study population.

The education process

Cross-disciplinary problems again arose in defining the content of the education component of the project. Conceived as a process and not just a package of recommendations, the education component was designed to provide scientific explanation. This prompted the question, "Up to what point should an explanation be given?" For example, in a discussion of effects of pesticides on health, was it necessary to talk of the inhibition of cholinesterase or the accumulation of calories with respect to DNOC? In the case of forced tree blossoming, was it necessary to talk about dormancy and the breaking of dormancy produced by auxins and gibberellins stimulated by DNOC? These explanations demand a knowledge of chemistry, biology, and physics that the grower does not have and is unlikely to find much use for.

Faced with these obstacles, the research team had to be much more multidisciplinary and innovative in its approach, particularly when it was found that the growers' main interest was in learning more about the agricultural process in general and not the safe use of pesticides, as was the objective of the project.

The education component of the project aims to be an effective tool for reducing human intoxication resulting from the use of pesticides. It is directed at the growers and their families, but it also considers the health promoters and agricultural professionals in the study area.

The education process for the growers and their families consists of both intensive and extensive activities. Intensive interaction is facilitated through small groups and periodic meetings. Through this process, a systematic education program is developed consisting of three stages:

- Identification of the general knowledge and adequate and inadequate pesticide practices of the growers;

- Design and implementation of alternative experimental pesticide practices (the project is currently at this stage);

- Evaluation of the education process by the participants and support for the epidemiologic monitoring program.

The education process of the growers and their families is extended through the use of scheduled village meetings and the involvement of local organizations. Schools, in particular, play a principle role. Teachers are encouraged to educate and sensitize their students and involve the parents in this process. The education component also includes parallel programs for health promoters and medical and agricultural science professionals in the area. The purpose of these programs is both preventative and proactive, seeking to raise awareness and promote accountability to the epidemiologic surveillance program.

As an interactive process, the educational component also works to ensure the internal coherence of the project by:

- Gathering information from the participants to find out not only what they do and how they do it, but also why;

- Comparing and analyzing gathered information to formulate hypotheses;

- Transferring new knowledge in a systematic and usable manner; and

- Discussing and analyzing alternative options for action on the basis of new knowledge and understanding.

There is no set order in which to carry out these steps. During the gathering of information, the growers not only gave their perceptions but also sought comment on their conduct and actions. During the analysis of knowledge and practices, new information suggested new hypotheses to the researchers. While still in the process of gaining new understanding and knowledge, some of the participants proposed control programs and others began to adopt new practices, such as reducing the frequency or duration of DNOC application.

The education process, which draws information from knowledge and experience, is itself part of the research process. Through dialogue, the research strategy is adapted (this was especially true for the investigation into self-diagnosis) and results are confirmed and enriched (Table 1).

During the education process, the research team came to accept the need to stratify the sample population for the second phase of research and determined that the medical questioning and blood test should be done 24 h after application. This was also the process through which the problem of the growers' perception of "normal" was detected and the self-diagnosis research designed.

Table 1. Relation between information revealed in the research and education processes of the project.

Research component revealed	Education component revealed
Lack of protection of the face and feet because it is "bothersome."	The material used for gloves, boots, and edges of masks increases sweating and impedes work. Wiping away sweat spreads the pesticide even more.
Incorrect preventative or first-aid measures, e.g., not bathing immediately after finishing work.	The pesticide creates a new level of heat equilibrium that is dangerous to break until the body naturally drops to a lower level of equilibrium.
The custom of drinking lemonade or milk after spraying because "it makes me feel good" or "because I always do it."	The toxic pesticide causes harm to the stomach. Drinking promotes filtering and dissolving of the toxins.
Sources of information are most often, "I learned by myself," but occasionally, "I follow to the letter the instructions on the labels of the bottles."	Necessity for adequate labeling of recommendations on bottles. Exchange of experiences provides knowledge about blossoming, time of harvest, economics.
Few cases of complaints at the end of work days when pesticides were applied.	Most symptoms begin at night when subjects are at rest.

The epidemiologic monitoring program

The principal activities of the epidemiologic monitoring program involved:

- Education of growers' groups (base groups);

- Training of health promoters (parallel to or integrated with the base groups) to popularize the program, learn first-aid techniques and skills, understand the educational and organizational aspects of the program, correctly manage the registers, transfer cases of intoxication to other levels of the health service system, and periodically evaluate the program;

- Sensitize health and agricultural professionals and technicians to popularize the program, participate in the creation and management of the registers, receive and treat cases transferred by the health promoters and villagers, and periodically evaluate the program; and

- Create a register and transfer system.

Originally, these activities presupposed that a vertical chain of communication existed from the village level to the professional or formal level (villagers–promoters–professionals). Educational action was to be directed only at the base and intermediate levels of this vertical chain, although control of the system remained at the professional level. Before establishing the monitoring program, the problems inherent in such assumptions were revealed and activities were adjusted accordingly. Most importantly, it was understood that the information chain could not terminate with the professionals, but had to feed back to the lower levels (those most intimately involved and affected by the program), closing the circle of information.

Educational action is equally important for the higher level professionals. The information concerning cases of death by acute poisoning with DNOC over the past 5 years, revealed common treatment with atropine. Atropine, the antidote for poisoning by organophosphates and carbamates, has negative effects in cases of acute DNOC poisoning for which there is no antidote. This indicates that medical programs continue to impart information that is 20 years out of date. Worse, rural doctors do not recognize poisoning by pesticides whose effects and treatment are not known (e.g., pyrethroids and dinitros). Educational action at the highest level of the health system is necessary to make prevention and proper treatment possible.

Without validation and refining of the self-diagnosis research, the effort of installing the monitoring program would not have been warranted. Clearly, the grower uses judgement with regard to the degree of intoxication. Low levels are considered normal and medical assistance is sought only in severe cases.

The formation of groups consisting of villagers, promoters, and doctors began before the formal establishment of the monitoring program. It was initiated as part of the education process, which is not simply an intellectual process in the classroom but aims to promote action in the community. These groups provide an incentive (directly or indirectly by example) for the creation of other groups in the larger program-monitoring area.

Although the problem of pesticide intoxication is urgent, an education process that tries to modify practices and attitudes makes gains slowly. Therefore, it is not possible to wait until the education process has achieved all of its objectives before initiating the monitoring program. These two activities are mutually reinforcing and the education process must be continued if the monitoring program is to become an effective preventative instrument.

Conclusions

The problems encountered in this research project are not unique. Many research projects that attempt to use a multidisciplinary approach experience similar difficulties and challenges. The research strategy of the study is complex by virtue of its attempt to implement an essential feedback process among the research, education, and epidemiological monitoring components. It involves handling a large number of variables, operating at different levels of analysis, and engaging in various types of activities. In the process, tensions and difficulties created by competing perspectives and beliefs were clearly evident.

For example, epidemiology has as its central objective of study the prevalence of sickness in a population. Biological and other factors condition the manner in which sicknesses manifests itself in a given population. Occupation, earnings, education, norms, values, beliefs, and habits, for example, influence the prevalence of sicknesses. These variables can be interdependent, but analysis strives to determine the degree of association of each as an independent factor with respect to a particular health problem.

In epidemiology, the interaction of factors that determine the health–sickness process is perceived as multicausal. Epidemiology favours statistical analysis as an instrument, which requires the use of precategorized and precoded surveys to collect information.

In contrast, the social sciences have understanding of the social system as their central objective. Socioeconomic and cultural factors are perceived as a system in themselves and as a set of dynamic interconditioned elements. Such an orientation makes epidemiologic analysis more difficult, given that some variables (e.g., occupation and size of farm) are conditional factors and may be conditioned reciprocally. In seeking to understand all elements of the multicausal social relation, nonquantitative aspects (values, beliefs, interests, etc.) are considered equally important.

The joint participation of all researchers in the education process plays a decisive role in overcoming disciplinary barriers and prejudices. As a result, the successful demonstration of health surveillance or exposure-monitoring activities, as integrated functional elements of the services offered to farmers at the most peripheral levels of the health-care system, will be of considerable interest to other Third World countries searching for solutions to the problem of providing appropriate preventative health care to their rural populations.

Acknowledgment — This project (3-P-88-0186-02) is supported by the International Development Research Centre, Canada.

Worthing, C.R.; Walker, S.B., ed. 1987. The pesticide manual: a world compendium (8th ed.). British Crop Protection Council, Croydon, Surrey, UK. Pp. 326–327.

IOCU (International Organization of Consumers' Unions). 1986. The pesticide handbook: profiles for action (2nd ed.). IOCU, Penang, Malaysia.

Agrochemicals: a potential health hazard among Kenya's small-scale farmers

M.A. Mwanthi and V.N. Kimani

Department of Community Health, University of Nairobi,
Nairobi, Kenya

A community-based study in 10 villages in Githunguri, Kiambu District, was conducted to establish the extent of use of agrochemicals and to assess the attitudes, behaviour, and general awareness of the people of the health hazards posed by improper handling of agrochemicals. Another objective of the study was to design and test a health-education curriculum with active participation and feedback from the community and certain target groups. The use of pesticides and agrochemicals, in general, was found to be extensive. The health hazard posed by these chemicals is high because the community's awareness of the danger is limited. The only precaution taken was to ensure that the substances were not "within the reach of children." Cooperation was sought from chemical manufacturers, retailers, farmers, and community leaders to minimize the dangers of exposure through improper handling of pesticides.

Kenya's economy is dependent on agriculture and the agricultural sector uses enormous quantities of pesticides to improve crop yields. Small-scale farmers are encouraged to use pesticides, although often they neither follow instructions nor understand the potential hazards of careless handling of these agrochemicals. The use of insecticides in homes to combat disease vectors, such as mosquitoes, houseflies, and cockroaches, is also becoming increasingly common.

These chemicals are readily available in nearly every grocer's shop. They are purchased from retailers who measure them out from their original containers into all kinds of unlabeled packages and containers, compounding the problem of improper handling.

Kenya's Pest Control Products Board estimates that over 700 million KSH (30 million USD) is spent on pesticides annually (Table 1), mainly by farmers, for both large-scale cash crops such as coffee and tea and small-scale horticultural crops.

Table 1. Value (million KSH) of pesticides imported and used in Kenya, 1985–87.

Year	Insecticides and acaricides	Herbicides	Fungicides	Other	Total
1985	127.4	101.8	168.7	13.3	411.2
1986	134.9	121.3	281.3	42.6	580.1
1987	182.3	173.4	357.3	28.1	741.1

Source: Pest Control Products Board of Kenya 1987.

In the Third World, most of the human poisoning and environmental contamination caused by agrochemicals stems from indiscriminate use and mishandling, usually due to ignorance of the potential hazards posed by these chemicals. Labeling of the products and misleading advertisements by pesticide dealers intensify the problem. As a result, many small-scale farmers use pesticides that have not been recommended for their specific crops. Others use concentrated preparations in the belief that they will be more effective against pests; instead this has led to pest resistance and resurgence of new pests (Akhabuhaya 1989). The number of resistant arthropod species has risen from 137 in 1960 to 392 in 1980, an average rise of 13 species per year (Youdeowei 1983).

Improper handling and misuse of pesticides continue to cause untold numbers of poisoning cases in the developing world. Although improper handling may be due to carelessness, many other factors contribute to the problem: farmers' ignorance; poverty, which may force the farmers to continue to use poor and leaky equipment; improper or nonexistent protective clothing; use of wrong or cheap pesticides; use of the wrong antidotes or none at all; and poor handling of utensils such as measuring devices and storage containers. Empty chemical containers are commonly used in the home for storing sugar and salt and for fetching water or milk.

In developing countries, pesticides are usually stored inside the home, in rooms that are also used for sleeping and storing food and drinking water. Small-scale farmers, their families, and particularly children are exposed to a high risk of acute and chronic pesticide poisoning. Sometimes the chemicals are used for deliberate poisoning and suicide attempts.

All pesticides are biologically active. Their toxicity affects not only pests but also all other forms of life in the ecosystem. The situation is much worse in developing countries, where illiteracy, poverty, and general morbidity rates are high.

A study of mercury poisoning among children in a rural Kenyan community found elevated levels of mercury in their blood and chronic mercury poisoning (Brown et al. 1982). The blood of their family members also contained elevated levels of mercury. The incidence of mercury poisoning was higher during rainy seasons. These observations led to the conclusion that agrochemicals, used haphazardly, may be associated with the high incidence of mercury poisoning. Food, milk, and drinking water used by these families did not contain mercury; however, an analysis of chicken feed, skim milk, and other animal feeds revealed residues of pesticides.

In other developing countries, dichlorodiphenyltrichloroethane (DDT), lindane, aldrin, and endrin, among others, have been banned or are very highly restricted. In Kenya, they are still in circulation in one form or another. Lately, there has been much talk about banning DDT in Kenya, but the situation has not really changed. Farmers are still growing pyrethrum as an important cash crop and DDT is still in demand.

Records from the Kenyatta National Hospital reveal that at least two cases of pesticide poisoning are seen daily, i.e., 730 cases annually at this hospital alone. In 1981, 221 cases of organophosphate poisoning were treated at the Kenyatta National Hospital; 13 of the patients died. Of Kenya's estimated 5 million agricultural workers, 7% or 350 000 per year are poisoned by pesticides. The resulting economic impact is estimated at 336 million KSH annually (Choudhry and Levy 1988).

Much effort is needed at the community level to sensitize people to the dangers of improper handling of agrochemicals. To this end, a community-level study of awareness, attitudes, and behaviour with regard to handling pesticides was undertaken in Githunguri, Kiambu District. This study was unique in Kenya; previous studies have tended to concentrate on the clinical aspects of chemical poisoning. None appears to have assessed the socioeconomic behaviour of a community regarding pesticide handling.

The overall objective of the study was to assess people's perception of hazards posed by improper handling of agrochemicals, specifically pesticides at the community level. The specific objectives were:

- To identify the types of pesticides found in the community;

- To observe where and how these chemicals are handled and stored;

- To note the disposal methods used for empty containers and leftover chemicals;

- To determine whether the farmers understand and follow the instructions on container labels;

- To establish the use of protective clothing during handling as well as application;

- To measure awareness and assess practices;

- To develop baseline data about the types of pesticides used and determine the extent of use in the selected rural agricultural community of Kenya; and

- To develop a health-education package based on findings of the knowledge, attitudes, and practices (KAP) study and disseminate it in the community.

The study area

The study was carried out in the rural community of Githunguri in Kiambu District. It is about 50 km northwest of Nairobi and is 2 000 m above sea level. The 1988 projected population for the district was 1 012 438; the 1979 census reported a population of 119 548. These were the only official data available at the time of the study. The population is composed mainly of the Kikuyu ethnic group.

Githunguri is an agricultural area where coffee and tea are the only cash crops. Maize, pulses, potatoes, vegetables, and other horticultural crops are also grown. Most community members are small-scale farmers who use fertilizers, fungicides, pesticides, and other chemicals throughout the year to improve agricultural yields. Unemployed school leavers have turned to horticulture as an alternative income-generating activity. They use agrochemicals extensively to improve the crop yield from overcultivated land. The farmers find a ready market for these crops in Nairobi and other local towns.

Methods

The study population consisted of all households in Githunguri, Kiambu District. At the time of planning for the study, there were 6 351 households in the area. Because there was no systematic numbering of households, selection was based on villages. The villages were considered as clusters and households within them as listing units.

Of the 34 villages in the study area, 10 were randomly selected using cluster sampling with probability proportional to the number of listing units in the cluster. This procedure was chosen because the clusters had unequal numbers of listing units and this method improves reliability of estimates. Of the 2 454 households in the 10 villages, 1 797 were surveyed by interview. The minimum number of households needed was calculated to be 288. This sample size was obtained using the formula:

$$n = z^2 p(1 - p)/c$$

Where n is the minimum sample size; z is the chosen critical value from standard normal distribution tables; p is the estimated prevalence rate of pesticide use in the area; and c is the confidence interval (a measure of precision).

In a pilot study, we found that the prevalence rate of pesticide use in the area was 75%; c was set at 0.05 and z (from tables) at 1.96 for a probability of 5% in a two-tailed test. Therefore:

$$n = (1.96)^2(0.75)(0.25)/0.0025 \text{ or } 288$$

Data were collected using interviewing techniques with the aid of a questionnaire. Locally hired and trained research assistants visited each household in the selected villages daily from September 1987 to December 1987 under the supervision of the two principal investigators.

All questionnaires were in English, but, during the interview, they were interpreted in Kikuyu. Attempts were made to interview the head of the household. In his absence, the wife or an adult member of the household over 15 years of age was interviewed. Agrochemicals (mainly pesticides) were collected from every sixth household if available.

Results

During data collection, 1 797 households were surveyed by interview and 99 unknown agrochemical samples were randomly collected. To establish authenticity, these were qualitatively analyzed in the laboratory using gas chromatography. Among the 99 samples, only 21 pesticides were identified because of duplication, a situation that could not be established in the field. The most commonly used chemicals in the community were fertilizer, fungicides, insecticides, herbicides, nematicides, and rodenticides.

All households reported that they were currently using agrochemicals or had used them within the preceding 6 months (Table 2). The frequency of use ranged from continuously throughout the year to every crop season.

Members of the community other than farmers were also exposed to agrochemicals at a high level, e.g., coffee factory workers, career spray operators, and women who sold agrochemicals in the open markets. Career spray operators are young professionals who go from farm to farm, spraying coffee crops on a contract basis. During the peak of the spraying season, they sometimes work many hours every day of the week.

Table 2. Frequency of use of fertilizers and pesticides.

Frequency	Number of respondents	%
All year	922	51.9
Every crop season	840	46.7
Irregular	24	1.3
Total	1 797	99.9

Table 3. Storage of agrochemicals.

Storage site	Number of respondents	%
Living area (where family sleeps)	1 120	62.3
In food storage area	258	14.4
In separate shelter	172	9.6
Other places (underground, in cage, etc)	247	13.7
Total	1 797	100.0

Most families stored agrochemicals in their living quarters, particularly in the eaves of the roof (Table 3). Although such places are out of reach of children, the practice facilitates their spread as air-borne dust and fumes, especially when agrochemicals are loosely packed in secondary containers. Much exposure takes place during the night hours of sleep. Among the 1 797 households, only 35.3% had pesticides stored in their original containers; 64.7% could not produce an original container. This made it difficult for researchers to identify chemicals on sight.

Those who produced original agrochemical containers with chemicals in them claimed that they understood the instructions on the labels and followed them when spraying. However, when asked to read the labels, more than 60% of the respondents were unable to make sense of them. There is no doubt that many of the farmers were literate, but the instructions were sometimes too technical for them to understand.

Half the farmers used domestic containers, such as cooking pots or water containers, for mixing chemicals; the rest used empty agrochemical containers. The residues in these containers probably contaminate food and, in one way or another, come into contact with members of the family. Almost all empty agrochemical containers were converted to domestic use. Judging from the poor degree of protection against exposure, storage patterns, and the community's general attitude toward agrochemicals, most respondents were probably not fully aware that pesticides are dangerous to their health.

The level of education of the subjects was correlated with method of preparation of chemicals and attention to the instructions; use of protective clothing while spraying; and caution when handling chemicals (Table 4). About 69% of the respondents stated that they take precautionary measures when handling certain chemicals and pesticides, among them malathion, Difolatan, and Roundup.

In 1987, there were 35 cases of organophosphate poisoning at Kiambu District Hospital (Fig. 1). Of these victims, 21 (60%) were 21 years of age and over. Similarly, in 1988, organophosphates caused a large majority of poisonings treated at the district hospital. However, the number of cases decreased notably from 1987 to 1990. Most of the victims were adult males. Typically,

Table 4. Number and proportion (%) of respondents who prepared agrochemicals according to the instructions, wore protective clothing while spraying, and took precautions when handling chemical by level of education.

Level of education	Followed instructions	Wore protective clothing	Caution with — All chemicals	Certain chemicals	Sample size
None	64 (15.0)	96 (22.5)	116 (27.2)	259 (60.8)	426
Primary	424 (44.0)	342 (22.5)	217 (22.5)	699 (72.6)	963
High school and over	256 (62.7)	167 (40.9)	98 (24.0)	289 (70.8)	408
Total	744 (41.4)	605 (33.7)	431 (24.0)	1 247 (69.4)	1 797

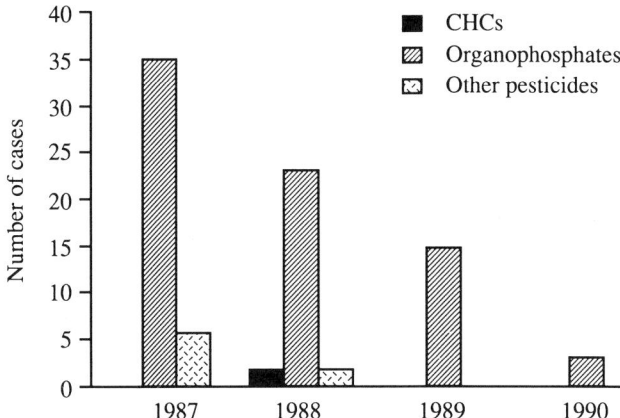

Fig. 1. Cases of pesticide poisoning reported to Kiambu District Hospital between January 1987 and April 1990.

career spray operators are male, and it is possible to speculate that ready access to agrochemicals is a factor in the considerable number of suicides by poisoning. From 1987 to April 1990, among poisonings of people 21 years and over, 44 (55.7%) were of undetermined causes, 33 (41.8%) were intentional, and 2 (2.5%) were accidental. During the period of the study, 133 cases of agrochemical poisoning were treated at Kiambu District Hospital alone; there were 8 (6%) deaths (Table 5).

Discussion

The community uses a variety of pesticides and fertilizers extensively almost all year. Community awareness of the environmental and personal health hazards posed by these chemicals, especially pesticides, is minimal. The number of poisonings treated at Kiambu District Hospital suggests that action

112

Table 5. Causes of agrochemical and other chemical poisoning treated at Kiambu District Hospital from 1987 to April 1990.

Age group (years)	Sex M	Sex F	Cause of poisoning Suicidal	Cause of poisoning Accidental	Cause of poisoning Undetermined	Outcome Alive	Outcome Dead	Total
0–5	15	6	1	6	14	21	0	21
6–10	0	2	1	0	1	2	0	2
11–20	16	15	12	2	17	31	0	31
21+	61	18	33	2	44	71	8	79
Total	92	41	47	10	76	125	8	133

is needed to remedy the situation. The belief of the majority of the population that there are no alternatives to pesticides calls for urgent steps to be taken to ensure safe handling of pesticides.

The investigators, together with the district multidisciplinary team, developed a health-education curriculum for different target groups. Teams from the health centres in the region were identified to assist in this activity and to integrate the curriculum into their community-health activities. Target groups included school children, women's groups, factory workers, and the career spray operators. Pesticide manufacturers and retailers have begun to cooperate, especially in packaging chemicals in smaller and safer containers.

The response of the community to the health-education curriculum has been encouraging because some of the problems are obvious. This study has motivated people to take responsibility for their own safety. Target groups, health centres, schools, coffee-factory workers, and all other members of the community have adopted safe handling practices recommended in the health-education program in their day-to-day activities.

Acknowledgment — We acknowledge the financial support of the International Development Research Centre of Canada, which made this study possible.

Akhabuhaya, J.L. 1989. Use of pesticides in agriculture. East Africa Newsletter on Occupational Health and Safety, 31 (December), 8–9.

Brown, J.D.; Meme, J.S.; et al. 1982. Mercury poisoning in Kenyan children: a further report on epidemiologic aspects. East African Medical Journal, 59(2).

Choudhry, A.W.; Levy, B.S. 1988. Morbidity estimates of occupational illnesses and injuries in Kenya: human and economic costs. Paper presented at the 9th Annual Medical Scientific Conference, 1–5 February 1988. Kenya Medical Research Institute and Kenya Trypanosomiasis Research Institute, Nairobi, Kenya.

Youdeowei, A.; Service, M. 1983. Pest and vector management in the tropics. Longmans, Harlow, Essex, UK.

Pesticide use in agriculture, public health, and animal husbandry in Cameroon

F.D. Mbiapo and G. Youovop

Faculty of Science, University of Yaoundé, Yaoundé, Cameroon

A pilot study was conducted in the western, central, and southern regions of Cameroon, West Africa, to define the problem posed by the use of pesticides. Cameroon imports organochloride, organophosphate, carbamate, and pyrethroid insecticides. Although imports amounted to 60 000 t in 1987, they were reduced by half in 1988 and 1989. Legislation reflects the standards recognized by the United Nations Food and Agriculture Organization and the United Nations Environment Programme. Pesticides are used mainly on cash crops, such as coffee and cocoa. However, food and vegetable producers commonly obtain pesticides on their own initiative. The main pesticides used in Cameroon are insecticides and fungicides. Producers rarely respect the recommended last date of application before the harvest and users generally are not adequately protected when spraying.

Pesticides are substances or mixtures of substances intended to repel, destroy, or combat pests that ravage plants or act as vectors for human and animal diseases. However, their use is not without danger to plants, animals, humans, and even the abiotic environment.

The aim of this study was to examine the conditions under which pesticides are used in Cameroon, including:

- The quantities and types of pesticides imported and the country's legislation governing use;

- The ways in which pesticides are distributed and circulated within Cameroon: agricultural, pastoral, and public uses; timing and duration of treatments; interval between the last application and harvest; dosage applied; methods of application; and how empty pesticide containers are handled.

Methods

A pilot field survey was undertaken to ascertain the status of pesticide use in Cameroon and to generate baseline information, enabling the team to carry out a more substantial study of pesticide poisoning. A questionnaire was administered to 360 respondents, 30 from each of the 12 districts of the west (Bafoussam, Bandjoun, Dschang, Foumbot, Galin, and Mbouda), central (Mfou, Monatélé, Obala, and Yaoundé), and south (Ebowala and Zoétélé) provinces of the country. Further information was gathered from official government records.

Results and discussion

Cameroon imports insecticides, fungicides, herbicides, rodenticides, and acaricides. Between 1985 and 1987, imports amounted to some 60 000 t of pesticide per year. In 1988 and 1989, this amount decreased considerably, to almost half.

Legislation on pesticide use is still in the draft stage. The standards followed are those set out in the decree on the regulation of poisonous substances and in the standards and recommendations of the United Nations' Food and Agriculture Organization (FAO) and the United Nations Environment Programme (UNEP).

Pesticides used on cash crops (coffee and cocoa) are distributed through an established system. On the basis of a list of pesticides tested and recognized as effective by the Institute of Agronomic Research, the Ministry of Agriculture purchases the pesticides abroad through agrochemical companies (such as Rhône Poulenc, ADER, Shell Cameroon, etc.) with operations in Cameroon. The substances are then distributed to coffee growers via the Ministry's regional branches, consisting of provincial plant-health stations, departmental brigades, and agricultural stations in the districts and villages.

In the cocoa sector, a state-owned enterprise is responsible for protecting plantations. It orders pesticides and is responsible for their use in the field or their distribution to cocoa growers in the producing areas.

However, pesticides are being diverted from the agricultural outstations. Food and vegetable producers buy pesticides for use in production and conservation. They are purchased from agrochemical companies and resold on the open market or by one user to another. Domestic agrochemical companies place their orders directly with private industry. The pesticides used in public-health and animal protection are also purchased on the open market or from private industry.

Two groups of pesticides make up the bulk of those used: insecticides and fungicides. Insecticides include organochlorides, organophosphates, carbamates, and especially pyrethroids. The fungicides are primarily copper based. In the west, both groups are used on coffee and vegetable crops to conserve seeds and food products. In the south, they are used only on cocoa crops and, in the central province, by vegetable and cocoa growers.

The insecticides used in homes have a pyrethroid base. In the west, centre, and south, 70%, 62%, and 76%, respectively, of people surveyed used household insecticides. Unlike northern Cameroon, raising livestock and small animals is not common in the three regions under study. However, to kill poultry ectoparasites, dog ticks, human chiggers, and head lice, 29%, 29%, and 18% of the people interviewed in the west, centre, and south, respectively, used pesticides of the organochloride, organophosphate, and pyrethroid groups.

Insecticides were used against flying insects (tse-tse flies, mosquitos, black-flies, etc), crawling insects (cockroaches), and other household pests by 44%, 43%, and 17% of respondents in the west, centre, and south, respectively. People used organochloride, organophosphorous, and especially pyrethroid insecticides as well as rodenticides.

Timing and duration of treatment were usually not according to recommen-dations, because the insecticides were received late or because of seasonal variations in rainfall patterns. A high percentage (86%, 62%, and 83% of respondents in the west, centre, and south, respectively) admitted that they did not allow for any time lag, or no more than 1 week, between the last application of insecticide and the harvest, increasing the likelihood of pesti-cide residues in harvested food.

In antifungal treatments administered by users, 100%, 70%, and 80% of respondents in the west, centre, and south, respectively, increased the dosage beyond that indicated on the product package. Recommended dosages were respected when insecticide treatments were applied by employees of govern-ment agencies or state-owned enterprises. The number of treatments was, however, not usually respected in applications to vegetable and cash crops in the three provinces; they may be increased or decreased, depending on the user.

Application methods included spraying by users or thermonebulization or misting by the staff in agricultural stations. None of the users had adequate protection, even though some reported that they were aware of the risks they incurred in handling pesticides. Many of them drank milk after applying pesticides, but were unaware that this practice has been discredited.

In many cases, pesticide containers were not properly disposed of — 66%, 76%, and 75% of users in the west, centre, and south, respectively, discarded their empty containers in water, on the plantation, or in the bush. In rare instances, drums and barrels that had contained pesticides were reused for other purposes or sold.

Conclusion

The practices followed by pesticide users in the regions studied are not adequate to minimize harmful effects on humans, animals, plants, and the environment. The use of pesticides under prevailing conditions constitutes a danger to the entire ecological chain. In general, the most commonly used pesticides in Cameroon are the organochloride insecticides, organophosphorous insecticides, carbamates, and, above all, pyrethroids. Copper-based fungicides are also widely used. Herbicides are not as widely used, but they are used in vegetable and cash-crop production, in the livestock- and human-health sectors, and — especially in the western region — the conservation of food stocks. Further study of the use of pesticides and their effects on human health in Cameroon is recommended.

Acknowledgment — This research was supported by a grant from the International Development Research Centre, Canada.

Health profile of workers exposed to pesticides in two large-scale formulating plants in Egypt

Mahmoud M. Amr, M. El Batanouni, A. Emara, N. Mansour, H.H. Zayat, G.A. Atta, and A. Sanad

Kasr El-Aini Faculty of Medicine, University of Cairo, Cairo, Egypt

Large quantities of pesticides used in Egypt are manufactured or formulated locally. Growing concern about the number of unrecorded cases of pesticide intoxication prompted a study of 300 workers in the industry. Cases of both acute and chronic intoxication were found with a direct relation between the severity of signs and symptoms of chronic intoxication and exposure. Peripheral neuritis was confirmed by electromyography in 25% of the exposed workers compared with 2% of the control population; pulmonary function was impaired in 30% of the subjects. Optic-nerve abnormalities, psychiatric disturbances, and respiratory (including asthmatic) problems, atherosclerosis, hypertension with cardiac asthma, and enlargement of the liver were also found among workers exposed to pesticides in the workplace. About 90% of the exposed group presented with abnormal liver function, and abdominal sonography revealed a significant number of abnormalities.

Since 1945, some 1 500 compounds and more than 35 000 formulations have come into use as pesticides. In developing countries, the adverse effects of pesticides have become a serious public-health problem (Davis et al. 1986). Workers engaged in pesticide manufacture and formulation are among those at greatest risk (Levine 1986).

Epidemiological data on the effects of pesticides are limited in most parts of the world (Hayes 1982). In developing countries, where about 99% of pesticide-related poisonings occur (Xue 1987), few studies have been conducted because of financial and technical constraints.

Concern about the environment and health hazards of pesticides has been expressed in Egypt where 30 000–60 000 t of pesticides are used annually in agriculture or for public health. Large quantities of these pesticides are manufactured or formulated locally, in both urban and rural areas. Although the last two decades have seen continuing growth in the number of these industries, safety measures are generally poorly applied and workers lack proper knowledge or training in the safe handling of these chemicals.

Acute, as well as chronic, pesticide poisonings have been reported among the workers. Although acute health effects are reasonably well documented in case studies or reports on animal experiments, the manifestations of chronic pesticide intoxication are less well recognized or understood (Hayes 1982).

The health profiles of workers in two large-scale pesticide-formulating plants in Egypt were studied in a 3-year, cooperative project, between the International Development Research Centre and Cairo University, that began in April 1987. The objective of the study was to determine the prevalence of pesticide intoxication among workers in the formulation industries and to correlate epidemiologic data with pesticide exposure.

Materials and methods

Participants were a group of 200 employees from a rural plant located in the middle of the Nile Delta about 100 km from Cairo and 100 employees from a semiurban plant 5 km from Cairo. The rural plant, which is well established and large, employs 1 200 people in the formulation of organochlorine, organophosphorous, carbamate, and pyrethroid pesticides, as well as sulfur compounds. The urban plant, with 600 employees, formulates pyrethroids, organophosphorous compounds, and carbamates, as well as detergents and other intermediate chemicals. The workers studied at these plants were directly exposed to pesticides for at least 2 years.

A control group of 300 subjects consisted of 200 rural and 100 urban employees from textile factories. The control groups were matched for age and socioeconomic status with the subjects. They were not exposed to pesticides directly or in their work.

A health profile of all participants was developed through the administration of a questionnaire and a medical examination, in which special attention was paid to liver pathology and impairment of the central and peripheral nervous systems. In addition, 120 rural and 88 urban plant workers were examined by three psychiatrists and the diagnoses recorded according to Egyptian Psychiatric Association (1979). Other tests included electroencephalograms, electromyograms, and sonography of the liver.

Clinical and biochemical laboratory investigations included liver function tests, determination of acetylcholinesterase (AChE) activity, and definition of the types and levels of pesticides found in blood samples. Urine and stool samples were examined for the presence of bilharzia and other parasites. Air samples were taken using personal samplers to assess the concentrations of pesticides present in the working environment. Associations between study findings, the rural or urban location of workers, the duration of exposure, and the presence of bilharzia were examined.

Results

All participants in the groups exposed to pesticides were male, almost all were married, and 55% were smokers. Duration of exposure ranged from 2 to 35 years for the rural group and 2 to 25 years for the urban group.

Medical histories revealed that there were no statistically significant differences between the exposed groups of plant workers and the control groups with regard to the diagnosis of bilharzia, hepatitis, and diabetes (Table 1). Hypertension was present in a significantly higher number of pesticide-plant workers of both groups. No diagnosis of acute poisoning was reported.

Table 1. Health profile of rural and urban groups exposed to pesticides compared with unexposed subjects.

Condition	Control (%) ($n = 300$)	Rural plant (%) ($n = 200$)	Urban plant (%) ($n = 100$)
Past illnesses			
Acute poisoning	0	0	0
Bilharziasis	48	51[a]	47[a]
Hepatitis	5	8.5[a]	5[a]
Diabetes	3.3	3[a]	3[a]
Hypertension	1.7	8[b]	7[b]
Current complaints			
Neuropsychiatric	10	48[c]	47[c]
Abdominal	30	25[a]	32[a]
Chest and URT[d]	16.6	21[a]	31[b]
Topical	5	20[c]	10[a]
Sexual	1	4[e]	15[c]
Others	6.3	22.5[c]	23[c]
Examination findings			
Neurological	5	45[c]	47[c]
Psychiatric	20	50	51[c]
Abdominal	11	23.5[b]	17[a]
Cardiorespiratory	30	26.5[b]	16[b]
Topical disorders	0	5.5	4
Without signs	40.7	26[c]	36[a]

[a] Not significantly different from control group.
[b] Significantly different from control ($p < 0.01$).
[c] Significantly different from control ($p < 0.001$).
[d] URT, upper respiratory tract.
[e] Significantly different from control ($p < 0.05$).

However, based on individual complaints, there were significant differences between exposed and control populations with regard to neuropsychiatric and certain other problems. Complaints regarding skin conditions were more commonly reported in the exposed rural group and chest and upper respiratory tract dysfunction was more common among the urban exposed group. Abdominal complaints were similar for all groups.

Medical examinations revealed significantly more neurological, psychiatric, and liver disorders among the exposed groups.

Polyneuropathy was significantly more common among exposed workers (37.33%) than in the control group (7%). The typical clinical presentation was a worker with several years' exposure, complaining of general ill-health, paresthesia (30.33% of all exposed workers), impotence (18.67%), depressed tendon reflexes, stock and glove hypoesthesia with muscle twitches and hypotonia occurring in some of the workers. Optic neuropathy and diminution of visual acuity were found, especially among the rural exposed workers. The polyneuropathy was generally mild and mainly sensory, although there was evidence of more severe neuropathy with distal wasting and weakness of the muscles of both the upper and the lower limbs in some cases. All neurological findings were more prominent among older workers and those exposed for a longer period. Neurobehavioural symptoms and signs in the exposed workers appeared to be related to duration of exposure.

Abdominal abnormalities were more common in the rural exposed group. Enlargement of the liver was frequent and associated in some with tenderness, more characteristic of pesticide intoxication than bilharziasis. On the other hand, signs related to the cardiorespiratory system, including atherosclerosis and hypertension with cardiac asthma, were more common among the urban workers ($p < 0.01$). Topical (dermal) manifestations were observed only among workers in the exposed groups.

Abnormal electrocardiograms (ECG) and ventilatory function were observed in both groups exposed to pesticides compared with the control groups (Table 2). More electroencephalographic (EEG) changes were observed in exposed groups than in the controls ($p < 0.001$). Electromyographic changes among rural exposed workers were about three times the number observed among urban workers and control subjects (Table 3).

A higher proportion of abnormal sonography was observed in the rural group and more abnormal electrocardiograms in both exposed groups compared with the control group. Impairment of ventilatory function was greater in both exposed groups than in control subjects, but the differences were not statistically significant.

Laboratory investigations by high-performance liquid chromatography revealed pyrethroids, chlorinated hydrocarbons (dichlorodiphenyltrichloroethane, DDT) and organophosphate pesticides in significantly higher amounts in blood samples from rural workers. The average concentration of

Table 2. Selected results of the medical examinations (values in parentheses are %).

Condition	Rural plant (n = 200)	Urban plant (n = 100)	Control (n = 300)
Smokers	110 (55.0)a	55 (55)a	180 (60.0)a
Abnormal liver	27 (13.5)a[a]	11 (11)a[b]	13 (4.3)b
Abnormal spleen	5 (2.5)a[c]	1 (1)b	11 (0.3)b
Ascitis	3 (1.5)a	1 (1)a	2 (0.6)a
Lower limb edema	8 (4.0)	0	2 (0.6)
Abnormal ventilatory function	65 (32.5)a[d]	34 (34)a[d]	27 (9.0)b
Abnormal ECG	67 (33.5)a[d]	33 (33)ad	34 (11.3)b
Low red blood cell count	131 (65.5)a[e]	14 (14)b	60 (20.0)b

Note: Within a row, values followed by the same letter are not significantly different.
[a] Significantly different from control ($p < 0.01$).
[b] Significantly different from control ($p < 0.05$).
[c] Significantly different from urban group ($p < 0.01$) and from control ($p < 0.05$).
[d] Significantly different from control ($p < 0.001$).
[e] Significantly different from urban group and control ($p < 0.001$).

Table 3. Proportion (%) of subjects showing abnormalities in special investigations (sample sizes are given in parentheses).

Test[a]	Control group (%)	Rural plant workers %	p	Urban plant workers %	p
EEG	2 (100)	24.5 (200)	< 0.001	27 (100)	< 0.001
EMG	10 (10)	36.7 (60)	n.s.	7.5 (40)	n.s.
ECHO	10 (300)	22.0 (200)	< 0.01	2 (100)	< 0.05
ECG	5 (300)	33.5 (200)	< 0.001	33 (100)	< 0.001
Ventilatory function	27 (100)	32.5 (200)	n.s.	34 (100)	n.s.
AChE activity < 50%	7 (300)	62.0 (180)	< 0.001	24 (100)	0.001
Liver function	20 (300)	90.0 (200)	< 0.001	86 (100)	< 0.001

[a] EEG, electroencephalograph; EMG, electromyograph; ECHO, abdominal ultrasonograph; ECG, electrocardiograph; AChE, acetylcholinesterase.

DDT in blood plasma of exposed rural subjects was about six times that in the controls. Pyrethroids, organophosphates, and carbamates were also detected in the blood of the urban workers.

All pesticides being formulated were detected in air samples from work areas in both factories. The concentrations in air were two to five times the permissible levels.

Severe depression of AChE activity was found in both exposed groups, but was marked in the rural group (Table 4). Health profiles of those with AChE activity below 60% of normal were reviewed (Table 5). Exposed workers had significantly more psychiatric and cardiorespiratory problems. Neurological

Table 4. Acetylcholinesterase activity in rural plant workers ($n = 159$).

Level of activity	Number of cases	%	Maximum	Minimum	Average	SD
High activity	6	3	9.50	6.30	7.22	1.13
Normal	40	25	5.90	2.10	3.62	1.19
Moderate inhibition	34	21	1.90	1.00	1.43	0.29
Sever inhibition	40	25	0.99	0.51	0.76	0.13
Extremely severe inhibition	39	25	0.48	0.03	0.25	0.13

Table 5. Health profile of subjects with acetylcholinesterase activity below 60% of normal ($n = 159$).

System affected	Control (%) ($n = 21$)	Rural plant ($n = 114$) %	p	Urban plant ($n = 24$) %	p
Neurological	52.4	48.2	n.s.	83.3	< 0.001
Psychiatric	14.3	69.3	< 0.001	100	< 0.001
Cardiorespiratory	9.5	39.5	< 0.01	45.8	< 0.01
Hematological	4.8	71.1	< 0.001	4.2	n.s.
Hepatic	19.0	35.1	n.s.	—	n.s.
Urinary	4.8	9.6	n.s.	—	n.s.
Gastrointestinal	9.5	2.6	n.s.	12.5	n.s.
Topical	0	3.5		4.2	n.s.
Total	57.2	98.2	< 0.01	100	< 0.01
Subjects without health problems	42.8	1.8	< 0.001	0	

symptoms in the urban group and hematological manifestations in the rural group were also significantly related to AChE depression. Further investigation showed that neurological symptoms were significantly associated with blood contamination by pesticides in both exposed groups; cardiorespiratory and hematological abnormalities were associated with blood contamination by pesticides in the rural group of workers. The latter may be due to heavy pesticide exposure, socioeconomic status, and a higher prevalence of endemic diseases.

Examination (physical, clinical laboratory, and sonography) of liver function and pathology revealed a burden of liver disease attributable to the effects (additive or synergistic) of both pesticides and bilharzial infestation (Tables 6 and 7). There was a significant difference in liver pathology between rural workers exposed to pesticides and the control subjects, clearly demonstrated by a combination of findings: hepatosplenomegaly with ascites, edema of the lower limbs, abnormal sonography, and abnormal liver-function tests (Tables 6 and 7). A significant difference was also found between exposed urban workers and control subjects with respect to hepatosplenomegaly and ascites.

Table 6. Summary of liver function tests and pathology.

Measure	Rural workers		Urban workers	
	Control ($n = 200$)	Exposed ($n = 200$)	Control ($n = 100$)	Exposed ($n = 100$)
Abdominal complaint	55	50	45	32
History of bilharziasis	98	102	54	47
History of hepatitis	10	17	5	5
Hepatosplenomegaly and ascites	15	34[a]	1	13[a]
Tender epigastrium	4	4	1	4
Lower limb edema	2	8[b]	0	0
Positive sonography	20	4[a]	10	24[b]

[a] Significantly different from control ($p = 0.01$).
[b] Significantly different from control ($p = 0.05$).

Table 7. Hepatic and renal function derangement among workers exposed to pesticides.

Measure[a]	Rural workers		Urban workers	
	Control ($n = 200$)	Exposed ($n = 200$)	Control ($n = 100$)	Exposed ($n = 100$)
Reduced AChE (< 50%)	15	104	6	24
LDH	6	35	2	—
GPT	36	28	14	14
YGT	20	136	9	86
Total protein	2	52	2	1
Albumin	24	16	11	0
Bilirubin	40	42	10	6
Alkaline phosphatase	42	—	17	26
Sugar	6	80	2	13
Urea	5	68	1	3
Creatinine	3	4	5	57

[a] AChE = acetylcholinesterase activity; LDH = lactic dehydrogenase; GPT = glutamic pyruvic transaminase.

Hepatocellular damage, indicated by liver-function tests, significantly affected both rural and urban pesticide workers (Table 7). Bilirubin and alkaline phosphatase were not affected significantly in either group, suggesting that cholestasis was not a factor in the liver damage. Urea and creatinine levels were significantly affected in exposed groups. Total protein values were significantly lower in exposed groups, a finding compatible with chronic liver dysfunction.

The influence on hepatorenal function of duration of exposure was investigated by comparing people exposed to pesticides in the work place for more than 10 years with those exposed less than 10 years. Rural workers exposed more than 10 years had significantly higher blood sugar levels ($p < 0.01$), and AChE activity was significantly reduced with longer exposure ($p < 0.001$).

Age did not appear to be an important factor except for the occurrence of certain respiratory conditions and anemia that affected younger workers (under 40 years) exposed to pesticides. Abnormal ECGs, although significantly more frequent among the younger rural exposed workers, were also associated with older workers (over 40 years) in the urban exposed group.

The clinical laboratory investigations, including the hepatorenal function tests, were also reviewed in relation to the presence of bilharzia in the pesticide workers. Comparison with the control groups suggested that bilharzial infestation was associated with significantly more severe functional changes, particularly among the rural pesticide workers.

Discussion

Of special interest in this study was the significantly more frequent occurrence of neurologic, psychiatric, and liver disorders among the workers exposed to pesticides (Albert 1976; Gupta et al. 1978; Amr et al. 1984a,b,c; El Samra et al. 1984). The high prevalence of polyneuropathy supports two hypotheses: that organophosphates can produce delayed neuropathy and that prolonged low exposure to these substances can induce nervous-system symptoms (Lotti et al. 1984). However, chronic and cumulative effects of other compounds used in pesticide formulation cannot be excluded. Sridhar (1986) postulated that changes in the concentration of trace elements in people exposed to pesticides might explain the observed neuromuscular changes. The neurobehavioural symptoms and signs in the exposed workers seemed to be related to duration of exposure.

Behavioural effects are now a recognized outcome of exposure to many industrial and environmental chemicals, especially pesticides (Weiss 1988) and the EEG changes observed in both exposed groups corroborate this. EEG abnormalities, which are destructive or irritative in nature (Klimmer 1969; Mannaioni 1960; El Samra et al. 1984; Savage et al. 1988), might reflect toxic encephalopathy, cerebral anorexia, brain edema and osmotic permeability, or metabolic disturbances prevailing among pesticide-exposed workers.

Chronic liver pathology, found by examination and confirmed by abdominal ultrasonography and liver-function tests, was a common, but not unexpected, finding, as liver problems are common in Egyptians (Amr et al. 1984a,b,c; Kundiev et al. 1986; Moses 1989). Impaired liver function was confirmed in 90% of rural and 86% of urban workers. Establishing conclusive correlations between liver function and pesticide exposure is confounded by the high prevalence of bilharzia and hepatitis.

Liver injury results in failure or delay in the body's detoxification mechanism causing earlier development of cumulative pesticide effects. Subnormal metabolism and elimination of neurotoxins, pesticides, and their metabolites is evident among susceptible and sensitive subjects (Couri and Milks 1982).

Humans, in general, have a limited capacity for metabolizing, for example, chlorinated hydrocarbons (Roan and Morgan 1971). Pesticides cause mild acute and subacute liver-cell damage and affect renal function and possibly B cells in the pancreas (Amr et al. 1984a,b,c). Hepatic susceptibility to pesticides is also related to low dietary protein as well as parasitic infestations and other demographic factors.

The high degree of abnormal liver function in workers not suffering from bilharzia may indicate a toxic effect of the pesticides; among those with bilharzia, the bilharzial infestation and the chemical effects of pesticides may have exerted effects in an additive or synergistic manner. However, the chronic hepatocellular damage among those heavily exposed to pesticides cannot be discounted. Our findings focus attention on the importance of initial and periodic medical examinations to identify those with liver fibrosis or with liver-function abnormalities.

There was some evidence of renal pathology in subjects exposed to pesticides. Toxic compounds may damage the kidneys directly and release renal antigens, eliciting the production of antiglomerular basement-membrane antibodies. This may be followed by antigen–antibody or autoimmune reactions. Susceptibility to this autoimmune reaction might be inherited (Kleinknecht et al. 1980).

The occurrence of hypertension among pesticide workers is thought to be attributable to the possible arteriosclerotic effects of pesticides (Hill and Carpenter 1982); their abnormal ECGs might be due to either normal aging processes or myocardial pesticide effects. However, abnormal ECGs were significantly more prevalent among younger workers in the rural group, but in older workers in the urban group. A possible explanation is repeated infestation by bilharzia in the former and the cumulative effect of pesticides in the latter. The high prevalence of bilharzia and the extensive exposure to pesticides in the rural plant may explain the higher number of cardiorespiratory disorders (e.g., bilharzial corpulmonale and cardiomyopathy) in the rural workers (Wegever et al. 1982). Young workers exposed to zinc phosphide are also more frequently found to have abnormal ECGs (Amr et al. 1984a,b,c).

Zipf et al. (1976) reported evidence of cardiac muscle damage, with abnormal ECG tracings associated with chest pain, palpitation, and dyspnea in pesticide-exposed subjects. In spite of the larger number of abnormal ECGs among exposed workers in our study, there were no significant differences between the two exposed groups and the control subjects except for ill-defined pathological effects.

Anemia among Egyptians is usually a result of parasitic infestation, but may also be attributable to pesticide exposure (Klimmer 1969). McConnell et al. (1980) noted a dose-related, progressive anemia due to exposure to pentachlorophenol; pentachlorophenol is also associated with leukocytosis (Clemmer et al. 1980). The most likely explanation for blood dyscrasia associated with pesticide exposure appears to be idiosyncratic bone-marrow reactions to the pesticides in occupationally exposed individuals. Among workers exposed to zinc phosphide, number of red blood cells and percentage of hemoglobin were significantly higher (Amr et al. 1984a,b,c), probably due to the irritative stimulant effects on the homeopathic system of small doses of zinc phosphide.

Impairment of ventilatory function observed in both exposed and control groups was probably related to pesticide exposure in the former and hazards in the textile-industry environment in the latter. Although respiratory problems have been linked to pesticide exposure (Lehnigk et al. 1985; Mechkow et al. 1987), the significantly more prevalent symptoms observed among young workers in the rural plant compared with urban workers may also be linked with smoking habits and parasitic infestations.

Cases of asthma have been linked with exposure to organophosphate pesticides (Hayes 1982), but it is uncertain whether the active ingredient or other components in the formulation produce the effects. Carbamates may cause fibrotic changes noticed among survivors of suicide attempts (WHO 1984). A high rate of impairment of ventilatory function (70%) was observed among workers exposed to zinc phosphide (Amr et al. 1984a,b,c); Koilpakov (1987) reported significant deviation from standard functional parameters among workers exposed to pesticides. Asthma and impaired ventilatory function among those exposed to organophosphates may have a biochemical basis: accumulation of endogenous acetylcholine with consequent cholinergic stimulation after AChE inhibition. However, among those exposed to DDT, the effects might be due to either the interaction of DDT with beta-receptor adenylate cyclase complex (Dudeja et al. 1980) or interaction with adrenal cytochrome P-450 and inhibition of endogenous steroid formation.

Reduction in AChE activity is an indicator of exposure to organophosphate and carbamate pesticides. It is a measure of the biochemical effect of the pesticide rather than of the pesticide itself. Cholinesterase inhibitors (organophosphates and carbamates) prevent the breakdown of AChE, which consequently builds up in the brain and in the peripheral synapses (Risch and Janowsky 1984). Significant depressive (Bowers et al. 1964), neurologic, and cardiorespiratory symptoms are often displayed as a result. These clinical manifestations may be due to the direct effect of the pesticides on the nervous system, myocardium, and bone marrow, or indirectly due to the metabolic changes that coincide with the high pesticide level in blood, e.g., effects of hormones, trace elements, and enzymes. The degree to which blood AChE activity was reduced in both exposed groups seems to reflect insufficient awareness of acute pesticide poisoning by attending physicians as well as a lack of appropriate safety measures and ventilation systems in the working environment.

The long retention time of pyrethroids, organophosphates, and carbamate compounds in the blood of the exposed urban workers might be due to heavy exposure, liver damage, or both. Blood contamination with chlorinated hydrocarbons (e.g., DDT) suggests that neuropsychiatric changes and liver damage (detected by ultrasonography or biochemical parameters) in the rural group of workers might be due to this group of pesticides.

Contact dermatitis and allergic sensitization have been observed frequently in pesticide workers after exposure to a variety of pesticides (Adams 1983). Photoallergic reactions have been reported as well (WHO 1989). The number of cases in exposed workers in this study differed significantly from the control group, but the numbers were small.

It was difficult to determine the degree or duration of exposure to specific pesticides. Workers were shifted from department to department as vacancies occurred, and formulation of pesticides in both establishments depended on market factors. However, the overall results demonstrated the existence of hazardous working conditions in the two plants.

Signs and symptoms of acute and chronic pesticide exposure were prominent, particularly for organophosphate and carbamate insecticides and for the exposed group of rural workers, for whom a reduction in AChE activity of more than 50% was recorded in 62% of workers. Liver dysfunction was observed in 90% of exposed workers, but the significance of this in terms of acute and chronic toxicity effects remains unclear because of the endemicity of bilharzia. There was good evidence of significant chronic toxicity in the form of neurologic disorders.

The results of this study indicate the need to improve safety in the working environment and to train workers in the safe handling of pesticides.

Acknowledgment — This work was supported by a grant from the International Development Research Centre, Canada.

Adams, R.M. 1983. Occupational skin diseases. Grun and Stratton, New York, NY, USA.

Alber, G.V. 1976. Zbravookhkirg, 2, 38.

Amr, M.M.; El-Batanouni, M.M.; El-Gendy, M.; Gaafer, T. 1984a. Some cardiorespiratory and haematological manifestations due to exposure to zinc phosphide (ZnPO$_4$). Egyptian Journal of Occupational Medicine, 8(2), 131–141.

Amr, M.M.; Abbas, E.Z.; El Samra, G.H.; Ayed, F.A. 1984b. Clinical and biochemical evaluation of hazards in workers occupationally exposed to zinc phosphide pesticide. Medical Journal of Cairo University, 52(2).

Amr, M.M.; et al. 1984c. Some metabolic studies among workers exposed to zinc phosphide (ZnPO$_4$) during its preparation in Egypt. Journal of Occupational Medicine, 8(2), 211–222.

Bowers, M.B.; Goodman, E.; Sim, V.M. 1964. Some behavioral changes in man following anticholinesterase administration. Journal of Nervous and Mental Disease, 138, 383.

Clemmer, H.W.; et al. 1980. Clinical findings in workers exposed to PCP. Envonmental Contamination and Toxicology, 1980, 715–725.

Couri, D.; Milks, M. 1982. Toxicity and metabolism of the neurotoxic hexacarbons n-hexanone and 2,5-hexanedione. Annual Review of Pharmacology and Toxicology, 22, 145–166.

Davis, J.E.; et al. 1986. Changing profile in human health effects of pesticides. Paper presented at the 6th international congress of pesticide chemistry, Ottawa, Canada, 10–15 August 1986.

Dudeje, P.K.; Sanyal, S.N.; Agarwal, N.; Rao, T.N.; Subrahmanyam, D.; Khuller, G.K. 1980. Acute exposure of rhesus monkeys to DDT: effect on carbohydrate metabolism. Chemico-Biological Interactions, 31(2), 203–208.

Egyptian Psychiatric Association. 1979. Diagnostic manual of psychiatric disorders. Egyptian Psychiatric Association, Cairo, Egypt.

Gupta, S.K.; Jani, J.P.; Saiyed, H.N.; Kashyap, S.K. 1984. Health hazards in pesticide formulators exposed to a combination of pesticides. Indian Journal of Medical Research, 79, 666–672.

Hayes, W.J., Jr. 1982. Pesticide studies in man. Williams and Wilkins, Baltimore, MD, USA.

Hill, E.F.; Carpenter, J.W. 1982. Response of Siberian ferrets (Mustles cuersmanni) to secondary Zn_3P_2 poisoning. Journal of Wildlife Management, 46(3), 678–685.

Kleinknecht, D.; Morel-Maroger, L.; Callard, P.; Adhemar, J.P.; Mahieu, P. 1980. Antiglomerular basement membrane nephritis after solvent exposure. Archives of Internal Medicine, 140(2), 230–232.

Klimmer, O.R. 1969. Beitrog zur Wickung des phosphorwosser H3 peffesot. Archives of Toxicology, 24, 164–187. [Cited in Hayes 1982.]

Koilpakov. 1987. The possibility of early detection of disorders of the respiratory system in persons working with pesticides. Fiziologicheskii Zhurnal, 33(6), 505.

Kundiev, Y.; Krasnyuk, E.P.; Viter, V. 1986. Specific features of the changes in the health status of female workers exposed to pesticides in green houses. Toxicology Letters, 33(3), 85–89.

Lehnigk, B.; Thiele, E.; Wosnitzka, H. 1985. Respiratory function disorders caused by agromedical exposure. Zeitschrift fuer Erkrankungen der Atmungsorgane, 164(3), 267.

Levine, R.S. 1986. Assessment of mortality and morbidity due to unintentional pesticide poisonings. World Health Organization, Geneva, Switzerland. WHO/VBC/869.929, p. 24.

Mannaioni, P.F. 1960. Clinical toxicological considerations in some cases of acute poisoning by zinc phosphide. Minerva Medica, 51, 3 721–3 724.

McConnell, E.E.; et al. 1980. The chronic toxicity of PCP in cattle. Toxicology and Applied Pharmacology, 52, 468–490.

Mechkov G.; Petkova, V.; Zakariev, N. 1987. Bromsulphatein clearance in persons in occupational contact with pesticides. Problemi na Khigienata, 12, 99–102.

Moses, M. 1989. Pesticide-related health problems and farm workers. AOHN Journal, 37(3), 115–130.

Risch S.C.; Janowsky, D.S. 1984. Cholinergic–adrenergic balance in affective illness. *In* Post, R.M.; Ballenger, J.C., ed., Neurobiology of mood disorders. Williams and Wilkins, Baltimore, MD, USA. Pp. 652–663.

Roan, C.C.; Morgan, D.P. 1971. Absorption, storage and metabolic conversion of ingested DDT and DDT metabolites. Archives of Environmental Health, 22(3), 301–308.

El Samra, G.H.; Amr, M.M.; Osman, A.L.; Nahlah, M.A. 1984. Neuropsychiatric manifestation due to zinc phosphate exposure. Egyptian Journal of Occupational Medicine, 8(2), 177–196.

Savage, E.P.; Keefe, T.J.; Mounce, L.M.; Heaton, R.K.; Lewis, J.A.; Burcar, P.J. 1988. Chronic neurological sequelae of acute organophosphate pesticide poisoning. Archives of Environmental Health, 43(1), 38–45.

Sridhar, M.K. 1986. Pesticide usage and poisoning in Nigeria. Journal of the Royal Society of Health, 1986, 182–184.

Wegever, N.R.; et al. 1982. Myocardial disease. *In* Hurst, J.W., ed., The heart (5th ed.). McGraw Hill, New York, NY, USA.

Weiss, B. 1988. Behavior as an early indicator of pesticide toxicity. Toxicology and Industrial Health, 4(3), 351–360.

WHO (World Health Organization). 1985. Environmental pollution control in relation to development. WHO, Geneva, Switzerland. Technical Report 718.

_____ 1989. Dichlorvos. *In* Environmental and health criteria. WHO, Geneva, Switzerland. P. 157.

Xue Shou-Zheng. 1987. Health effects of pesticides. American Journal of Industrial Medicine, 12, 269.

Zipf, K.E.; Arndt, T.; Heint, R. 1976. Klinisch seoboehtunger beieiner phostoocin-Vergiftung. Archives of Toxicology, 22, 209–222. [Cited in Hayes 1982.]

Extent of exposure of farm workers to organophosphate pesticides in the Jordan Valley

R.M. Sansur,[1] S. Kuttab,[2] and S. Abu-Al-Haj[1]

[1]Center for Environmental and Occupational Health Sciences and
[2]Chemistry Department, Birzeit University, West Bank, via Israel

The aim of this study was to determine the extent to which farm workers were exposed to organophosphate pesticides and, based on the results, to develop a program to teach farmers safe practices in the handling and use of pesticides. Low acetylcholinesterase (AChE) levels were found among participating farmers; recovery was dependent on the dose. AChE levels were higher during the spraying season than at the end of the season possibly due to enzyme induction resulting from long-term exposure. Legs were the parts of the body most exposed during spraying. A knowledge, attitudes, and practices (KAP) survey revealed lack of knowledge about the safe handling of pesticides and the dose required for spraying. An intensive long-term training program on the safe use of pesticides is urgently needed.

When used properly, pesticides have been of tremendous benefit to humans and the environment, but, when used carelessly, they have caused considerable and, in some instances, inestimable harm. In developing countries, the increasing use of pesticides in agriculture and the absence of adequate worker education and effective control measures have given rise to a growing concern about the magnitude of the health risk to farm workers exposed to these chemicals. It is suspected that many illnesses and deaths go unrecorded.

In the Jordan Valley of the West Bank (Al Ghor), spraying of crops with pesticides has increased tremendously over the last 20 years. The increased use of dangerous organophosphate insecticides as substitutes for chlorinated hydrocarbons, which degrade more slowly and have a higher environmental impact, has added a new dimension to a serious occupational health problem in this area.

Al Ghor is on the west side of the river Jordan. It is a warm, semitropical area, 100–200 m below sea level and is unique in that farm produce can be cultivated during the winter months. Intensive agriculture is thus practiced from October to May. Although now diminished in size, the Palestinian area being cultivated amounts to 40 km^2. The Arab population is about 30 000, two-thirds of whom are involved in agriculture. Agriculture involves all members of the farming families, including pregnant women and children. About half the farmers are landowners; the others are tenants.

Organophosphates are potent inhibitors of acetylcholinesterase (AChE), an enzyme responsible for the hydrolysis of acetylcholine, which is synthesized at nerve endings and is involved in transmission of impulses from nerve to muscle fibres. A 40% decrease in blood AChE is associated with the first symptoms of pesticide poisoning and an 80% decrease is associated with severe neuromuscular effects. Absorption of organophosphate sufficient to inactivate all red blood cell AChE may result in death. Absorption may occur through the skin, by inhalation, or by ingestion.

Farmers are exposed to organophosphates either through continuous spraying of these pesticides or from contact with their residues on treated plant surfaces. No statistical information was available concerning poisoning incidents related to crop-spraying activity among the farmers of the region. However, a preliminary survey of a small number of farming operations revealed unacceptable mixing and spraying habits. Chronic exposure, poor working conditions, and lack of education about potential hazards could have serious health and economic implications for the region.

The objective of this study was to learn the extent of exposure to organophosphate pesticides of farmers in the Jordan Valley during and after the spraying season and to determine the knowledge, attitudes, and practices (KAP) of the farm workers regarding the use of pesticides. The information will be used to design a program for farmer education and training in the safe use and handling of pesticides.

Methods

Population and sampling

All of the participants were tenant farmers from the Jiftlic region of the Jordan Valley, who leased land and shared some of the expenses and the profit with the landlord. The participants were divided into groups:

- Group I consisted of 14 farmers in three subgroups (4, 6, and 4), each spraying 1 000 L of pesticide;

- Group II consisted of 10 farmers in two subgroups (5 and 5), each spraying 2 000 L of pesticide;

- Group III consisted of 31 farmers whose blood was sampled just before the beginning of the spraying season and at the end of the summer to determine AChE levels during the off-season; and

- Group IV consisted of 24 farmers whose blood was sampled after a minimum of 3 days abstention from handling pesticides (Group IVa) or randomly, regardless of their handling of pesticides (Group IVb).

Two farmers from the same region, with junior college or vocational training, agreed to act as supervisors to ensure that participating farmers complied with the experimental protocol. Methamidophos (available under the trade names Tamaron, Prodex, and Monitor) was chosen by the farmers and was used at a concentration of 1.2 g/L. A tractor-driven 500-L tank, fitted with a pump and spray gun, was used for spraying. In each subgroup, farmers alternated every 20–30 min between carrying the hose, operating the spray gun, and driving the tractor. Thus, each farmer was exposed to about the same dose of pesticide.

Blood analyses

The level of AChE activity in 10-μL samples of whole blood, obtained by finger-pricking, was measured spectrophotometrically using a method developed for field use (WHO 1984).

Dermal exposure

A standard protocol (WHO 1982), developed for field surveys, was used to measure the extent of exposure to the pesticide during spraying operations. Seven polyethylene-backed filter paper pads (Schleicher and Schuell, catalogue no. 295 PE), cut to 10×10 cm, were attached to different parts of the body with tape before spraying began. At the end of the work period, the pads were removed, a 5×5 cm piece was cut out and placed in a plastic envelope in a cooler for later analysis. As a control and to determine recovery, a known concentration of the organophosphorous pesticide was placed on a pad that was treated in a similar fashion.

Each 5×5 cm pad was cut into strips and placed in a beaker containing 5 mL ethyl acetate and extracted in an ultrasonic bath for 5–7 min. The ethyl acetate extract was concentrated and analyzed by gas chromatography using a 30-m capillary column SE 54. Column temperature was 180 °C, injector temperature was 200 °C, and TSD detector temperature was 300 °C.

Knowledge, attitudes, and practices

The survey to determine KAP was based on a standard protocol (WHO 1982) with certain modifications to elicit information about knowledge regarding the toxicity of pesticides.

Data analysis

Because of the many factors that influence AChE levels, nonparametric statistical tests were used to evaluate the significance of differences in activity. These included Spearman and Wilcoxon ranked matched-pair tests. Significance tests were based on Spearman's matched-pairs rank correlation coefficient.

Results and discussion

Choice of pesticide

The choice of methamidophos, the organophosphate pesticide used in this study, was left entirely to the farmers. To achieve uniform results, we opted to work with farmers using the same pesticide throughout the study. Methamidophos was not the best choice for our purposes, as it is a relatively weak AChE inhibitor in spite of its good insecticidal activity (Quistad et al. 1970). In addition, it is highly toxic, with a Class I toxicity rating; in rats, the acute oral LD_{50} (lethal dose to 50% of animals tested) is 30 mg/kg in rats (WHO 1988).

The concentration of pesticide after dilution was 1.2 g/L, which is double that recommended for the type of equipment the farmers normally use. Again the choice of concentration was left to the farmers.

AChE levels in blood samples

Mean AChE levels for Group I were significantly lower (at the 95% confidence level) 2 h after completion of the spraying operation compared with levels before spraying (Fig. 1). The difference was not significant on the following 2 days, indicating a recovery in AChE level. Exposure time for this group, which sprayed 1 000 L per subgroup of 1.2 g/L methamidophos, was 1.5–2 h per subgroup. At no time did AChE levels fall below normal range.

Group II also displayed a significant decrease in AChE levels in whole blood during the first 2 h after spraying, but their AChE levels were also significantly below those measured before spraying on the following 2 days. This group sprayed 2 000 L of 1.2 g/L methamidophos per subgroup. We observed that 2 000 L (four tanks) was the minimum a farmer would use on a particular spraying day; hence this level of exposure is realistic. Total exposure time was 3–3.5 h per subgroup.

No clinical symptoms of intoxication with methamidophos were evident among the participants, except for fatigue.

AChE levels in whole blood were measured twice: during the spraying season in March and again at the end of the season in September. For the early test,

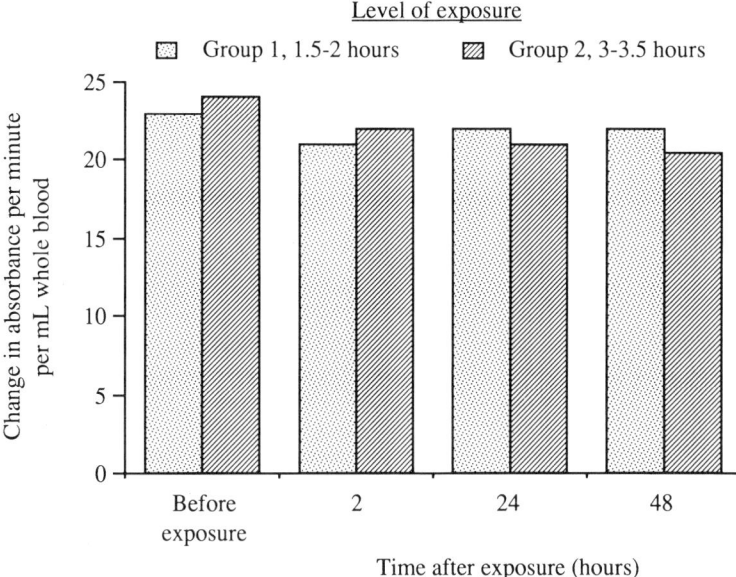

Fig. 1. Acetylcholinesterase levels in whole blood of participating farmers at selected times after they had finished spraying pesticide.

farmers were asked not to handle pesticides for at least 3 days before blood samples were taken.

Mean AChE activity (measured as change in absorbance per min per mL of whole blood) was 19.13 ± 0.394 (mean $\pm$ standard error) at the end of the season compared with 23.34 ± 0.470 during the spraying season, a significant difference of 22% for Group IVa and 15% for Group IVb. This result was opposite to what we expected, i.e., lower AChE activity during the spraying season due to the effects of organophosphate and carbamate pesticides on cholinesterase enzymes.

The seasonal differences in AChE activity may be the result of excessive contact with pesticides resulting in enzyme induction (Burgess and Roberts 1980). Roberts (1980) mentions the possibility that persistent exposure to pesticides may cause "cumulative biological changes." Individual differences in AChE activity are common (Loosli 1980). The farming population in the Jordan Valley has been using pesticides for at least the past 20 years.

Dermal exposure

Dermal exposure generally accounts for 90% of total exposure to pesticides by farmers. Seven pads were placed on each subject to determine which areas of the body were most exposed (Table 1). As expected, Group II, which operated spraying equipment twice as long as Group I, experienced a higher degree of

Table 1. Concentration of methamidophos (g/cm^2) on pads
collected from Groups I and II.

Pad no. [a]	Group I		Group II	
	Mean ± SE	Range	Mean ± SE	Range
1	0.70 ± 0.40	0–1.87	3.15 ± 1.41	0.20–13.86
2	26.62 ± 6.30	13.03–41.51	117.59 ± 34.21	10.26–303.60
3	10.78 ± 2.35	6.61–16.17	37.55 ± 16.98	2.91–174.68
4	1.09 ± 0.43	0.21–2.26	1.67 ± 0.33	0.43–2.86
5	0.34 ± 0.16	0–0.76	1.53 ± 0.57	0–5.08
6	1.19 ± 0.82	0–3.63	2.93 ± 1.22	0–12.51
7	0	0	0.46 ± 0.20	0–1.37
Total	40.72 ± 1.74		164.88 ± 7.85	

[a] 1, left upper arm; 2, left lower leg; 3, left mid-thigh; 4, chest over sternum; 5, back between shoulder blades; 6, forehead; 7, belly under clothing.

exposure. The longer duration of exposure resulted in the pesticide penetrating the clothing and becoming absorbed on the skin pads in more of the participating farmers. For both groups, the area of the body that received the highest concentration of pesticide was the leg (pads 2 and 3). A sizable reduction in dermal exposure could be achieved by covering legs and feet with boots made of rubber or other inexpensive but pesticide-proof material.

Knowledge, attitudes, and practices

The pesticide chosen by the farmers is very toxic. However, the farmers took no special precautions during mixing or spraying. Although methamidophos is classed as "highly hazardous," i.e., complete protective gear must be worn during spraying operations, participants wore only their everyday clothing and some sprayed without footwear.

After an argument and some discussion, the farmers decided to prepare the pesticide at a concentration of 1.2 g/L, double the recommended level. Lack of knowledge and inability to read the package label, which was in Hebrew, contributed to the error.

The legs of farmers were most exposed and became wet from the pesticide spray. This was especially evident when it was windy. No metering of the pump, volume delivered by the spray gun, or distance walked was done to optimize the amount of pesticide used.

Except for fatigue, no gross toxic symptoms were observed. However, most of the farmers indicated that methamidophos had a "bad" effect on them without clarifying what they meant.

We did not observe government or other extension agents during the course of the study.

Pesticides were stored everywhere covered only by a plastic sheet: in homes, in animal or food sheds, and among plants. Children of all ages were present on the farms at all times.

Most of the participants (54%) were between the ages of 20 and 29 years, indicating that a young population is involved in farming (Table 2). In this age group, we found a number of people with vocational diplomas or college-level education. Most family members (68%) who participated in spraying and other farming activities were below the age of 20; 55% were age 16 years or younger.

Knowledge about the choice and use of pesticides was gained mostly from personal experience and, to some extent, from pesticide sales outlets. Government extension agents played a minor role; 60 serve the whole West Bank and they do not have enough vehicles or petrol to move around.

Although 8% of the farmers indicated that they had had some training in the use of pesticides, no participating farmer could substantiate that claim by giving a source for the training.

Some farmers believed that, after using pesticides for a long time, they became immune to them. Although this belief was false, our results indicated a possible adaptive mechanism involved in raising AChE levels during the spraying season.

Although 88% of the participants said that they washed or changed their clothes after spraying, our observations indicated the opposite; we saw many farmers wearing the same clothes on a number of consecutive days. In addition, 75% indicated that they did not shower or bathe after spraying.

Eating and smoking while spraying were widespread among the farmers and the responses to this question were more or less accurate.

We did not observe anyone burying empty containers in the field although 33% of those surveyed said that this was their practice. Empty containers were scattered everywhere including around homes.

Conclusion

This study revealed a number of areas that must be addressed in the design of a program for training farmers in the safe handling of pesticides:

- Farmers lack knowledge about the proper choice of pesticides for pest management;

- Farmers are unaware of the correct amount to be sprayed for a given concentration of pesticide;

- Spray equipment is never checked for proper dose delivery;

Table 2. Results of the survey on knowledge, attitudes, and beliefs.

Question	Answers	
Age of participants (years)	< 20	8.3%
	20–24	37.5%
	25–29	16.7%
	29–33	0
	> 33	37.5%
Sex	Male	100%
Type of farm work	Mixed	100%
Average duration of work in the field, with emphasis on spraying		5 h
Proportion of smokers		41%
Proportion helped by other family members		100%
Age group and sex of family members assisting in farm work	< 20 years	68.0%
	20–24 years	12.0%
	25–29 years	5.3%
	30–33 years	4.0%
	> 33 years	10.7%
	Males	50.7%
	Females	49.3%
Source of information about pesticides (some respondents indicated mixed sources)	Government extension agents	42%
	Pesticide sales outlets	25%
	Other farmer	13%
	Experience	88%
Awareness about pesticide toxicity	Positive	33%
	Negative	67%
Training in the proper use of pesticides	Yes	8%
	No	92%
Harvesting produce after spraying	Same day	46%
	Next day	46%
	Few days later	8%
Reentry into the field after spraying	On same day	96%
	1 day later	4%
	2-7 days later	0%
	After 7 days	0%
Are pesticides dangerous to your health?	Yes	50%
	No	4%
	Do not know	46%
Should farmers wear protective apparel?	Yes	88%
	No	0%
	Do not know	12%

Continued

Table 2. Concluded.

Question	Answers	
Do you wear protective clothing?	Yes	4%
	No	96%
Do you believe you are immune against pesticides?	Yes	12%
	No	21%
	Do not know	67%
Do you change or wash your clothes after spraying?	Yes	88%
	No	12%
Do you shower or bathe after spraying?	Yes	25%
	No	75%
Do you smoke while spraying?	Yes	46%
	No	54%
Do you eat while spraying?	Yes	63%
	No	37%
Have you ever spilt pesticides on your body?	Yes	92%
	No	8%
How do you dispose of empty containers?	Throw in the field	63%
	Bury	33%
	Sell for recycling or other uses	4%
Do you have any unusual symptoms after spraying?	Yes	33%
	No	67%

- Adequate information about pest and crop management is unavailable to farmers;

- Spraying is done casually without proper planning or safety precautions;

- There is no regard for reentry periods or length of time between spraying and harvest;

- There is little awareness of the long-term effects of pesticides on farmers or their families as evidenced by young children participating in spraying and the casual way in which pesticides are handled; and

- Government agencies, the media, and chemical companies advise farmers to spray prophylactically.

In conclusion, we believe that a training program for farmers and others in the safe handling of pesticides coupled with pest management is urgently needed. Such a program, if run efficiently, will benefit a wide sector of society and reduce health hazards associated with exposure to pesticides.

Acknowledgments — This study was partly funded by the International Development Research Centre, AMIDEAST, and Birzeit University. We are indebted to Khaled El Hidmi of the Union of Agriculture Work Committees, who assisted us in organizing the farmers. We thank the farmers of the Jiftlic region who helped make this study a success. Aqel-Abu-Qara, Joumana Naser El Din, and Hanan Sa'adeh of the Center for Environmental and Occupational Health Sciences (CEOHS) assisted us in the field and laboratory. We also thank Mrs Zreikat for typing the report. This study would not have been possible without the generous grant from the Shoman Foundation enabling CEOHS to purchase equipment. The Catholic Committee against Hunger and for Development and the Ford Foundation funded some of the pilot projects that gave impetus to this study.

Burgess, J.E.; Roberts, D.V. 1980. *In* Tordoir, W.F.; van Heemstra-Lequin, E.A.H., ed., Field worker exposure during pesticide application. Elsevier, New York, NY, USA. Studies in Environmental Sciences 7, pp. 99–103.

Loosli, R. 1980. *In* Tordoir, W.F.; van Heemstra-Lequin, E.A.H., ed., Field worker exposure during pesticide application. Elsevier, New York, NY, USA. Studies in Environmental Sciences 7, pp. 93–98.

Roberts, D.V. 1980. *In* Tordoir, W.F.; van Heemstra-Lequin, E.A.H., ed., Field worker exposure during pesticide application. Elsevier, New York, NY, USA. Studies in Environmental Sciences 7, pp. 173–176.

Quistad, G.B.; Fukuto, T.R.; Metcalf, R.L. 1970. Journal of Agricultural Food Chemistry, 18, 189.

WHO (World Health Organization). 1982. Field surveys of exposure to pesticides standard protocol. WHO, Geneva, Switzerland. VBC/82.1.

_____ 1984. Spectrophotometric kit for measuring cholinesterase activity. WHO, Geneva, Switzerland. WHO/VBC/84.888.

_____ 1988. The WHO recommended classification of pesticides by hazard and guidelines to classification 1988–1989. WHO, Geneva, Switzerland. WHO/VBC/88.953.

PART II

RELATING PESTICIDE DEVELOPMENT,

MANUFACTURING,

APPLICATION TECHNIQUES,

AND REGULATORY CONTROL

TO USER SAFETY

Development of safe chemical pesticides

W.F. Tordoir

Occupational Health and Toxicology, Health, Safety and Environment
Division, Shell International Petroleum,
The Hague, Netherlands

Pesticides must meet many requirements, of which efficacy is fundamental. Product selection, therefore, always begins with efficacy testing; if these results are promising, additional work is carried out to check for human and environmental safety. However, commercial and technical requirements must also be met. Progress has been made in both increasing the selectivity (i.e., restricting the pesticidal activity to target species) and in improving efficacy. This has made dramatically lower application rates possible. In a number of cases, a reduction in intrinsic toxicity for mammals has also been achieved. Guided by a better understanding of intrinsic toxicity and good data on personal exposure during application, product development and product stewardship can contribute to safety through improved formulations, packaging, and advice on handling and personal protection. A comparison of three insecticides as representatives of organochlorines, organophosphates, and pyrethroids shows that, with the development of the pyrethroids, a major step forward has been made in the creation of safe pesticides. Similarly, the example of a highly toxic rodenticide demonstrates that a high efficacy and sophisticated product development can compensate for high toxicity.

It is only logical that in the history of the development of chemical pesticides, nearly all attention was focused initially on the ability of a substance to kill pests; concern for safety was secondary and followed somewhat later. The drive to develop safe pesticides is reflected in the introduction of pesticide registration and the increasingly stringent requirements to meet registration criteria. In 1950, only acute toxicity tests and a 30- to 90-day feeding study using rats were required for registration. By 1960, a 2-year study of effects on rats and a 1-year study using dogs were necessary. By 1970, most current requirements were in place (Table 1). The need for registration, combined with requirements for toxicity testing and restrictions for marketing certain pesticides, have, in turn, had a positive effect on the development of safe pesticides. ("Safe" is used here in relation to effects on human health.)

Table 1. Requirements and subrequirements met by
the "ideal" pesticide.

Test	Ideal result
Toxicity	
LD_{50}	
oral, rodents	High
rodent	High
dermal, rodents	High
Skin irritation, rabbit	None or slight
Eye irritation, rabbit	None or slight
Skin sensitization, guinea pig	Absent
Mutagenicity screen for predicting carcinogenicity	Nongenotoxic
Subchronic	
oral 14 days, rats, dogs	High NOAEL
oral 5 weeks, rats	High NOAEL
dermal, 14–21 days, rabbits	High NOAEL
feeding, 90 days, rats, dogs	High NOAEL
Carcinogenicity in rats and mice	None
Chronic toxicity, rodent	High NOAEL
Additional mutagenicity studies	Nonmutagenic
Teratogenicity in two species	No potential
Multigeneration reproduction, rats	No reproductive effects
1-year feeding study, dogs	High NOAEL
Neurotoxicity, chicken	Not neurotoxic
Other properties and aspects	
Dermal penetration	Very low
Metabolism	Fast, not leading to toxic metabolites
Accumulation	None
Excretion	Fast, mainly via kidneys (to allow for easy biological monitoring)
Behaviour in environment	Not persistent, breaks down into innocuous substances
Movement into groundwater	None
Residues in food crops	None
Residues in food and mother's milk	None
Exposure monitoring	Sampling and analytical procedures easy and available
Biological monitoring	Available, urinary metabolite(s) that can be related to an oral dose of the pesticide
Cases of overexposure	Warning properties
Product design	Contributing to safety
Antidote	Treatable with specific antidote

Note: LD_{50}, lethal dose to 50% of test animals; NOAEL, no adverse effect level, i.e., the findings in these studies do not generate any specific concern and there are no qualitative discrepancies between species.

Finding a molecule with specific pesticidal properties and low toxicity for nontarget species is still the first step in a long and complicated process. Today, however, modern pesticides must satisfy so many requirements that it seems easier to find a needle in a haystack than to discover a new molecule that meets the full specifications and still yields a profit. The essential requirements of a new pesticide are:

- Specific pesticidal activity;
- Safe for the crops on which it is applied and not seriously affecting nontarget species;
- Not inducing resistance in the target pests;
- Safe to health;
- Safe to the environment;
- Good market opportunities;
- Able to be produced in sufficient quantities in an acceptable manner (with respect to health, environment, and economics);
- Good formulation possibilities; and
- Stable when stored.

Failure to meet one of these requirements will jeopardize successful development of a project. Each of these core requirements can be split up into many individual subrequirements; even one unfulfilled subrequirement can be decisive in not pursuing product development.

Finding a pesticide

The usual process of selecting new chemicals for pesticide use is by elimination. The first step must be testing for pesticidal activity, thereby eliminating inactive candidates. Screening a range of molecules that are structurally similar to a known pesticide is often the approach adopted, although sometimes fundamental studies of the physiological properties or the biochemistry of the target pest provide clues for identifying the class of chemicals that would interfere with its essential biological processes.

Agrochemicals are also identified through a method of widespread screening — the testing of any new chemical against species representing target pests. Discovering a new pesticide often requires luck and scientific curiosity, combined with perseverance (Fig. 1).

Early testing of pesticidal activity is carried out in the laboratory; testing for plant toxicity and for the induction of resistance in pests follows shortly. Promising results in laboratory tests lead to small-scale field studies and limited toxicity data; initially, only ranges of oral and dermal acute toxicity are

144

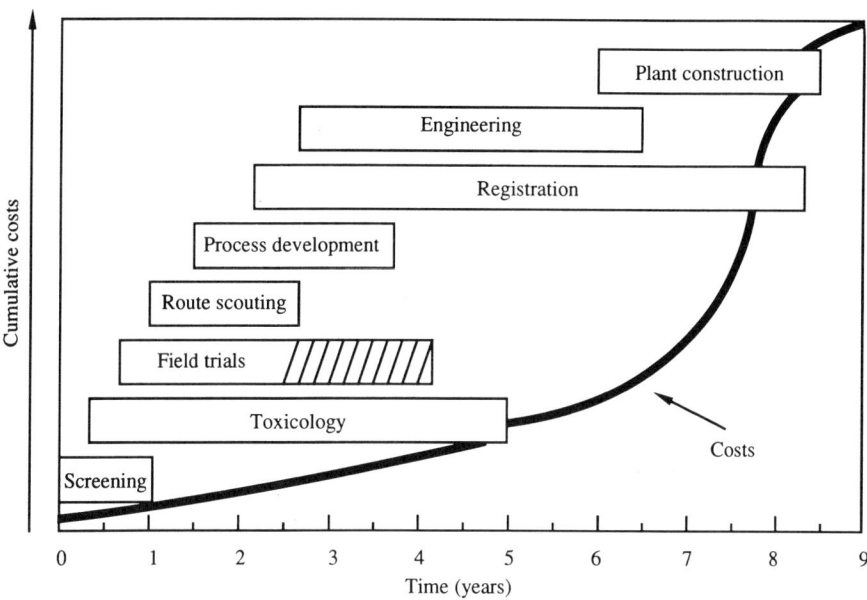

Fig. 1. Stages and costs of agrochemical research and development.

identified. If a new chemical survives to this stage, a market survey is carried out to obtain preliminary information about sales prospects. If a commercial future seems reasonable, further efficacy testing in the field and toxicity testing will be undertaken. The results of efficacy and toxicology testing are reviewed regularly and route scouting is used to determine the most economic, safe, environmentally sound, and technically feasible process of manufacture.

Assuming that all goes well (i.e., successful field-testing, no unpleasant surprises from toxicity testing, promising market prospects, and successful route-scouting), the first preliminary registration for experimental use in larger field studies is applied for (Fig. 2). These tests provide the opportunity for monitoring exposure of those applying the pesticides under realistic conditions. Exposure data are needed, in combination with animal toxicity information, for risk assessment. Early generation of exposure data facilitates the development of safe products and helps avoid unpleasant surprises (Tordoir 1980).

At this stage, process development begins. Laboratory and, later, pilot-plant production, using the most promising process found during route-scouting, are established. While toxicity and efficacy testing continue and the registration procedure expands, product development enters an engineering phase, in which a full-scale plant is designed. Experts are consulted to ensure an optimum design with respect to health, safety, and environment.

When the engineering phase is well under way, a crucial point is reached in the continuous process of reviewing the feasibility of the project. A construction site must be chosen, which may require obtaining permits, and

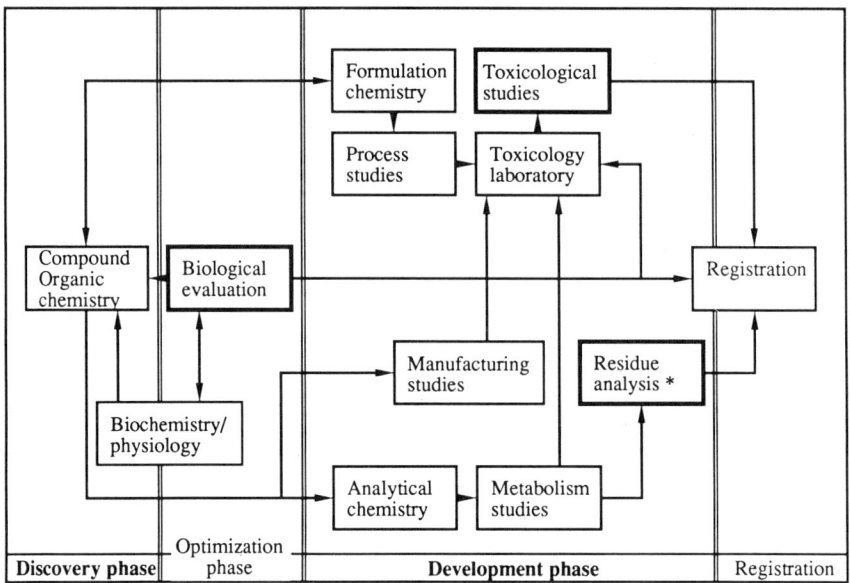

Fig. 2. Stages in developing and registering a new agrochemical.
(* Field evaluation in various parts of the world.)

construction begun. Once construction starts, the costs of the project rise exponentially; a thorough review of all aspects of the project is essential. All core requirements (see above) must be sufficiently met to justify the decision on plant construction.

The last phase, which is actually open ended, is the product development phase. Much research is needed to determine the best formulations to meet efficacy, application, health, safety, and environment requirements without becoming too costly. Another aspect to be considered is packaging. The amount of material per unit and the nature of the packaging can contribute to the safety of the product sold. Commonly, only a few of the many thousands of chemicals tested ever reach the manufacturing and marketing stage; the industry average is 1 in 25 000.

To confirm the safety of pesticides, field studies must be carried out once a product is to be marketed. Such studies assess and measure personal exposure and check for any biological or clinical effects. In many situations, biological monitoring is the method chosen for personal exposure measurement as it provides data on the total absorbed dose via all routes of entry (van Sittert 1986).

Ideally, regular health surveillance, including biological monitoring, should be carried out on workers engaged in pesticide manufacturing, formulation, and application. These data should also be used for epidemiologic evaluations of possible chronic adverse health effects.

146

Progress in improving safety

Pesticides in use at the turn of the century typically had a broad spectrum of biological activity and a low level of pesticidal activity. They were not particularly toxic to humans, but had to be applied in large quantities (compared with current application rates) to compensate for their low potency. The resulting human health hazard could, therefore, become significant.

Research has focused primarily on improving selectivity, i.e., enhancing the pesticidal activity on target species while leaving nontarget species, including man, more or less untouched. In the process, a higher efficacy is usually also obtained.

Qualitative selectivity is achieved when a pesticide interferes with a process or processes that occur only in the target species. For example, inhibition (by an acylurea pesticide) of chitin formation, which is essential for the formation of the legs, shield, etc., of insects, does not affect mammals. Interfering with photosynthesis in weeds does not harm birds and mammals as these organisms do not photosynthesize. Herbicides may, therefore, have a favourable toxicity profile (Table 2).

Quantitative selectivity, on the other hand, is based on differences between the target and nontarget species that determine susceptibility, e.g., skin-penetration rate, metabolism (toxification or detoxification), and excretion rate. A drawback of quantitative selectivity is the possibility that the target species becomes resistant. The properties that make nontarget species less susceptible to the pesticide may, in a number of generations, be acquired by the target species through genetic selection. Oppenoorth and Besemer

Table 2. Comparative toxicity of several herbicides, aspirin, and caffeine.

Product	LD_{50} in rats (mg/kg)
Caffeine	200
Alachlor	1 200
Cyanazine	1 200
Aspirin	1 750
Metribuzin	2 200
Terbutryne	2 380
Flamprop-isopropyl	>3 000
Glyphosate	4 320
Terbacil	5 000
Asulam	> 5 000
Imazapyr	> 5 000
Sulfometuron-methyl	> 5 000
Chlortoluron	> 10 000

Source: Graham-Bryce (1990).

147

(Inauguration speeches, Agricultural University, Wageningen, 1976, unpublished) point out that selectivity can also be improved by choosing the optimum time, place, and method of application; by improving the formulation; and by mixing a pesticide with other chemicals.

In certain cases, higher pesticidal activity can also be achieved by removing the less-active optical isomers (e.g., in pyrethroids) or less-active and nonactive components (e.g., removing 1,2-dichloropropane from a mixture of chlorinated C_3 hydrocarbons that is sold as a soil fumigant).

Progress in improving the selectivity and efficacy of pesticides is demonstrated by the fact that application rates have decreased several orders of magnitude since the beginning of the century (Fig. 3). Lower application rates also mean less exposure for those applying the pesticides in the field. Insofar as the increase in pesticidal activity has been matched by an increase in target species selectivity, an overall reduction in the risk to nontarget organisms (including humans) has been achieved.

Although the differences in toxicity between the target species and humans is of paramount importance, other factors can influence the health hazard. As skin contamination is the most important route of exposure during pesticide application, the degree of skin penetration of a substance has a significant bearing on the risk of intoxication. The amount of active ingredient used, the

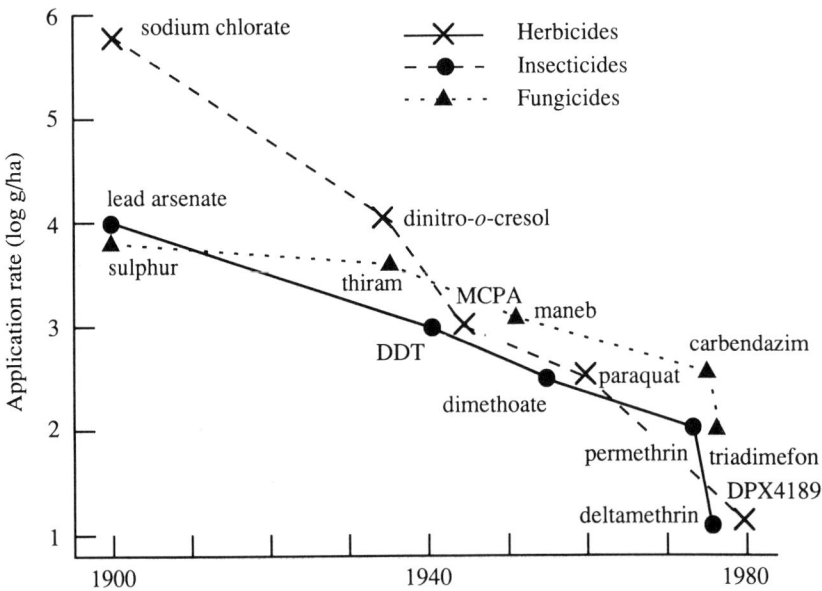

Fig. 3. Application rates (logarithmic scale) for crop protection chemicals plotted against the date of their introduction to illustrate the progressive increase in activity. Rates are representative values from the range employed in commercial practice for each compound (I.J. Graham-Bryce, Royal Society, London, UK, 1981).

type of formulation, and warning properties (odour and taste) are other important factors.

Key toxicity parameters and other features of representatives of four groups of pesticides that have been developed since the 1940s are compared to illustrate the progress made in improving the safety of pesticides (Table 3). Dieldrin represents organochlorines, a highly effective group of insecticides that includes dichlorodiphenyltrichloroethane (DDT) and has been in use since the early 1940s. Dichlorvos belongs to the large group of organophosphates that became available in the 1950s (it is not possible to find a true representative of this group because of the large differences in toxicity among the related substances). Cypermethrin is a pyrethroid, a group that was developed in the 1960s and 1970s. The rodenticide, flocoumafen, represents the new generation of anticoagulants, the so-called "super-warfarins," developed in the past decade.

Dieldrin has oral and dermal LD_{50} values (dose lethal to 50% of animals tested) that place it in the highly hazardous class IB in the World Health Organization's (WHO) classification system. Low NOAELs ("No adverse effect level") in the 90-day and chronic feeding studies also point to a high systemic toxicity. In spite of its toxicity and rapid skin penetration, lethal intoxication from occupational exposure has been rare over 40 years of use; only one case has been reported (IPCS 1989; van Raalte, unpublished paper presented at the Conference on occupational health, Caracas, Venezuela, 1965). Clinical recovery from intoxication, occurring during manufacture and formulation, was always complete (Jager 1970).

The persistence of dieldrin in the environment was initially viewed as an advantage because it was associated with long-lasting pesticidal activity. However, in light of ecotoxicological consequences, it is now considered a major disadvantage. Although dieldrin and the other organochlorines have good records of safety in regard to human health, their use has been restricted because of ecotoxicological considerations and, in North America, because of a perceived cancer hazard.

Dichlorvos should also be classified as highly hazardous, because of its acute toxicity in laboratory animals and its skin penetrating properties. The 90-day and chronic feeding studies show NOAELs 10 and 20 times higher, respectively, than dieldrin, although observed carcinogenic effects in laboratory animals are not relevant to man. No reports of lethal work-related intoxications with dichlorvos are known, but other organophosphates, in particular parathion and malathion, have caused many deaths despite the availability of specific antidotes. Certain organophosphates, now withdrawn, cause irreversible neurological defects in patients recovering from intoxications (IPCS 1986). Not all organophosphates represented a step toward safer pesticides. However, their rapid degradation in the environment represented an improvement from an ecotoxicological point of view.

149

Table 3. Toxicity and other characteristics of three insecticides and one rodenticide.

Characteristic	Dieldrin	Dichlorvos	Cypermethrin	Flocoumafen
LD$_{50}$ (mg/kg)[a]				
Oral, rat	37	30	251	0.25
Dermal, rat	60	75[b]	>1 600	0.54
Irritation, rabbit				
Skin	Slight	n.a.[c]	Moderate	None
Eye	Severe	n.a.	Moderate	None
Skin sensitization, guinea pig	Negative	Negative	Very weak	Negative
Mutagenicity				
In vitro	Negative	Positive	Negative	Negative
In vivo	Negative	Negative	Negative	Negative
NOAEL[d], rat in diet				
90 day	1 ppm	10 ppm (105 days)	100 ppm	0.02 ppm
Chronic	0.5 ppm	10 ppm	100 ppm	n.a.
Carcinogenicity	Mouse liver tumours only	Some, but not relevant to humans	Negative	n.a.
Teratogenicity	Negative	Negative	Negative	Negative
Multigeneration reproduction	Increased mortality in preweaning pups	No specific effects	No specific effects	n.a.
Dermal penetration	Rapid	Rapid	Very slow	Rapid
Excretion	Very slow, retention in fat	Rapid	Rapid	Slow, retention in liver
BM method	Available	Available	Available	Available
Treatment	Symptomatic	Atropine, cholinesterase reactivators	Symptomatic	Vitamin K

Source: Reviews of Toxicology, HSE Division, Shell International Petroleum Maatschappij B.V., The Hague, Netherlands.
[a] Lowest values are given; LD$_{50}$ = lethal dose to 50% of animals tested.
[b] Female.
[c] n.a. = not available.
[d] NOAEL, no adverse effect level, i.e., the findings in these studies do not generate any specific concern and there are no qualitative discrepancies between species.

The pyrethroid, cypermethrin, is moderately hazardous (class II) orally and slightly hazardous (class III) when absorbed through the skin. NOAELs in the 90-day and chronic feeding studies are 10 times those of dichlorvos and 100 times higher than those of dieldrin. Metabolism and excretion are rapid and other toxicity parameters are also favourable. The lower toxicity of pyrethroids, combined with their slow skin penetration make them safer to use. Skin sensations (tingling) that occur with overexposure usually disappear within a few hours and leave no aftereffects. These sensations act as a warning signal that exposure should be better controlled.

Reported cases of occupational systemic poisoning with pyrethroids are rare. However, He et al. (1989) reported 229 cases in China, 2 of which were fatal. The explanation was a failure to observe basic principles of industrial hygiene (He et al. 1988, 1989). With respect to health and safety, however, the development and use of the pyrethroids was an improvement over the more toxic organophosphates.

Flocoumafen, a representative of the latest generation of anticoagulants, was included in this overview to illustrate that extreme toxicity according to the WHO classification, does not necessarily mean a high risk to human health. Although flocoumafen is classified as extremely hazardous (class IA), it is also an extremely efficient rodenticide. The oral and dermal LD_{50} values of 0.25 and 0.54 mg/kg, respectively, and a 90-day NOAEL apply to the pure substance, but the formulation used to poison bait is almost nontoxic because it contains only 50 ppm (w/w) of the active ingredient. In addition, during product development, a deliberate effort was made to make poisoned bait unattractive to nontarget species by incorporating it into wax blocks with a deterrent colour (blue) and a bitter taste. These measures have rendered a potentially hazardous product safe for practical use. In the unlikely event of an intoxication, a specific antidote (vitamin K) is available. No cases of human intoxication have been reported.

Conclusion

Pesticides must meet many requirements, the most important of which is efficacy. Therefore, selection always begins with efficacy testing. If results are promising, additional work will be carried out to test for human and environmental safety. Commercial and technical requirements must also be met. Progress in increasing specificity and improving efficacy has made possible dramatically lower application rates, resulting in a major reduction in exposure of people applying the substances. In a number of cases, a reduction in intrinsic toxicity for mammals has also been achieved.

Guided by better understanding of intrinsic toxicity and data on personal exposure during application, product development and product stewardship can contribute to safety through improved formulations and packaging and advice on handling and personal protection. The comparison of the major groups of pesticides shows that with the development of the pyrethroids, a major step forward has been made in the creation of safe pesticides. Similarly, rodenticide is an example of a substance whose high efficacy and sophisticated product development compensate for high toxicity, allowing a final product that is safe for humans.

Graham-Bryce, I.J. 1990. Shell Agriculture, 6, 20–22.

He, F.; Sun, J.; Han, K.; Wu, Y.; Yao, P.; Wang, S.; Liu, L. 1988. Effects of pyrethroid insecticides on subjects engaged in packaging pyrethroids. British Journal of Industrial Medicine, 45, 548–551.

He, F.; Wang, S.; Liu, L.; Chen, S.; Zhang, Z.; Sun, J. 1989. Clinical manifestations and diagnosis of acute pyrethroid poisoning. Archives of Toxicology, 63, 54–58.

IPCS (International Programme on Chemical Safety). 1986. Organophosphorus insecticides: a general introduction. World Health Organization, Geneva, Switzerland. Environmental Health Criteria 63.

_____ 1989. Aldrin and dieldrin. World Health Organization, Geneva, Switzerland. Environmental Health Criteria 91.

Jager, K.W. 1970. Aldrin, dieldrin, endrin and telodrin: an epidemiological and toxicological study of long-term occupational exposure. Elsevier, Amsterdam, Netherlands.

van Sittert, N.J. 1986. Report of rapporteur. Toxicology Letters 33, 205–213.

Tordoir, W.F. 1980. Field studies monitoring exposure and effects in the development of pesticides. In Tordoir, W.F.; van Heemstra-Lequin, E.A.H., ed., Field worker exposure during pesticide application. Elsevier, New York, NY, USA. Studies in Environmental Sciences 7, pp. 21–26.

Studying the effects of pesticides on humans

M. Maroni

International Centre for Pesticide Safety, Busto Garolfo, Milan, Italy

Assessment of the health risks of pesticides to humans and regulation of their use are based on experiments on laboratory animals. Concern over the applicability of such studies prompted discussion at the 9th International Workshop of the International Commission on Occupational Health (ICOH). Highlights of the discussion and recommendations made at the workshop are presented in this paper, along with a review of difficulties associated with studies on people exposed to pesticides.

The use of human-exposure and health data for improving assessment of the toxicological risk of pesticides and setting regulations to govern their use was a topic of discussion at the 9th International Workshop of the International Commission on Occupational Health (ICOH 1990). There was concern that the regulation of pesticides is based mainly on data generated from animal-toxicology studies and only to a limited extent on information about effects on human health.

The aim of the workshop was to discuss studies on human exposure to pesticides and the resulting health effects, identify advantages and limitations on the use of such data by regulatory bodies compared with animal data, and to formulate recommendations for ICOH. Participants were 28 experts from 9 countries representing international organizations, governmental agencies, academia, and industry.

Presentations included: four on the generation of human-exposure data in field studies, using dermal-deposit measurements and biological monitoring; four on health effects in fieldworkers applying pesticides, using biological and clinical tests; an epidemiologic study of a large farming population in Canada, investigating possible associations between the use of pesticides and health effects; three on the development of protocols for conducting epidemiological studies; the limitations of data on animal carcinogenicity for predicting carcinogenic risk in humans; and the activities of the International Centre for Pesticide Safety (ICPS).

Participants at the meeting endorsed five recommendations:

- Preliminary registration, registration, and reregistration of pesticides is frequently, if not exclusively, based on the evaluation of comprehensive animal data. To enhance and improve assessment of risk to human health, more use should be made of information based on human exposure and health effects.

- In addition to generating hypotheses, epidemiological studies should be designed and used to verify hypotheses. Particular attention should be paid to verification of specific exposures, selection of controls, and occurrence of mixed exposure. Whenever possible, prospective records on the use of pesticides should be kept, preferably linked to individuals or groups of individuals.

- Various techniques and methods for assessment of exposure, such as analyses of data on use, measurement of external exposure, and biological monitoring, each have a specific contribution to make. Combining the various techniques and methods will increase the accuracy of an assessment.

- Data on early biological effects may serve as indicators of possible health effects. Case-by-case evaluation will be helpful in assessing the clinical significance of a biological effect.

- Data on health or biological effects, obtained in situations with multiple exposures, should be used to stimulate measures to reduce pesticide exposure.

General discussion during the meeting highlighted several points. In assessment of toxicological risk of pesticides, the experimental model by which single compounds are tested is limiting. In practice, agricultural workers are commonly exposed to formulations and mixtures of pesticides. Although it is not feasible to test all possible mixtures or associations of compounds encountered by users, when there is reason to suspect that an interaction between two pesticides can occur (e.g., based on the mechanism of action), specific experiments with mixtures should be carried out. However, in most cases, mixed exposure in real life will remain an unprecedented experiment, emphasizing the importance of human observations and epidemiology in detecting additive or synergistic effects.

Because of lack of statistical power or other methodological flaws, experiments to investigate carcinogenicity in humans may produce false-negative results. On the other hand, animal experiments to investigate carcinogenicity of chemicals are carried out in extreme conditions, i.e., a very high dose is administered over a prolonged period. This method is appropriate to detect the carcinogenic potential of a molecule even through testing a limited number of animals, but it generates information that is more qualitative than quantitative. For pesticide users, who are exposed to doses four or five orders of magnitude smaller than those used on experimental animals, the assessment of the risk is a quantitative issue; lifetime risks of cancer of 10^{-2} and 10^{-8} are entirely different and would produce markedly diverging consequences in terms of risk management. Furthermore, high–low dose extrapolation with mathematical models has very weak scientific support and tends to be an administrative procedure rather than a real scientific evaluation of the risk.

Discussion highlighted the importance of data on personal exposure doses. In terms of information, an epidemiological study in a small cohort with accurate personal-exposure data may be equal or even better than one in a large cohort with poorly defined personal exposures.

Currently, health-risk assessment of human exposure to pesticides is mainly based on data from experiments on animals. Little information is available about effects on humans because of the lack of "strong" data from epidemiological studies. This problem may be overcome by applying a more scientific approach in health-risk assessment. Knowledge of the pharmacokinetics and pharmacodynamics of a compound in different species must be increased by carrying out studies on the mechanism of action.

To obtain "good" human data, investigators should critically review the protocol for future epidemiological studies and address the question of whether the effort required in such a study is proportional to the possible outcome. Specific questions should be addressed, such as the possibility of accurate exposure measurements, the selection of a suitable control group in case–control studies, and the feasibility of collecting accurate data on health status of exposed and unexposed people.

Cohort studies of factory workers who manufacture pesticides can minimize these problems. In these studies, controls can be selected among unexposed workers in the same factory or nearby factories where workers are exposed to different chemicals or processes. Employment records can be used to obtain dates of employment and job titles and, thereby, identify those who had the opportunity to be exposed and the duration of exposure. If subgroups of workers are exposed to only a few pesticides or chemicals used in production, then it is possible to relate excess incidence or mortality to a particular exposure. Such studies often suffer from having a very small sample size; evaluation of workers in more than one factory and in more than one country where the same pesticides are produced may be necessary.

In choosing a reference group, using two similar industrial populations for comparison eliminates the healthy-worker effect, which may appear when using the whole general population. One should, in such cases, select a second industry with workers of similar social and economic backgrounds so that diet, physical demands, and other life-style factors are likely to be similar.

One way to handle the comparison problem is the selection of the unexposed or lowest exposed group within the cohort as an internal reference. When this is done, care should be taken to assure similarities in duration of follow-up and in the size of the groups. In case–control studies, controls may come either from the general population, often referred to as community controls, or from hospitals where cases were identified. Three criteria have been proposed for the selection of controls:

- Controls should come from the same registry (e.g., hospital, mortality records) as the index cases.

- Control illness should be unrelated to the risk factors under study (i.e., a given risk factor under study should not predispose one to getting the control illness).

- Control illness should be similar to the illness under study, bearing in mind the factors influencing its appearance (or referral) in the registry.

One might want to add a second control if these criteria are not met.

If farmers are under study, then ideally, the control group should also be farmers. Farming is a unique way of life, not only in terms of chemical exposure, but also with respect to diet, physical activity, exposure to viruses, habits, and life-style. Opportunity for exposure to viruses, for example, may be affected by the presence of intermediate hosts, including livestock or even particular crops.

In carrying out epidemiological studies, accurate measurement of exposure is important because errors may reduce estimates of relative risk and dampen dose–response gradients. Misclassification of toxic substance increases the chance of attributing elevated cancer risk among exposed people to the wrong agent. Employment records are commonly used to assess exposure in cohort studies and interviews are used to determine exposure in case–control studies. However, these techniques are of use in both types of study and should be used, when possible, to test and improve the reliability of exposure evaluations.

Information can be obtained from different sources to increase confidence in exposure estimates. Sources include suppliers who sell pesticides to the subjects under study, records of applicator companies, interviews about the type of equipment used for application, and duration and frequency of application coupled with biochemical monitoring data. For populations, estimates of exposure may be developed by a panel of experts, including agricultural scientists, entomologists, and industrial hygienists familiar with patterns of pesticide use. Biochemical factors can provide better measures of delivered dose than estimates based on concentration of substances in patches and ambient air or those derived from job descriptions or interviews. However, biochemical measures are typically only available for a sample of study subjects, for only a few pesticides, and at only specific times.

Ideally, to allow epidemiological surveillance of a farming population using pesticides, it would be necessary to implement a system of registration that would identify the exposed subjects and specify and quantify exposure. This registry could be linked to the licence required in some countries to buy and use some classes of pesticides.

An experiment on the feasibility of this approach over a large geographical area is currently under way in Lombardy, Italy. The ICPS intends to establish an epidemiological observatory through which people and pesticide usage will be registered and specific epidemiological investigations carried out on selected subgroups of the farming population. This project will include about 500 000 farmers and cover an area equal to one-eighth of the country.

ICOH (International Commission on Occupational Health). 1990. Proceedings of the 9th International workshop, 2–4 May 1990, International Centre for Pesticide Safety, Busto Garolfo, Milan, Italy. ICOH, Geneva, Switzerland.

Safer packaging and labeling
of pesticides

L.S. Dollimore

Shell International Chemical Company Ltd,
London, UK

During the past 5 years, major developments in the area of pesticide packaging and labeling, particularly the use of pictograms, have improved safety for the user and the environment. Pictograms are symbols indicating essential advice and warning messages to the user. Introduced only a few years ago, they are now being used in at least 66 countries, mainly in the developing world. The most significant advances in packaging have been in the use of new plastics and container designs. New products, such as polyethylene terephthalate and fluorinated high-density polyethylene, are able to withstand aromatic solvents, allowing many new safety features, such as simple-to-use dispensers, to be built into containers.

In the past 5 years, major changes in pesticide labeling and packaging have contributed significantly to safeguarding not only end-users and those in the supply chain, but also the general public and the environment. A major stimulus was the *International code of conduct on the distribution and use of pesticides* (FAO 1986), which outlines rules of responsible practice for both governments and the crop-protection industry.

Labeling

One of the recommendations of the Food and Agriculture Organization of the United Nations (FAO) was to "include appropriate symbols and pictograms whenever possible, in addition to written instructions, warnings and precautions." A pictogram is a symbol designed to convey a message, e.g., the no-smoking symbol. The message should be recognizable at a glance.

In the late 1970s, a specialist on information graphics, Shirley Parfitt, was working on an FAO project in Bangladesh when her illiterate maid was poisoned through mishandling of an agrochemical product. The incident convinced Parfitt of the need for pesticide labels to contain easily understandable symbols that convey safety warnings and advice. Using her artistic skills, she designed her own symbols and approached Shell International for sponsorship and agrochemical expertise. As a result of this collaboration and field-testing in six developing countries, a labeling system was developed (Fig. 1).

In 1984, these pictograms were submitted to FAO for consideration and endorsement for worldwide use. FAO requested more exhaustive field-testing to ensure that the symbols were understandable to people of diverse cultures and with low levels of literacy. In response, the International Association of Agrochemical Manufacturers established a pictogram working group comprising labeling, safety, and environmental specialists and an FAO representative.

The working group reviewed previous work on agrochemical labeling systems using pictograms and examined general international and national warning and advice symbols, such as road and factory signs. Two decisions had to be made: which messages to portray and where to locate the symbols. Because of space limitations on labels, only the most important messages could be portrayed and these were to be selected on the basis of incident experience. For example, not wearing gloves when handling pesticide concentrates has led to far more incidents than smoking during product use. In fact

Fig. 1. Pictogram labeling system developed by Parfitt and Shell International.

159

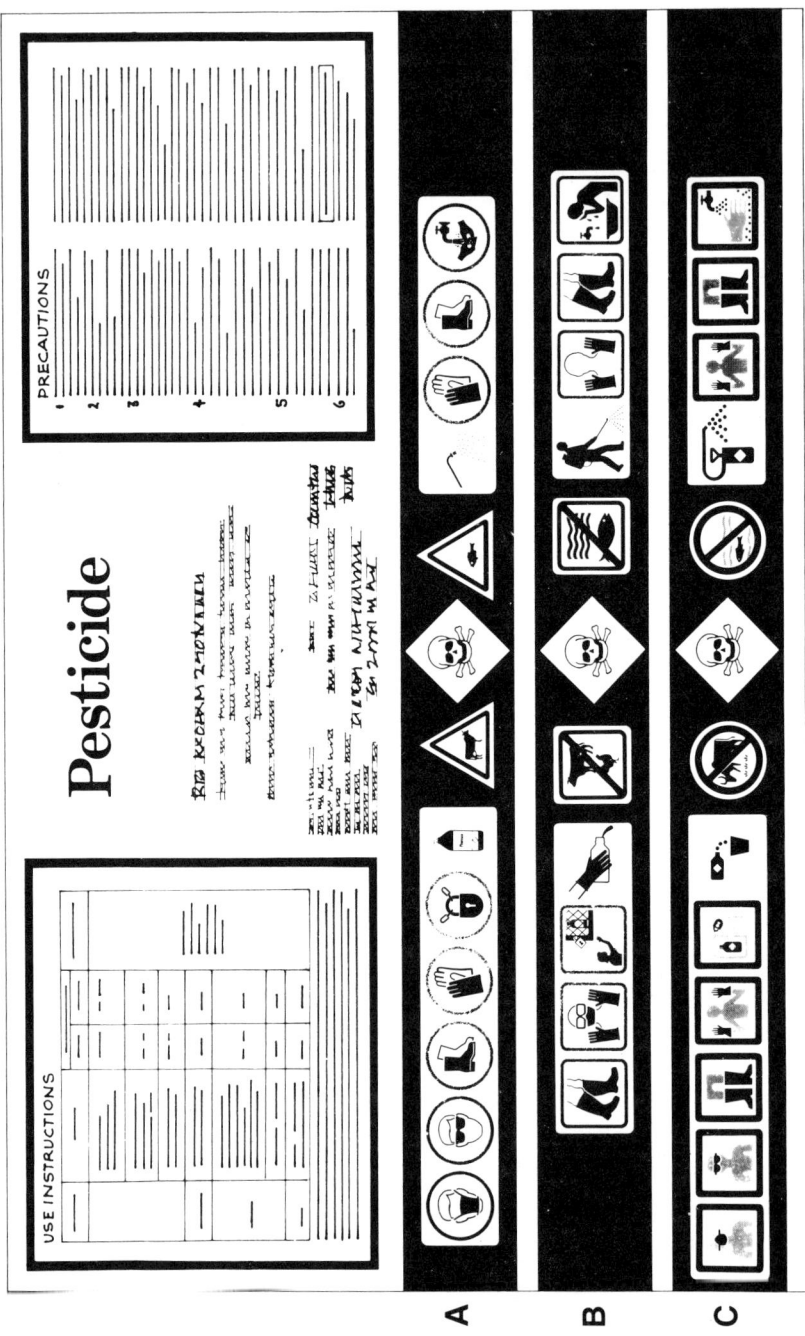

Fig. 2. Sample label with three sets of pictograms presented by designers to the working group. (Original size of label = 18 × 28 cm.)

no documented incident could be found relating to the latter. Space limitations also led the working group to recommend that the pictograms be placed in the hazard warning band recommended by the FAO.

Three graphic designers, including Parfitt, were given full background information, and asked to produce a set of pictograms of their own design (Fig. 2: bands A, B, and C). An important aspect of the designs was the linking of activity pictograms with advice pictograms, e.g., "when handling the concentrate — wear gloves."

Table 1. Regional distribution of pictogram survey.

Region	Number of completed questionnaires	Percentage
Far East	237	25
Latin America	214	23
Africa	153	16
Indian subcontinent	127	14
Europe	105	11
Middle East	69	7
Australasia	41	4
Total	946	100

Table 2. Characteristics of the population surveyed ($n = 940$).

Characteristic	Percentage of respondents
Occupation	
Farmer	66
Worker	28
Not stated	6
Age	
< 18 years	1
18-34 years	40
35-54 years	49
> 54 years	10
Level of education	
No formal schooling	12
Primary	43
Secondary	24
Advanced	19
Literacy	
Literate (claimed)	91
Illiterate	9

An international survey was carried out by field staff of the FAO, extension services, industry, and other organizations, using a questionnaire and interview procedure. The length of the interview ranged from 40 min to over 2 h in one case. Of the 3 000 questionnaires distributed, about 1 000 were returned completed from 42 countries (Table 1).

The questionnaires were analyzed according to such factors as age, educational background, literacy, and occupation (Table 2). The most surprising finding was that only 80% of those interviewed understood the meaning of the skull-and-crossbones symbol, which has been in use for more than 50 years. It was reassuring, therefore, to find a high level of spontaneous understanding of the proposed pictograms; immediate understanding of the symbols ranged from 60% to 85% (Fig. 3). The most difficult concept to convey without explanation was the linkage of advice pictograms to activity pictograms. Ot those interviewed, 90% considered pictograms a useful way to provide safety instructions on labels.

Both the International Association of Agrochemical Manufacturers and the FAO began promoting the use of the recommended pictograms in 1988. A survey of 96 countries in 1990 revealed that 69% were using them (Table 3).

It is reassuring to observe the high level of use in Africa and the Middle East and, to a lesser degree, in the Far East and Latin America, although an impediment to early use in some countries has been the need to change national legislation. Use of pictograms has been rejected in the USA, Canada, and many European countries, where they are thought to be for developing countries only. However, these countries have both illiterate and immigrant

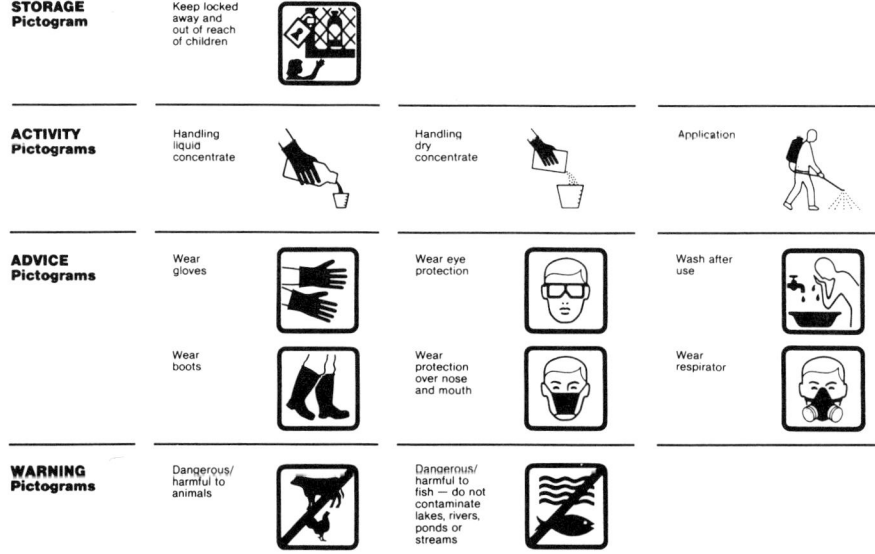

Fig. 3. Pictograms preferred by the majority of people surveyed.

Table 3. Use of using pictograms, 1990.

Region	Countries surveyed	Countries using pictograms	
		No.	%
Africa	26	26	100
Middle East	15	14	93
Far East and Australasia	17	13	76
Latin America	20	12	60
Europe	16	1	6
North America	2	0	—
Total	96	66	69

groups. They also make widespread use of pictograms in road signs and factory warnings. Moreover, many people do not read label instructions and warnings before using chemicals. Developed country arrogance is suspected in these cases, although once again the daunting task of changing legislation is no doubt a factor.

Packaging

Pesticide containers provide a safety barrier between those handling the container and its chemical contents. Container design has a major effect on minimizing the risk of exposure in transferring the product to a spray tank. Liquid agrochemical products usually present the most difficult packaging problems and the greatest potential risk of exposure to the applicator.

Major changes in agrochemical packaging occurred during the 1980s and further progress is expected during the 1990s. These developments are a result of a number of factors:

- Introduction of package-performance tests by the United Nations;

- Introduction of high-activity and high-value pesticides leading to lower application rates and smaller package sizes;

- Advances in plastics and moulding technology; and

- Collaboration within the industry to improve user safety and convenience and product shelf-life.

Changes in small packages for the small-scale farmer have probably had the greatest impact on safety. In the past, most small containers were glass or tin plate: the former were too fragile and had narrow necks that caused spattering during pouring; the latter dented easily and had leaky seams and a metal insert under the cap, which was difficult and often hazardous to pry out. Both materials were severely limiting in terms of design flexibility. Furthermore, people frequently used the caps as measures. Although the common plastics (high-density polyethylene, polyvinyl chloride, and polyethylene) offered design flexibility and were suitable for aqueous products, they could not be used for aromatic solvents, which are the base for most liquid agrochemical formulations.

A major breakthrough was achieved with the discovery of polyethylene terephthalate (PET). This plastic is widely used for carbonated soft drinks and, more recently, for alcoholic spirits sold by airlines. When this polymer is biaxially stretched during blow-moulding, it produces a bottle that is compatible with aromatic solvents. The potential benefits of this plastic for the agrochemical industry were seen immediately by three companies, who collaborated to carry out the basic testing. PET bottles are now widely used for agrochemicals. They are transparent and very tough. New bottles have design features such as wide necks to avoid spattering.

However, farmers continued to use the cap of the PET bottles as a measuring container, thereby contaminating the outside of the bottle. Shell overcame this problem by designing a standard volume dispenser that is welded into the neck of the bottle during manufacture. Not only does this prevent the cap being used, it also allows for precise measurement and keeps the contents from spilling if the bottle is accidentally knocked over when the cap is off. A further safety benefit is that the bottles cannot be reused without cutting the neck off. PET bottles with dispensers have proven to be popular with small-scale farmers in Africa. The principal limitations of PET bottles are their unsuitability for ketones or products that are susceptible to water degradation because water vapour can permeate the bottle wall.

Two other recent developments in container technology have been the use of fluorinated high-density polyethylene (HDPE) and the design of multilayered containers. In the former, gaseous fluorine is included either during or after the blow-moulding process. This forms an extremely thin layer of fluorinated HDPE, which is highly resistant to the solvents commonly used in agrochemicals. A layer of only 150 Å is usually considered sufficient.

Containers with walls comprising two or more polymer layers are produced by coextrusion. Bottle strength is attained from a support layer, usually HDPE, and resistance to the chemical product is provided by a layer of barrier material such as polyamide (nylon). The various layers are bonded together using adhesives.

Plastics permit considerable flexibility in packaging agrochemicals. For example, 5-L tins are not easy to handle when wearing protective gloves, whereas easy-to-grip handles can be incorporated into plastic containers of this size. A common problem with early plastic containers was the retention of product in the hollow handles. This problem has been overcome by pinching off the base of the handles during blow-moulding to prevent liquid ingress. Similarly, wider necks on plastic bottles prevent splashing of contents during pouring. Drainability of containers has also been improved to reduce risk if the container is reused for drinking water or foodstuffs. This is a particular advantage of fluorinated HDPE because fluorination makes the inner surface slippery and facilitates draining.

A common constraint to the use of new plastic containers is that many developing countries require agrochemical containers to be manufactured locally to minimize foreign-currency expenditures. Although PET blow-moulding is becoming common because of the high-volume needs of the soft drinks industry, the technology needed to produce HDPE and coextrusion bottles restricts their manufacture to developed countries.

Speculating on future developments in agrochemical containers and packaging for developing countries, it is useful to consider recent advances in the USA and Europe. Returnable containers, probably the most important innovation in terms of safety, are designed to have a long life. Made of stainless steel or plastics, they can be returned to the supplier for refilling. This eliminates the problems of container misuse and disposal. Returnable containers are also linked to a closed system designed to minimize the risk of exposure during transfer from the container to the tractor or aircraft. However, this approach requires a good supply and transport infrastructure, often lacking in developing countries.

A resurgence of interest in water-soluble sachets has also occurred as this technology has improved. Although these sachets were originally designed to contain powders, they are now being used for liquid formulations. Finally, there are signs that rigid plastic containers may be replaced in many cases by "bag-in-the-box" packaging, which has proved successful for the supply of wine. This will certainly facilitate disposal and prevent unauthorized reuse.

FAO (Food and Agricultural Organization of the United Nations). 1986. International code of conduct on the distribution and use of pesticides. FAO, Rome, Italy. 31 pp.

Improving pesticide regulation in the Third World: the role of an independent hazard auditor

M.E. Loevinsohn

Consultant in Applied Ecology, Butare, Rwanda

Central to the environmental and health hazards created by the expanding use of pesticides in developing countries is the weakness of national regulatory agencies. International efforts to support these institutions include the establishment of a Hazard Audit Organization to assess the pesticide industry's adherence to accepted standards of health and environmental protection. An independent evaluation by a hazard auditor may be attractive to all parties in the long-standing confrontation over the control of pesticide technology: the industry, public interest groups, developing and developed countries, and international agencies. One approach to implementing the concept is proposed and initial responses to the proposal are reported.

The papers presented at this symposium join a growing body of evidence of the effect on human health and the environment caused by the rapid increase in pesticide use in developing countries. Chemicals of sometimes extreme human or environmental toxicity are transported, stored, used, and discarded in ways that expose people and other nontarget organisms to significant hazard.

The urgent need for effective regulation of these hazards, however, contrasts starkly with national capacity, which is limited in much of the Third World. More than 50% of developing countries have no legislation enabling government to regulate the marketing of pesticides or limit their availability to particular areas or users; in Africa, the proportion is 76% (FAO 1989). Even where adequate legislation exists, regulatory agencies are often unable to assess pesticide hazards in light of local conditions or to enforce the decisions they reach, because of a lack of qualified personnel, inadequate resources, or interference.

Current initiatives, at the international level, aimed at improving this state of affairs are discussed in this paper. A novel approach, the concept of a "pesticide hazards auditor," who would build upon and supplement these initiatives has been promoted over the last year with support from the International Development Research Centre (IDRC). Initial reactions to this proposal from developing and developed countries, international agencies, the pesticide industry, and consumer, environmental, and labour groups are described.

International initiatives

Technical assistance and information exchange

A number of bilateral and multilateral aid organizations have launched programs aimed at increasing the skills and resources available to pesticide regulatory agencies in developing countries. The Food and Agriculture Organization (FAO), United Nations Environmental Program (UNEP), the World Health Organization (WHO), the US Agency for International Development (USAID), and Germany's Agency for Technical Cooperation (GTZ), among others, are providing training, analytical equipment, information systems, continuing support in the evaluation of risks and benefits, and advice on legislative reform. In several cases, as in FAO's programs in the Far East and Africa, this assistance is organized on a regional basis. The task, however, is immense; fewer than one-quarter of developing countries claim to have received any technical assistance (FAO 1989).

A major cause of concern is the international trade in highly toxic pesticides, particularly the export to developing countries of products banned or severely restricted in the country of manufacture. The USA, UK, and the European Community have instituted schemes to notify importing countries of shipments of unregistered or severely restricted pesticides. In practice, notifications are often received well after the pesticides have arrived and do little to enable importing countries to control hazardous imports (Pallemaerts 1988).

Nongovernmental organizations (NGOs), with support from many developing countries, have mounted a determined lobbying effort within the governing councils of FAO and UNEP in favour of more restrictive schemes based on the principle of "prior informed consent" (PIC), whereby a designated authority in the importing country must explicitly agree to the import before it can take place (Anon. 1990a). Late in 1989, both organizations adopted complementary PIC procedures, after first refusing to do so. The Commission of the European Community is currently considering a draft directive that would incorporate PIC into European law (T. Casey, Consultant to the Directorate-General for the Environment, Commission of the European Community, June 1990, personal communication) and the proposed *Pesticide Export Reform Act* would do the same in the USA (Anon. 1990b).

Although PIC, as operated by FAO and UNEP, will extend to many of the pesticides that have been implicated most often in human poisoning, the degree to which it will actually improve the regulation of such hazards is open to question. Pesticides banned or severely restricted by 10 or more countries will be the first to be covered by the scheme, followed, probably in late 1990, by those so treated by five or more countries. Thereafter, substances labeled "banned or severely restricted" in a decision by any additional country will be included. A working group will determine whether formulations based on WHO class 1A (extremely hazardous) compounds should be covered as well (Anon. 1990a). A recent report by the British-based Pesticides Trust (1989) contends, however, that several class 1B (highly hazardous) pesticides that have frequently been involved in poisoning incidents may escape the informed consent provisions.

The scheme hinges on a government's ability to evaluate and act on the notices it receives, and it is precisely this capacity that is deficient in many instances. As well, PIC begins with the decisions industrialized countries have taken to protect health and the environment within their own jurisdictions. Industry has often claimed that a different balance of risks and benefits may lead developing countries to judge acceptable a number of pesticides strictly controlled in industrialized countries (Willis 1986).

The argument works as well, however, in the other direction, e.g., the application methods and worker protection typical in much of the Third World may result in operators being dangerously exposed when using products not subject to any significant restriction in industrialized countries. These determinations can only be made in the light of local conditions, emphasizing once again the need for effective national regulation.

The drafting of the *International code of conduct on the distribution and use of pesticides* (FAO 1986) is another major initiative that addresses the weakness of pesticide regulation in developing countries. The code calls on the pesticide industry at all levels, as well as exporting nations, international agencies, and public-sector organizations to assume a share of responsibility for ensuring safety in the use of pesticides. The code's provisions are entirely voluntary and there has been considerable controversy over the extent to which they are respected in practice.

Two reports (ELC 1987; Pesticides Trust 1989) prepared for the Pesticides Action Network (an international group of NGOs) allege widespread infringements of the Code, for the most part by industry, in all developing countries investigated. Evidence is presented regarding misleading advertising, inappropriate packaging, poor quality control, and marketing of banned and dangerous products. Governments in developing countries also cite widespread failure by industry, as well as other parties, to abide by the Code's provisions (FAO 1989). In response to these findings, the FAO Conference has asked the Director-General to report by next year on the feasibility of transforming the Code into a convention that governments could make legally binding within their jurisdictions. Effective enforcement of such legislation, however, would come up against both the vague wording of many of the Code's provisions and, once again, the limited resources available to Third World governments.

The industry perspective

At Ciba-Geigy Ltd, a principal pesticide manufacturer, the FAO Code is accepted apparently without reservation and has been incorporated into the Agriculture Division's quality policy (Anon. 1988a). It is seen as being consistent with the principle of "product stewardship," which entails continual monitoring and periodic internal audits (Anon. 1988b).

The true measure of corporate commitment to these policies and principles is in their application, particularly in cases where there may be conflict with short-term profitability. Ciba-Geigy claims, in several instances, to have voluntarily refrained from marketing products where evidence suggested that they could not be safely used (as required under section 5.2.3 of the Code). For example, chlordimeform (Galecron) was removed from the Latin American market following reports of poisoning; dichlorvos (Nuvan) and phosphamidon (Dimecron) were considered too toxic for agricultural application in the Philippines and Burkina Faso, respectively.

Officers of the company point to a range of initiatives aimed at reducing risks to health and the environment, including improvements in formulations and packaging and an increased emphasis on safety training. Progress is slow but continual, they say, yet little credit is given to these efforts by the company's critics.

Pesticide hazard auditor

Crisis and opportunity

The pesticides industry finds itself under increasing pressure from national authorities, international bodies, and environmental and consumer groups (GIFAP n.d.). Its public image has suffered from a series of widely publicized disasters (Seveso, Bhopal, and the Rhine), as well as from more localized crises, such as the contamination of groundwater in Italy's Po valley.

An alternative to confrontation may be found in an historical analogy. By the late 18th century, there had emerged in Britain a large number of common-law corporations engaged in commerce and manufacturing. A highly speculative and unregulated market in corporate stocks developed, leading to several spectacular financial failures. Investors and creditors led the resulting public demand for investigation, which required the services of independent accountants. By the early 19th century, it had become common practice to call upon such skilled outsiders to assist in settling disputes and bankruptcies and, increasingly, to attest to the soundness of enterprises seeking investment or credit. It is to these developments, given legal support in 1844, that the Anglo-American tradition of independent financial auditing can be traced (Anderson 1984).

The recent trend in the United States toward "environmental auditing" appears to have a similar history. A growing number of firms whose activities may give rise to pollution and occupational-health hazards have retained independent environmental auditors to help ensure compliance with regulatory standards and to oversee internal auditing procedures. Once again, the need of companies to maintain investor and creditor confidence and to safeguard their public images appears to have been as crucial in this decision as court-sanctioned or regulatory requirements (Palmisano 1989)

Companies producing and marketing pesticides in developing countries should have their practices, with regard to impact on health and the environment, examined by an independent pesticide-hazards auditor. To the extent that a company's good name or image has value in a competitive environment, a hazard auditor might help create a market-based mechanism for ensuring compliance with accepted standards that would reinforce official regulation. For the system to gain acceptance in the industry, it must embody certain characteristics:

- Independence — the auditor must be seen to have no link, direct or indirect, with the company being examined.

- Authoritativeness — the audit must be based on explicit and recognized standards, as have been codified for financial auditing in the form of generally accepted accounting principles. The FAO Code (FAO 1986) is subscribed to by all parties and might provide one of the bases for defining acceptable corporate practice with respect to pesticide hazards.

- Expertise — the individuals performing the audit must inspire confidence by their demonstrated technical knowledge and mastery of the standards underlying the hazard audit.

- Openness — while respecting proprietary and commercial information whose disclosure might prejudice a company's interests, the detailed and material conclusions of the auditor must be made public if its function is to be fulfilled. Similarly, the company must be prepared to make available to the auditor all relevant documents and records.

- Service — beyond assessing a company's compliance with accepted standards, a financial auditor often provides advice on internal auditing procedures. Similarly, the hazard auditors would make a more useful contribution (and not only to the company) if they suggested changes in, say, a company's environmental- and health-monitoring programs that would enable problems to be identified earlier.

Benefits of a hazard auditor

From the company's perspective, a positive and unqualified attestation from the hazard auditor would provide authoritative confirmation that the company was acting on the high standards to which it laid claim. This would help reassure the increasingly restive society within which agrochemical companies operate and at the same time serve to differentiate the firm from less responsible competitors.

Among developing countries, those whose national regulation is the weakest would stand to benefit most from a hazard audit. The audit would provide an immediate form of control of pesticide hazards, based on the application of broadly accepted principles to the local context in which the products are marketed and used. In no sense, however, should the hazard auditor be seen as substituting for national regulation over the longer term.

A financial audit, in most industrialized countries, is sanctioned by law and backed by administrative and legal measures that ensure compliance with accepted norms. Either form of audit, financial or hazard, relies on market forces and corporate self-interest to raise and maintain an industry's standards. Internal and external audits may lessen the requirements for government enforcement, benefiting developing countries with operational, if constrained, regulatory systems. However, public supervision is still essential to ensure that these mechanisms function efficiently.

For NGOs and their allies on one hand and the pesticides industry and its supporters on the other, the hazard audit may represent one element of a solution to a long-running conflict that, for both, has absorbed considerable energy and resources.

Implementing the concept

Initial steps

A description of the hazard audit (Loevinsohn 1989) was sent to some 150 organizations on all sides of the debate. The concept was further discussed at two scientific conferences and in meetings with some of the major organizations. The response has been generally positive. A meeting of representatives of the main sectors has been suggested to explore in greater detail whether a consensus is attainable and to chart further action.

Participants at such a meeting could discuss its outcome in their respective constituencies and, if general agreement is obtained, working groups could be formed to define "accepted standards," develop procedures for the audit teams, and prepare a draft charter for a Hazard Audit Organization. The output of the working groups would be reviewed at a further meeting involving all major actors. At the same time, the concept would be given wider circulation through print and other media.

Several case studies could be conducted to build confidence and gain experience. These would take place in developing countries whose governments support the aims of the audit.

If the studies were judged successful, the Hazard Audit Organization might be established by a substantial portion of companies in the industry, the major NGOs engaged in campaigning, and other groups representing the public. The support of influential governments in the North and South, key professional associations, and leading international bodies would also be essential.

Structure and function

An autonomous, nonprofit Hazard Audit Organization would have, as its primary task, external hazard audits of companies involved in the manufacture and sale of pesticides in developing countries. Financing would be provided by participating companies, the members, and the industry association, Groupement international des associations nationales de fabricants de produits agrochimiques (GIFAP), as well as firms outside this body. Companies would be charged on a cost basis for each audit, but would also make annual contributions toward the organization's administrative expenses.

General supervision, policy formulation, and the further development of "accepted standards" would be the responsibility of a governing council whose members would be drawn from four broad sectors: the pesticides industry; national regulatory agencies and international bodies (e.g., FAO, WHO, and UNEP); research institutions and professional associations; and consumer, producer, and environmental organizations. Relative proportions remain to be negotiated, but no sector should be allowed to dominate. A technical subcommittee would be responsible for planning and setting terms of reference for individual audits, selecting team members, and reviewing their reports. A small secretariat would also be required. Well-qualified auditors would be drawn from professional associations, international agencies, and national regulatory bodies in the North and South. Retained initially as consultants or on secondment, auditors might eventually be hired by the Hazard Audit Organization.

Standards

The FAO Code of Conduct (FAO 1986) may provide a framework of generally accepted principles on which to base the hazard audit, but in many respects the Code's provisions lack specificity. What, for example, constitutes "safe use" or an "unacceptable hazard"? An operational definition of these terms might be based on the practice of well-established regulatory agencies. A residue concentration or exposure level that falls within the range of what different agencies take to be permissible can be said to be "generally acceptable." The variation in national tolerances to health hazards appears to be greatest with respect to chronic effects which, in statistical terms, are often weak and uncertain. The consensus is generally clearer for acute effects in spite of the preeminent threat to populations in the Third World (Jeyaratnam 1985).

Procedures

Rather than focus on one company's operations worldwide, the hazard audit might be conducted in one developing country at a time and involve all participating companies that do business there. In this way, it should be possible to cover several countries each year. Given the number of firms and the range of their activities, the auditors would have to rely on sampling techniques, as do financial auditors. The hazards entailed in different aspects of companies' operations might be stratified by severity and risk on the basis of published information, reports from government agencies and NGO groups, and the experience of the technical committee and auditors. Giving greatest weight to the most severe and probable hazards, a sample of practices would then be drawn and assessed in relation to the standards that had been defined.

The audit team would examine company documents and facilities, interview employees, and investigate the distribution of products and the manner in which they were employed. Auditors would also consider information from regulatory bodies, research institutions, and producer, consumer, or environmental groups. Where it is deemed necessary, the team might undertake or commission research that would enable it to reach an informed opinion.

Every effort would be made to ensure the active support of governments of the countries in which the audits are performed. The Hazard Audit Organization and host governments might work out different relations, according to the latter's needs and desires. Following their investigations, auditors would be well placed to report to the government on the effectiveness of national regulation and provide some advice on remedies, possibly focusing on aspects that the government had identified beforehand as problematic. An audit that covered perhaps several months would not, however, provide an opportunity for extensive technical assistance, although the team might make a useful contribution by identifying critical needs for other agencies to follow up.

Reporting

The auditor's report would express a considered opinion regarding the extent of a company's adherence to accepted standards of conduct. Where deficiencies were noted, the report would detail how practices should be improved to meet standards. This might entail, for example, changes in labeling, packaging, promotional material, educational programs, or restrictions on the availability of the product in that market.

As suggested above, the auditor's report would be made public, except for commercially sensitive or proprietary information. At the company's request, dissemination might be delayed a few months to permit it to bring its practices into line with the recommendations.

It is conceivable that, at some point, the conclusions of the auditor may conflict with the judgement of the national regulatory agency. For example, the former may find that a company should not be marketing a certain product, given pesticide practices in that country, even though the agency might have recently renewed the product's registration. The audit is of a company; it is not intended to limit a government's prerogative to evaluate risks as it sees fit. In the face of an auditor's public report, however, a decision to permit continued use would call for an alternative interpretation of the evidence or a demonstration of overriding benefits. In this way, the hazard auditor might serve to raise the standards of risk assessment and to open it to public scrutiny.

Initial reactions

Several dominant themes emerged among 58 written replies to the hazard auditor proposal (Loevinsohn 1990). Of the opinions expressed, 5 (9%) were negative and 53 (91%) were positive in varying degrees.

Increase support to national regulatory authorities

The most widely voiced view (39% of the 49 detailed responses) was that greater emphasis should be placed on evaluating and assisting regulatory agencies. Some industry respondents felt that the focus on industry alone was unfair and others, from several sectors, thought the auditor's recommendations would more likely be acted on if government was more closely involved in the process.

An audit that puts developing country governments on the same footing as industry makes no sense; the weakness of national regulation is universally acknowledged and is the underlying rationale for initiatives such as the Code of Conduct and the pesticide hazard auditor. Although some assistance to national authorities might take place within or along with the audit, supported from nonindustry sources, there is a danger of overlap with existing or planned programs of agencies, such as FAO, were this function to take on a much greater significance. Closer integration with FAO is indeed a possibility and, in that context, the audit might extend to other sectors addressed by the Code of Conduct.

Ensure the auditor's independence from industry

The second most frequent comment (22% of the 49 detailed responses) concerned the danger of the hazard audit being dominated by the pesticides industry and of its serving to legitimize pesticide use. Several respondents believed that these risks could only be avoided by complete financial autonomy and by excluding industry representatives from the governing and technical bodies of the audit organization.

Reasonable safeguards against domination should be built into the scheme. There would be justification, for example, in excluding company representatives from the technical committee where audits would be planned. Sanctions should be available against companies that, in their advertising or labeling, misconstrue audit results to imply an endorsement of their products. Funding from other sources could be sought to dilute the dependence on industry, but the self-financing character of the proposal is one of its chief attractions. Any attempt to exert undue influence would lead other sectors to withdraw their support from the scheme and lose companies the commercial benefit they derive from an independent audit.

Furthermore, financing by industry is economically rational. An independent audit can legitimately be seen as part of the regulation required to minimize the external costs to which pesticides give rise when they damage human health or the environment. Outside financing of the audit would amount to a subsidy, leading to greater use than if real costs were reflected in market price (Repetto 1985; Brader 1990) and distorting the choice between chemical-based pest control and alternative techniques.

Emphasize the incentives for industry compliance

A number of respondents suggested measures to increase the benefits to a company that agreed to be audited and implement the audit's recommendations. These include proposing that multilateral and bilateral aid organizations make a satisfactory audit report a requirement in their procurement programs. Developing country governments could similarly agree to purchase only from manufacturers who have received such an evaluation from the auditor. Governments might also make the external hazard audit a legal requirement, as is the case for financial audit in many countries.

Measures such as these might indeed usefully increase incentives for compliance and penalize companies who remain outside the scheme to gain, for example, a price advantage. Additional incentives may be particularly important to smaller manufacturers based in Third World countries where public opinion is often poorly informed. Many of these firms produce hazardous pesticides and are often not affiliated with national or international industry associations.

Increase collaboration with FAO and other United Nations agencies

Several respondents wrote that, as the hazard auditor aims at improving compliance with accepted standards, particularly those embodied in the FAO Code of Conduct, a closer relation with FAO should be sought. Some questioned the need for an independent audit organization.

I investigated the possibility of an association with FAO and other United Nations (UN) agencies. Senior FAO officials recognized the value of the auditor, particularly as a possible means to improve compliance with the Code should it become a convention made binding under national law. However, two difficulties were mentioned. FAO, the officials declared, would not accept the financial link with industry that the proposal envisages. The other problem involves openness in reporting, which is crucial to the functioning of the audit mechanism. Because FAO is responsible to its member governments, problems might arise if it were to publish reports that were critical of national administrations.

Similar concerns were voiced by officials of UNEP and WHO. These constraints might be loosened if there were widespread support for the innovation at the highest levels. However, the independence required of an external auditor would be more readily assured within an organization that is itself independent of the major actors.

Conclusions

The hazard auditor concept holds promise for improving the regulation of pesticide hazards in developing countries by creating a new market-based mechanism complementary to, and supportive of, national structures and international programs. It has already attracted widespread, if still provisional, support, but further progress toward implementing the concept will require collaboration among parties grown accustomed to confrontation. The proposal does not assume an congruency of interests among these parties, only that each side believe its interests are served by independent evaluation. Individual actors may conclude that the risks attendant on creating and operating this novel mechanism outweigh its benefits. It is not possible to predict the outcome of what will be a long process of negotiation. The prudent option, for all concerned, is to judge at each step whether the hazard auditor as it is emerging represents an improvement on what currently exists.

Anderson, R.J. 1984. The external audit (2nd ed.). Copp Clark Pitman, Toronto, ON, Canada.

Anonymous. 1988a. FAO code of conduct: 17 important messages. *In* Product quality manual (Appendix 3). Agriculture Division, Ciba-Geigy Ltd, Basel, Switzerland.

_____ 1988b. Product quality concept. Agricultural Division, Ciba-Geigy Ltd, Basel, Switzerland.

_____ 1990b. White paper proposals. Pesticides News, 8, 7–9.

_____ 1990a. London guidelines "prior informed consent." International Register for Potentially Toxic Chemicals Bulletin, 10(1), 4–6.

Brader, L. 1990. Integrated pest management: successes and constraints. Paper presented at the FAO/UNEP workshop on integrated pest management, 12–15 June 1990, Kishniev, Moldavia, USSR. 8 pp.

ELC (Environment Liaison Centre). 1987. Monitoring and reporting the implementation of the international code of conduct on the use and distribution of pesticides (the FAO Code). ELC, Nairobi, Kenya.

FAO (Food and Agriculture Organization of the United Nations). 1986. International code of conduct on the distribution and use of pesticides. FAO, Rome, Italy. 31 pp.

_____ 1989. International code of conduct on the distribution and use of pesticides: analysis of responses to the questionnaire by governments. FAO, Rome, Italy. AGP: GC/89/BP.1.

GIFAP (Groupement international des associations nationales de fabricants de produits agrochimiques). n.d. The role of GIFAP. GIFAP, Brussels, Belgium.

Jeyaratnam, J. 1985. Acute pesticide poisoning: a Third World problem. World Health Forum, 6, 39.

Loevinsohn, M. 1989. The hazard auditor and improved pesticide regulation in the Third World. International Development Research Centre, Ottawa, Canada. Consultant's report, 10 pp.

Loevinsohn, M. 1990. Evaluation of responses to the pesticide hazard auditor proposal. International Development Research Centre, Ottawa, Canada. Consultant's report, 9 pp.

Pallemaerts, M. 1988. Developments in international pesticide regulation. Environmental Policy and Law, 18, 62–68.

Palmisano, J. 1989. Environmental auditing: past, present and future. Environmental Auditor, 1, 7–20.

Pesticides Trust. 1989. The FAO code: missing ingredients. Pesticides Trust, London, UK.

Repetto, R. 1985. Paying the price: pesticide subsidies in developing countries. World Resources Institute, Washington, DC, USA. 18 pp.

Willis, G.A. 1986. Pesticide residue problems in developing countries in Asia: some contributions of industry. Paper presented at the 2nd session of the working group on pesticide residue problems in Asia, 2–5 April 1986, Chiang Mai, Thailand.

Safer pesticide-application equipment for small-scale farmers in developing Asian countries

Md. Jusoh Mamat,[1] A.N. Anas,[2] and S.H. Sarif Hashim[2]

[1]Fundamental Research Division and [2]Agriculture, Engineering and Water Management Division, Malaysian Agricultural Research and Development Institute, Kuala Lumpur, Malaysia

Many tropical Asian countries produce lever-operated knapsack sprayers locally. The quality of the sprayers ranges from poor to moderate, yet large numbers of these sprayers are bought and used in the region. Some of their unsafe features are: sharp edges on the tank body, poor positioning of the pump body and linkage system, leak-prone connections, and poor quality materials. Factors leading to the continued production of inferior sprayers include the lack of incentives for local manufacturers to produce good quality sprayers, farmers' poor appreciation of quality and safety features, and the lack of national minimum product standards. Overcoming these problems requires the establishment of research programs in national agricultural research institutes to design safer and more efficient knapsack sprayers that can be produced locally, upgrading of farmers' training programs to improve their appreciation of good quality sprayers and safer spraying practices, and establishing minimum standards for knapsack sprayers.

Unlike North America or Europe, the greatest concern relating to the escalating use of pesticides in developing Asian countries is the personal contamination of those applying the chemicals. Ironically, in cases of occupational poisoning or failed pest control, farmers put the blame on the pesticide. Seldom, if ever, is the method or apparatus used for application blamed. Yet, the design and condition of the equipment owned and used by farmers can contribute to the safe and efficient application of pesticides (Anas et al. 1987; Jusoh Mamat and Anas 1988).

In many developing Asian countries, the lever-operated knapsack sprayer is still the most commonly used pesticide applicator among small-scale farmers

(Adam 1976; Fraser and Burrill 1979; Heinrichs et al. 1979; Litsinger et al. 1980; Prasadja and Ruhendi 1980; Lim et al. 1983; Jusoh Mamat et al. 1985). Major agricultural countries in South Asia produce lever-operated knapsack sprayers locally. However, the quality of these sprayers, in terms of efficiency and safety, ranges from very poor to moderate. In Malaysia, for example, the lever-operated knapsack sprayers (96% locally manufactured) used by rice farmers were not well designed and did not satisfy the minimum product standards set by the World Health Organization (Anas et al. 1987).

Unsafe features of locally produced sprayers

In using a lever-operated knapsack sprayer, the operator is exposed to two kinds of hazard: direct physical harm such as wounds and bruises caused by the equipment and contamination by the pesticides being applied. Hazardous features of knapsack sprayers have been discussed in detail (Fisher and Deutsch 1985; Anas et al. 1987; Thornhill 1987).

Sharp edges on the tank body — Sharp edges are found at the bottom of the tank body, the body skirting, and the threaded knob where the pump lever is attached. These sharp edges can rub against and injure the operator's back or buttock area (if contaminated by pesticides, wounds can be even more serious). Sharp protrusions also cause injuries if the operator falls.

Narrow straps made of unsuitable material — Knapsack sprayer straps are often too narrow (less than 5 cm) and made of hard, coarse materials. When a full load of 15–20 L of pesticide is carried, a hard narrow strap will dig into the shoulder muscles causing skin bruises. Because of the constant movement of the tank on the operator's back, bruises can appear after only two or three rounds of spraying with a full load. Straps may snap under a heavy load and cause the tank to fall, wounding the operator. If the straps give way early in the work period, farmers may replace them with even less appropriate materials, such as plastic rope, which is very uncomfortable, or soft cloth rope, which can absorb spilled or splashed pesticide and become a persistent source of contamination.

Tank weight and balance — Most locally produced sprayers are made of brass and are heavy even when empty (over 6.5 kg). Carrying such a heavy sprayer often taxes the energy of slightly built Asian workers. This is especially true for women workers in the rubber and oil-palm plantations of Indonesia, Malaysia, and Thailand. The tank's centre of gravity, often located in the upper half, will cause the operator to be unbalanced and fall easily when walking on difficult terrain, such as wet and muddy paddy fields.

Small tank port and shallow basket strainer — A tank port smaller than 12 cm and basket strainers less than 5 cm deep can create an air lock. As a result, diluted pesticide can overflow and splash onto the operator's hands or legs as it is poured into the tank, particularly if this is done in haste.

Leak-prone spraying components — Leaky sprayer components are a major cause of pesticide contamination among small-scale farmers. Pump cylinders, air chambers, and cut-off valves may leak because of worn-out parts, such as the piston head, ball bearings, and washers, or because they are not greased properly. Leaks at hose connection points are mainly due to the use of crimp ferrule; continual use and flexing of the hose causes cracking.

Positioning of components — If the pump body and the air chamber are not firmly secured to the tank or are awkwardly positioned at the side of the tank body, misalignment of the piston head in the cylinder and leakage at weld points in the air chamber can easily result from knocks and falls during spraying operations. Misalignment was found in 64% of the knapsack sprayers owned and used by farmers in the rice-bowl area of Malaysia (Anas et al. 1987).

Large mesh and absence of strainers and filters — A common cause of contamination during spraying operations is blockage of the nozzle opening by solid foreign matter when strainers and filter mesh are too large, i.e., larger than 1 mm for strainers and 0.5 mm for filters, or completely absent. In Southeast Asian countries, water used to dilute the pesticides comes from irrigation canals, small streams, or rain-water storage tanks, which contain much foreign debris. As a result, nozzle blockage is common. Farmers often attempt to free blockages with sharp hard objects, such as pins or steel wire (possibly damaging the nozzle opening) or by blowing into the nozzle, contaminating both their hands and mouths. In a survey in Malaysia's rice-bowl area, 89% of 193 knapsack sprayers had strainers with a mesh size larger than 1-mm and none had filters at the cut-off valve and the nozzle (Anas et al. 1987).

Absence of an agitator — Not all lever-operated knapsack sprayers produced locally are fitted with an agitator. When wettable powder (WP) formulations are used, agitators keep the powder suspended in the solution, thus preventing it from settling or becoming aggregated and blocking the nozzle opening during spraying.

Length and design of the spray lance — All locally produced knapsack sprayers currently available are equipped with a 50-cm spray lance. This is too short to prevent drifting spray droplets from contaminating the operator. A 1-m lance, curved slightly at the front end, could make directing the spray much easier and decrease contamination due to dirt (Jusoh Mamat and Anas 1988).

Poor quality brass used for tank body — According to Anas et al. (1987), 50–58% of sampled sprayers had body indentations and evidence of corrosion. These were caused by knocks and falls, testifying to poor user habits as well as the poor quality brass sheets used by manufacturers. In Malaysia, local manufacturers produce sprayers with a price range from 60 to 120 MYR (2.6875 Malaysian ringitts (MYR) = 1 US dollar (USD) in 1991) according to brass content or thickness of the brass sheet used for the tank body. Sprayers in the lower price categories tend to be unsafe because they are easily dented and corroded and are thus too prone to leaking.

Improving local production of sprayers

The continued production of poor-quality sprayers in developing Asian countries cannot be blamed solely on the profit motive of the manufacturers. The lack of incentives for local manufacturers to produce good-quality knapsack sprayers, the attitudes of farmers, and the lack of government standards are equally responsible.

There are few economic incentives for local manufacturers to improve quality. In Malaysia, for example, local manufacturers, who produce about 300 000 sprayers annually, have virtually no foreign competition. Even though their sprayers are inferior to imported models in quality and safety, the overwhelming majority of farmers choose to buy cheaper, locally produced sprayers. Any improvement would merely add to the cost of production without any market advantage. Moreover, improvements of certain components of the sprayer system can involve producing an entirely new model. This may require changes in manufacturing equipment and possibly retraining workers, which may result in an initial decrease in the efficiency of production.

On the demand side, farmers' poor appreciation of good quality and safety features also discourages improvement. In Malaysia, farmers and plantation workers pay little attention to the proper use of pesticides especially with regards to safety (Zain 1977; Zam 1980; Basri 1981; Heong 1982; Normiya 1982; Ooi et al. 1983; Heong et al. 1985, 1987; Hussein et al. 1985; Anas et al. 1987; Jusoh Mamat et al. 1987; Anon. 1990). Consequently, they fail to recognize or avoid sprayers with poor safety features. As long as the sprayers function cheaply, they will use them without concern for their own health and safety. Such attitudes cause complacency among local manufacturers about the quality of their sprayers.

The lack of national product standards for knapsack sprayers also contributes to the continued production of unsafe sprayers. In most Asian developing countries, monitoring quality or standardized testing of locally produced machines and equipment is left almost entirely to manufacturers. Generally, governments have made little effort to establish standards or enforce them.

In Malaysia, a positive step has been taken with the establishment of the Standard and Industrial Research Institute (SIRIM). Its role is to produce standards and approve products that conform. However, SIRIM has no authority to enforce these regulations, and even the product approval scheme has not yet been implemented for knapsack sprayers. Without such a scheme, local manufacturers have no incentive to alter their products to meet minimum safety standards.

Improving safety in design and use

Because manufacturers are not likely to produce better models, it is suggested that local government research institutions should accept this responsibility. Research institutions should develop a new sprayer design and provide blueprints, free of charge, to local manufacturers. Besides optimizing ergonomic and safety features, the new design must also take into account the production capability of the manufacturers and the availability of materials, which together determine the unit price of the product. Good, imported knapsack sprayers are available, but at a price double or triple that of locally manufactured sprayers. The challenge is to produce a new knapsack sprayer as good as the imported ones, if not better, at the same price as current local models.

Good equipment will not improve pesticide-application technology among small-scale farmers without training in its proper use and maintenance. A critical step will be to encourage the establishment of sprayers' clinics within existing farmers' cooperative centres. These clinics, besides selling approved knapsack sprayers, storing spare parts, and providing repair services, must also provide advice and training for farmers in the safe use of the sprayers.

Regulatory measures, such as setting local minimum standards for knapsack sprayers and mandatory use of protective clothing during spraying operations must also be considered. These measures are useless without effective enforcement.

Conclusion

Regardless of future improvements in spraying techniques and technology, there is much room for improvement in current conventional practices, especially in tropical developing countries. Although the inefficiencies of conventional foliar spraying are well recognized (Matthews 1983; Hislop 1988; Zeren and Moser 1988), its versatility makes it attractive to small-scale farmers of the Asian region. Any significant improvement in the current practice, be it in the equipment itself, the technique of spraying, safety attire, or the attitude and perception of the farmers, will, therefore, produce long-lasting benefits. The advent of a revolutionary new application technique is less likely to replace conventional spraying than to augment it in this part of the world.

Adam, A.V. 1976. The importance of pesticides in developing countries. *In* Gunn, P.L.; Steven, J.G.R., ed., Pesticides and human welfare. Oxford University Press, Oxford, UK. Pp. 115–130.

Anas, N.; Jusoh Mamat, Md.; Heong, K.L.; Ho, N.K. 1987. A field observation of lever operated knapsack sprayers owned by the rice farmers in the Muda Irrigation Scheme. *In* Proceedings of the international conference on pesticides in tropical agriculture, 23–25 September 1987, Kuala Lumpur, Malaysia. Malaysian Plant Protection Society, Kuala Lumpur, Malaysia.

Anonymous. 1990. Laporan ringkas bancian kegunaan racun makhluk perosak di kalangan peserta dan pekerja FELCRA Seberang Perak. Jabatan Pertanian, Kuala Lumpur, Malaysia.

Basri, M.W. 1981. Study on the use of and hazards posed by certain insecticides on tobacco and vegetables in Peninsular Malaysia. Crop Protection Branch, Department of Agriculture, Kuala Lumpur, Malaysia.

Fisher, H.H.; Deutsch, A.E. 1985. Lever-operated knapsack sprayers: a practical scrutiny and assessment of features, components and operation — implication for purchasers, users and manufacturers. International Plant Protection Center, Oregon State University, Corvallis, OR, USA.

Fraser, F.; Burrill, L.C. 1979. Knapsack sprayers: use, maintenance, accessories. International Plant Protection Center, Oregon State University, Corvallis, OR, USA.

Heinrichs, E.A.; Saxena, R.C.; Chellia, S. 1979. Development and implementation of insect pest management systems for rice in tropical Asia. *In* Sensible use of pesticides. Asian and Pacific Council, Taiwan. FFTC Book Series, 14, 208–247.

Heong, K.L. 1982. Pest control practices of rice farmers in Tanjung Karang, Malaysia. *In* Proceedings of the 3rd international meeting on the perception and management of pests and pesticides, Nairobi, Kenya, 21–25 June 1982. Insect Sci. Appl., 5(3), 221–226.

Heong, K.L.; Ho, N.K.; Jegatheesan, S. 1985. The perception and management of pests among rice farmers in the Muda Irrigation Scheme, Malaysia. Malaysian Agricultural Research Institute, Kuala Lumpur, Malaysia. Report 105.

Heong, K.L.; Jusoh Mamat, Md.; Ho, N.K.; Anas, A.N. 1987. Sprayer usage among rice farmers in the Muda area, Malaysia. *In* Proceedings of the international conference on pesticides in tropical agriculture, 23–25 September 1987, Kuala Lumpur, Malaysia. Malaysian Plant Protection Society, Kuala Lumpur, Malaysia.

Hislop, E.C. 1988. Electrostatic ground-rig spraying: an overview. Weed Technology, 2, 94–105.

Hussein, M.Y.; Jusoh Mamat, Md.; Azmi, M.A.; Monyvellu, S. 1985. Transfer of pesticide application technology to small farmers: practical aspects on approaches and constraints. Paper presented at the workshop and course on pesticide application technology, 21–26 October, Universiti Pertanian Malaysia, Serdang, Selangor, Malaysia. Malaysian Plant Protection Society, Kuala Lumpur, Malaysia.

Lim, G.S.; Hussein, M.Y.; Ooi, A.C.P.; Zain, M.B.A.R. 1983. Pesticide application technology in annual crops in Malaysia. *In* Lim, G.S.; Ramasamy, S., ed., Pesticide application technology. Malaysian Plant Protection Society, Kuala Lumpur, Malaysia. Pp. 13–41.

Litsinger, J.A.; Price, E.C.; Herrara, R.T. 1980. Small farmer pest control practices for rainfed rice, corn and grain legumes in three Philippines provinces. Philippine Entomologist, 4, 65–86.

Matthews, G.A. 1983. Problems and trends in pesticide application technology. *In* Lim, G.S.; Ramasamy, S., ed., Pesticide application technology. Malaysian Plant Protection Society, Kuala Lumpur, Malaysia. Pp. 163–170.

Jusoh Mamat, Md.; Anas, A.N. 1988. Herbicide application technology for irrigated rice in Malaysia. *In* Lam, Y.M.; Cheong, A.W.; Azmi, M., ed., Proceedings of the national seminar and workshop on rice field weed management, Penang. Malaysian Agricultural Research and Development Institute, Kuala Lumpur, Malaysia. Pp. 221–229.

Jusoh Mamat, Md.; Heong, K.L.; Rahim, M. 1985. Principles and methodology of pesticide application techniques. *In* Proceedings of the workshop and course on pesticide application technology, 21–26 October 1985, Universiti Pertanian Malaysia, Serdang, Selangor, Malaysia. Malaysian Plant Protection Society, Kuala Lumpur, Malaysia.

Jusoh Mamat, Md.; Anas, A.N.; Heong, K.L.; Chan, C.W.; Nik Mohd Nor, N.S.; Ho, N.K.; Zaiton, S.; Fauzi, A. 1987. Features of lever operated knapsack sprayer considered important by Muda rice farmers in deciding which sprayer to buy. *In* Proceedings of the international conference on pesticides in tropical agriculture, 23–25 September 1987, Kuala Lumpur, Malaysia. Malaysian Plant Protection Society, Kuala Lumpur, Malaysia.

Normiya, R. 1982. Problems in transfer, delivery and acceptance of rice technology. Malaysian Agricultural Research and Development Institute, Kuala Lumpur, Malaysia. Rural Sociology Bulletin 12.

Ooi, A.C.P.; Heong, K.L.; Lim, B.K.; Mazlan, S. 1983. Adoption of pesticide application technology by small-scale farmers in peninsular Malaysia. *In* Lim, G.S.; Ramasamy, S., ed., Pesticide application technology. Malaysian Plant Protection Society, Kuala Lumpur, Malaysia. Pp. 148–158.

Prasadja, I.; Ruhendi. 1980. Farmers' existing technology and pest control practices for food crops at three locations in Jogjakarta Province. Agency for Agricultural Research and Development, Central Research Institute for Agriculture, Bogor, Indonesia. 59 pp.

Thornhill, E.W. 1987. Lever-operated sprayer selection. *In* Proceedings of the international conference on pesticides in tropical agriculture, 23–25 September 1987, Kuala Lumpur, Malaysia. Malaysian Plant Protection Society, Kuala Lumpur, Malaysia.

Zain, M.B.A.R. 1977. Survey on the use of pesticides on tobacco in peninsular Malaysia. Crop Protection Branch, Department of Agriculture, Kuala Lumpur, Malaysia. Report 1.

Zam, A.K. 1980. Bancian pengurusan dan kawalan serangga perosak padi di rancangan perairan tanjung karang dan krian. Laporan Cawangan Pemeliharaan Tanaman, Jabatan Pertanian, Kuala Lumpur, Malaysia.

Zeren, Y.; Moser, E. 1988. Effects of electrostatic charging and vertical air current on deposition of pesticide on cotton plant canopy. Agricultural Mechanization in Asia and Latin America, 19(1), 55–60.

Field evaluation of protective equipment for pesticide appliers in a tropical climate

G. Chester,[1] A.V. Adam,[2] A. Inkmann-Koch,[3] M.H. Litchfield,[4] R. Sabapathy,[1] and C.P. Tuiman[4]

[1]ICI Agrochemicals, Fernhurst, Haslemere, Surrey, UK; [2]Food and Agriculture Organization, Rome, Italy; [3]Bayer AG, Monheim, Germany; [4]Shell International Petroleum, The Hague, Netherlands

To provide practical advice and information on personal protection used during pesticide application in tropical climates, a field study was conducted in Thailand. Items assessed included protective garments worn by workers mixing and loading the organophosphorous insecticide methamidophos and by sprayers applying the diluted formulation for several hours per day to a cotton crop using knapsack sprayers. Garments made of various materials were assessed for their acceptability to the workers, their comfort and durability, and the degree of protection they offered. Tyvek garments proved to be uncomfortable under tropical conditions. Kleenguard and cotton garments were acceptable to the workers; cotton proved more durable and comfortable over 6 days of use. Nitrile gloves and face shields worn by the mixer–loaders were also found to be suitable under field conditions. The effective use of protective equipment must go hand in hand with safe handling practices and good personal hygiene.

The heat and humidity in tropical countries often make it difficult for farmers to wear recommended protective clothing when handling and applying pesticides. They either suffer excessive heat discomfort when wearing such equipment or remove it to work more comfortably, thus risking increased exposure. To improve this situation, the Groupement international des associations nationales de fabricants de produits agrochimiques (GIFAP) and the Food and Agriculture Organization (FAO) undertook a field evaluation of personal protective equipment in Thailand (Working Group 1989). This paper describes the study and its findings.

Materials and methods

Assessment of protective equipment

The Working Group examined materials of potential use for protective garments, using criteria such as protection against pesticides, durability, heat-exchange properties, comfort, availability, and cost. Three materials were identified for field evaluation: polypropylene (Kleenguard EP), polyethylene (Tyvek S1422), and cotton. Two-piece protective garments were made from these materials. The upper garment was a double-apron design, the lower garment a pair of trousers.

A review of materials for protective gloves, using criteria such as protection against a range of pesticide formulations, cost, and availability, revealed that nitrile rubber was one of the most appropriate. Gauntlet-style gloves were made from this material and used by the mixer–loaders in the study.

A face shield of simple design was constructed with a transparent visor mounted with elastic so that the transparent vertical section rested about 2.5 cm from the face. These were worn by the mixer–loaders in the study.

Study site and crop

The study was conducted on five smallholder farms in the vicinity of Patananikom, Lopburi District, Thailand, in September 1988. The cotton crop under cultivation at the time was up to 1.5 m tall. Weather conditions, including temperature, humidity, and wind speed, were monitored during the study.

Pesticide and application equipment

The organophosphorous insecticide, methamidophos, was used in the study. The formulation, Tamaron 600 SL, contained methamidophos at 600 g/L; 30 mL was diluted to 17 L in each knapsack tank to give a methamidophos concentration in the spray solution of about 1.06 g/L. The application rate was 1 L/ha. The spray equipment consisted of semiautomatic knapsack sprayers of stainless-steel construction with a capacity of 17 L.

Workforce

Twenty pesticide workers were recruited locally and divided into two groups of 10. Each group consisted of two mixer–loaders and eight spray operators. All workers wore standard cotton T-shirts and shorts under the protective garments.

Study design

The study was carried out in two stages. A prestudy was conducted to evaluate the comfort, acceptability, and effectiveness of protective garments made of the nonwoven materials, Kleenguard and Tyvek, to select one for use in the main study. In the main study, we evaluated garments made of the selected material and cotton.

On day 1 of the prestudy, one group of workers wore Kleenguard protective garments, the other group Tyvek garments. All workers wore cotton sampling garments beneath their protective clothing to assess its permeability to the insecticide during work. Upon completion of the day's operations, the protective and sampling garments were removed carefully from each operator. Each garment was turned inside out and wrapped in aluminium foil. Each foil package was placed in a polyethylene bag labeled with the worker's number, day, type of garment, and date. All garments were kept in refrigerated storage before analysis.

On day 2, workers in both groups wore new Kleenguard protective garments without sampling garments underneath to assess comfort. On day 3, this procedure was repeated using Tyvek protective garments.

On days 1 and 6 of the main study, one group of workers wore Kleenguard protective garments and the other group wore cotton. All workers wore cotton sampling garments under their protective clothing to assess their protection against insecticide. Upon completion of the day's work, the protective and sampling garments were removed and stored as described for the prestudy.

On day 2 (main study), new protective garments were issued to the two groups of workers (Kleenguard for one group, cotton for the other) for daily assessment of comfort and durability. Each worker wore the same garment until either the study team agreed that it could no longer be worn because of deterioration or until the end of the study.

Regular assessments were also made of the performance and condition of protective gloves and face shields. The opinions of the mixer–loaders were obtained through the use of a questionnaire.

Daily operations

To standardize the daily work as much as possible, the two mixer–loaders in each group prepared Tamaron spray solution for the spray operators in their respective group. Each mixer–loader handled the same quantity of the insecticide formulation.

The spray operators applied the same volume of insecticide (seven knapsack tanks) each day. Operators worked sufficiently far apart to minimize the possibility of cross-contamination due to drift from other sprayers.

At the end of each day's operations, the protective garments were washed with detergent and water using a standard procedure and dried overnight.

Assessment of protection

Permeation was derived by measuring the amount of methamidophos found on the protective and sampling garments. Methamidophos was extracted from the garments using methanol for cotton and ethyl acetate for Kleenguard and Tyvek. The extracts were subjected to analysis by high-performance liquid chromatography (HPLC) using an ultraviolet detector for the determination of methamidophos. The results were corrected for losses of methamidophos through exposure to light, storage, and transit.

The permeation value, expressed as a percentage, was the amount of methamidophos found on a sampling garment, divided by the total amount found on the protective and sampling garments. Permeation data for each group were compared using Student's t-test after exclusion of outlying values (Tukey 1977).

Evaluation of comfort, thermal comfort, and durability was achieved through the use of a detailed questionnaire and from observations made by the study team.

Health surveillance

The workers were given a medical examination 1 day before the start of the prestudy and were kept under medical surveillance during both phases of the study. Blood samples were collected at regular intervals to test for cholinesterase activity. Measurement was carried out in situ using the tintometer method.

Results

Total working time on the 1st day of the prestudy was about 4 h. As the study proceeded, work rate increased. On the last day of the main study, working time had decreased to about 2.5 h. Because spray equipment broke down frequently, one person was employed full-time to look after the equipment and ensure continuity of spraying.

Over the 9-day study period, shade temperatures during work ranged from 26 to 33 °C. Relative humidity, measured on 3 days, was 52–81%. Wind speed was low, with gusts of up to 1 m/s recorded occasionally.

During the prestudy, all 20 workers said that they felt comfortable when working in the Kleenguard protective garments (Table 1). By contrast, 11 of the 20 workers said that the Tyvek garments were uncomfortable and this was reinforced by responses to questions about problems during spraying; 10 of the 16 sprayers complained that the material stuck to the skin and caused the

Table 1. Worker assessment of two types of protective garments:
Kleenguard and Tyvek.

Question	n	Kleenguard		Tyvek	
		Yes	No	Yes	No
Was it comfortable?	20	20	0	9	11
Did it feel hotter?	20	10	10	19	1
Any problem when spraying?	16	3	13	10[a]	6
Did you like the garments?	20	19	1	9	11

[a] The main problem was sticking and wrinkling.

Table 2. Comparison of Kleenguard (K) and cotton (C) protective garments
regarding comfort and acceptability (numbers represent positive answers).

Question	n	Day 2		Day 3		Day 5	
		K	C	K	C	K	C
Was it comfortable?	10	10	10	9	10	10	10
Did it feel hotter?	10	10	7	10	7	8	4
Any problem when spraying?	8	0	0	2	0	3	3
Did you like the garments?	10	10	8	10	10	10	9

garments to wrinkle. Most of the workers with Tyvek garments felt hotter as a result of wearing them. On the basis of questionnaire results and observations made by the study team, it was decided that Kleenguard garments would be used in the main study.

Workers wearing both Kleenguard and cotton stated that the garments were comfortable to wear every day of the main study (Table 2). Most workers in both groups experienced no problems. When problems did occur, they were due to the individual fit of the garment rather than general physical discomfort. Although most workers felt somewhat hotter when wearing the garments, this was not sufficient to cause undue stress or loss of working efficiency. By day 5, fewer workers wearing cotton garments felt hotter.

Most of the cotton protective garments were unaffected by wear over the first 3 days (Table 3). From day 4 on, an increasing number of cotton garments showed signs of slight chafing, mainly around the shoulders, lower back, around the buttock area, and knees. There was no evidence of tearing or splitting.

Kleenguard garments showed evidence of slight chafing on day 1. By day 3, the chafing was pronounced and tearing and splitting of the garments was also noted. This was more evident in the garments worn by workers who carried the knapsack sprayers, which rubbed against their shoulders and lower back. The damage was progressive; by day 6, all Kleenguard garments were affected, some producing severe chafing. The chafing was mostly on the shoulders and lower back.

Table 3. Number of garments chafing or showing wear over 6 days of use.

	n	Kleenguard 1[a]	2	3	4	5	6	Cotton 1	2	3	4	5	6
Upper garment													
Chafed		7	8	8	8	8	9	1	1	1	7	7	8
Torn or split		2	4	6	8	8	8	0	0	0	0	0	0
Total affected	10	9	10	10	10	10	10	2	2	2	7	7	8
Trousers													
Chafed		6	8	7	10	9	10[b]	1	0	1	4	6	10[c]
Torn or split		0	0	2	3[d]	1[d]	0[e]	0	0	0	0	0	0
Total affected	10	6	8	8	10	9	10	1	0	1	4	6	10

[a] Day of permeation assessment; new garments were issued on day 2.
[b] Mostly showed severe chafing.
[c] Mostly slight chafing.
[d] Two pairs trousers replaced at start of day.
[e] One pair trousers replaced at start of day.

The mean permeation value for the Kleenguard garments on day 1 of the main study (41.8%) was similar to that found during the prestudy (37.7%) (Table 4). The mean permeation value for the cotton garments (18%) was significantly lower than that of the Kleenguard garments ($p < 0.01$). After 6 days of wear, including daily washing, the permeation values of the two sets of protective garments had not increased and, in fact, were somewhat lower than on day 1. Permeation of the cotton garments was still significantly lower than that for the Kleenguard garments on day 6 ($p < 0.01$).

The permeation rate varied widely within each group of workers. This is consistent with findings from other worker-exposure studies, which show that the amount of contamination varies with individual workers (Wolfe et al. 1972).

Table 4. Permeation (%) of protective clothing by insecticide.

	Tyvek	Kleenguard	Cotton
Prestudy (day 1)			
Mean	17.3 ($n = 8$)	37.7 ($n = 7$)	—
Range	4.9–41.8	25.1–56.3	
Main study (day 1)			
Mean	—	41.8 ($n = 8$)	18.1 ($n = 7$)
Range		30.0–55.6	14.1–22.2
Main study (day 6)			
Mean	—	30.4 ($n = 8$)	10.4 ($n = 8$)
Range		13.3–48.1	8.7–15.5

Three of the four mixer–loaders found the protective gloves uncomfortable to wear. However, all four believed that their hands were being protected. Some experienced difficulty in gripping equipment when wearing the gloves, and two pairs of gloves had to be replaced because of abrasion or splitting. Workers found the face shields comfortable to wear; no misting of the visors occurred. The mixer–loaders had no difficulty in keeping the face shields on during the work periods. There were some complaints of reflection from the white garments onto the face shields.

Health surveillance

No clinical signs or symptoms of intoxication by methamidophos were reported or observed in anyone involved in the study. Although there were slight variations in cholinesterase activity in blood samples from workers during the study, all values were within normal ranges.

Discussion

Cotton protective garments of the design used in this study were comfortable to wear in hot and humid climates. They showed little deterioration over the period of use and provided as much or more protection from the organo-phosphorous insecticide than the synthetic materials.

When the benefits of availability and general low cost are taken into account, cotton has major advantages as a material for protective garments required by pesticide workers in tropical climates. Further investigation into the protective properties of cotton, i.e., variations in weight, thickness, and weave, would be worthwhile.

The design of the protective garments was intended to provide some degree of flexibility under field conditions. For example, workers are able to wear the top and trousers separately or together over their normal work clothing depending on the kind of work they are engaged in. Although this design is recommended by the Working Group, it may not be acceptable esthetically in all cultures and other designs may be more appropriate for some types of pesticide application. However, the general principle of the design proved to be successful in this study.

Nitrile rubber gloves, although somewhat uncomfortable, can be worn by mixer–loaders in hot and humid conditions and are, therefore, suitable for handling pesticide concentrates under these conditions. The simple face shield used in this study is also recommended.

The opportunity was taken during the study to seek the workers' opinions about acquiring and wearing gloves and face shields in normal practice. Commitment to wearing these protective items varied, even if they were made readily available. This type of response is common (Jeyaratnam 1982) and underlines the need to change attitudes through training and education programs.

Many pesticides can be handled and applied safely without the need for protective clothing; workers simply wear lightweight work clothing covering most of the body. However, where protective clothing is necessary for use with more hazardous pesticides, cotton garments of the type described in this paper are recommended for use in tropical conditions.

Finally, protective equipment cannot be regarded as the only answer for all aspects of personal protection. This equipment does not necessarily reduce pesticide contamination unless it is used and maintained properly. Its use must go hand in hand with other important protective measures: principally the avoidance of contamination, good personal hygiene, and the correct use and maintenance of safe application equipment.

Acknowledgments — The dedicated technical support provided by Mr Sant Kongsomboon (Bayer Thai Ltd), Dr Apichai Daorai (ICI Asiatic Ltd), and Dr Sujin Chantarasa-ard (Shell Thailand Ltd) is gratefully acknowledged. We also thank the Thai Pesticide Association for their encouragement and support of this project.

Jeyaratnam, J. 1982. Health hazards awareness of pesticide applicators about pesticides. *In* van Heemstra, E.A.H.; Tordoir, W.F., ed., Education and safe handling in pesticide application. Elsevier Scientific Publishing, Amsterdam, Netherlands. Pp. 23–30.

Tukey, J.W. 1977. Exploratory directory data analysis. Addison-Wesley, Reading, MA, USA.

Wolfe, H.R.; Armstrong, J.E.; Staiff, D.C.; Comer, S.W. 1972. Exposure of spraymen to pesticides. Archives of Environmental Health, 25, 29–33.

Working Group(Working Group on Protective Clothing for Hot Climates). 1989. Field evaluation of protective clothing materials in a tropical climate. Groupement international des associations nationales de fabricants de produits agro-chimiques and Food and Agriculture Organization, Bangkok, Thailand. 4 pp.

PART III

ROLES OF INFORMATION
AND HEALTH SERVICES
IN ACHIEVING USER SAFETY

Evaluation and management of pesticides on a global level: the role of the International Program on Chemical Safety

M.J. Mercier

International Program on Chemical Safety, World Health Organization, Geneva, Switzerland

In spite of the magnitude and severity of the health and environmental problems associated with pesticides, these chemicals will continue to be widely used in the foreseeable future, both in agriculture (to produce more food and fibres) and in public health (to control vector-borne diseases and pests). Two processes should take place before pesticides come into use. The first involves the scientific evaluation of any potential risk of a pesticide by virtue of its intrinsic properties to produce an adverse effect in humans and other biota. The second process deals with management of the risk, weighing it against benefit, cost of a product, and local socioeconomic conditions. The aim of this paper is to elaborate briefly on these two processes, especially in relation to developing countries. Particular attention is paid to activities developed and maintained by the International Program on Chemical Safety and other international bodies.

The use of chemicals in a broad range of human activities has increased considerably in recent decades. Society has undoubtedly benefited greatly from this trend, e.g., through the introduction of drugs for the prevention and treatment of many diseases; the development of pesticides to control diseases transmitted by vectors and to boost agricultural productivity; the introduction of food additives that allow food to be preserved where it would otherwise have been wasted.

Chemical substances play a key role in most industries and contribute significantly to society's technical progress by producing new materials for clothing, packaging, construction, and so on. It has now sadly become apparent that this progress carries a price — in terms of human health and the quality of our environment — and that, if adequate means are not found to control the use of chemicals, this price could increase dramatically to the point of becoming unacceptable.

With the exception of a limited number of certain types of chemicals (pesticides, food additives, some of the major air and drinking-water contaminants, and some hazardous industrial chemicals), few products have been tested appropriately for potential risks. The information available is, in most cases, inadequate to estimate the levels that can be tolerated safely by humans.

The constant increase in the number and overall volume of chemicals on the market increases the risk of exposure to humans and their environment throughout the various phases of the lifespan of these substances: production, storage, handling, transport, use, and disposal. Land-development activities, such as quarrying and dredging, also release naturally occurring chemicals from the soil in quantities that exceed those released in normal geological processes.

Sooner or later, many potentially toxic chemicals end up in the environment as residue or wastes that pollute air, water, food, and soil and, thus, affect the entire population. More and more chemicals are being added to animal feed and more and more veterinary drugs are being administered to livestock. Agricultural pesticides continue to increase, and their residues are found in food together with their various metabolites.

About half a million people die every year of serious accidents resulting from exposure to excessive doses of chemicals. The accidental leakage of methylisocyanate at Bhopal killed over 2 000 people in a very short time, but it also left more than 200 000 survivors severely incapacitated with painful and irreversible lung damage.

Although, by definition and in practice, pesticides are poisons designed to kill some forms of life, such as insects, nematodes, rodents, grasses, fungi, and so on, most people recognize the advantages inherent in their use because of the resulting benefits: controlling disease transmitted by vectors and increasing agricultural productivity. To assure safe and beneficial use of pesticides, two separate processes should take place before they are used in practice. The first step is a scientific evaluation of any potential risk of a pesticide connected with its intrinsic property to produce an adverse effect on humans and other biota. The second process deals with management of the risk arising from using a given pesticide, taking into account the balance of the risk against benefit, cost of a product, and even local socioeconomic considerations.

Toxicologic and safety evaluation of pesticides

Regardless of whether a pesticide molecule comes into contact with an organism during its use as an agricultural chemical, household pesticide, or vector-control agent or is consumed as a residue in food or water, it must reach a target and react with some crucial body constituent to produce an adverse effect. In most instances, the proper target is unknown and only for a minority of pesticides is the toxicologic mode of action in mammals and humans understood.

Before being allowed on the market, pesticides are extensively and meticulously tested to ascertain their biological potential in mammals regarding their carcinogenicity, mutagenicity, embryo toxicity, teratogenicity, and other specific toxic properties. These studies are normally performed by the manufacturer, using widely accepted procedures; the results are confidential, but are made available under specific circumstances.

At present, the only truly international evaluations of the risk of pesticides to humans and the environment are made by the International Programme on Chemical Safety (IPCS), a joint venture of the International Labor Organization (ILO), the United Nations Environment Programme (UNEP), and the World Health Organization (WHO). The two main roles of the IPCS are to establish the basis for scientific health and environmental risk assessments for the safe use of chemicals and to strengthen national capabilities for chemical safety. Because of their great importance and wide use, the IPCS has paid particular attention to pesticides and, consequently, has developed several activities.

Joint meeting on pesticide residues in food

The joint meeting on pesticide residues (JMPR), sponsored by the Food and Agriculture Organization of the United Nations (FAO) and WHO, provides member states with estimates of the levels at which various pesticides can be safely tolerated by the human body. The recommended levels are then used by national regulatory agencies and by the Codex Alimentarius Commission to establish safe levels of pesticides in foodstuffs. The meetings have been held annually since 1963. The WHO Group of Experts has described the procedures in the toxicologic evaluation processes (WHO 1990a).

WHO pesticides evaluation scheme

The JMPR does not evaluate pesticides that do not leave residues in food, such as household pesticides, and those primarily designed for public-health use. The latter usually undergo the WHO pesticides evaluation scheme (WHOPES) and are eventually reviewed by the WHO Expert Committee on the Safe Use of Pesticides, the results of which are published in WHO's *Technical Report Series* (e.g., WHO 1985).

WHOPES is a scheme for the evaluation and testing of new pesticides for public-health use that has existed since 1960. It was revised in 1982 to take into account new trends in the field of pesticide development and pest control. Specifically, WHOPES was established to evaluate new chemical compounds, submitted by the agrochemical industry, for their effectiveness against disease vectors and nuisance pests. It also includes evaluation of the toxicity of these compounds to man and beneficial species. It has four phases: laboratory evaluation, small-scale field evaluation on the natural population of pests, large-scale field trials, and establishment of specification for pesticides.

Pesticides in drinking water

Some of the pesticides that contaminate drinking water are specifically evaluated by a WHO ad hoc working group to provide appropriate information for the production or revision of WHO *Drinking Water Quality Guidelines* (WHO 1984).

Environmental health-criteria documents

The environmental health-criteria documents are designed for scientific experts who are responsible for the evaluation of risk to human health and the environment incurred by chemicals. They enable relevant authorities to establish policies for the safe use of these chemicals. The risk incurred by some physical factors is also studied. The information is detailed enough to allow the scientific reader to make his or her own validation. Over 30 documents dealing with pesticides have been published or are in press, and more are in preparation.

Health and safety guides

The health and safety guides are designed for the wide range of administrators, managers, and decision-makers in various ministries and governmental agencies, as well as in commerce, industry, and trade unions, who are involved in various aspects of safe chemical use. They summarize toxicity information in simple, nontechnical language and provide practical advice on safe storage, handling and disposal of the chemical, accident prevention and health protection measures, first aid and medical treatment in cases of overexposure, and clean-up procedures. So far, over 30 guides have been published and more are in preparation.

Pesticide data sheets

To provide member states, governments, and various institutions, public and private, with basic toxicologic information on individual pesticides, WHO, in collaboration with the FAO, issues data sheets on pesticides. Priority is given to compounds that are widely used in public-health programs and in agriculture or have a high or unusual toxicity record. Since the program began in 1975, 70 data sheets have been published in English and French. Data sheets are revised when appropriate. Draft versions are produced by a group of scientists on a contractual basis, then sent to independent referees and, through the Groupement international des associations nationales de fabricants de produits agrochimiques (GIFAP), to industry for comment. Publication follows only after mutual agreement of all involved.

International chemical safety cards

The international chemical safety cards summarize essential product-identity data and health and safety information on chemicals for use by workers and employers in factories, agriculture, and other workplaces. The cards are prepared using standard phrases, complemented, when appropriate, with information specific to the pesticide being used.

Risk management

After assessment of the risk arising from pesticides, the IPCS pays considerable attention to activities that reduce the risk to an acceptable level. This is done through three areas of activity: prevention of pesticide poisoning; promotion of safe use of pesticides; and education and training. The activities are implemented through joint programs with other international organizations.

Prevention of pesticide poisoning

Prevention of pesticide poisoning is a specific WHO objective and is defined as an outstanding activity within the IPCS. It includes the production of guidelines for poison control and the validation of antidotes (and their availability) used in the treatment of poisonings. The key role of IPCS in these activities is developing information systems for poison control, including harmonization and exchange of data. Particular attention is given to the medical response to chemical emergencies in the case of an accident.

Lack of information on the extent of pesticide poisonings is an immediate concern of WHO, industry, and national governments, particularly those of the developing world. Even where adequate data are available, the collection and collation of data on pesticide poisoning worldwide remains an on-going activity of WHO.

Safe use of pesticides

Among WHO's activities concerning the safe use of pesticides, special attention is paid to identifying sources of increased health hazards. Pest resistance to certain types of pesticides or a significant increase in cost requires a search for alternative compounds that are more effective or cheaper, but only occasionally safer. One of WHO's priorities is to encourage the development of safer, more effective compounds, specifically suited for integrated pest management.

Fully aware of the present-day need for pesticides, WHO tries to reduce pesticide risk to an acceptable level. One activity to support this effort is the regular updating of *Recommended Classification of Pesticides by Hazard and Guidelines for Classification* (WHO 1990b). Although the WHO classification system takes into account acute oral or dermal toxicity (whichever is higher), it also notes any irreversible effect that might be recognized. Classification of a pesticide is subject to regular revision based on scientific evidence.

Recognition of overexposure contributes significantly to the safety of pesticides used in agriculture and public health. WHO has produced a standard protocol for the assessment of exposure. Determination of erythrocyte cholinesterase activity is considered to be the most appropriate tool for field assessment of exposure to organophosphorous compounds. When used as a part of worker surveillance, it allows a person to be withdrawn from further exposure when the activity of his or her erythrocytic cholinesterase decreases significantly from an established preexposure value. In the past few years, WHO has developed a field method for measuring whole-blood cholinesterase activity. After several years of use in the field, the method is being improved and the field kit modernized.

Education and training

To support its efforts to promote education in the safe use of pesticides, several education and training programs have been developed within WHO. One program is designed to be used at several distinct education levels and is adaptable to local needs. It includes courses at the basic level, for supervisors, for health workers, and for physicians. The courses are divided into sections, each with specific objectives. Each section contains one or more subjects and each subject consists of a number of modules. Modules are supported by a visual aid in the form of a slide with key words or a photograph. A selection of modules from each section can be made according to the educational objectives and background of the participants.

The course manual is accompanied by an establishment manual designed to enable national authorities to create multilevel courses on the safe use of pesticides in their own country, in the local language, and with their own resources. Its format allows for easy modification to suit national needs and different levels of audience.

The FAO, in consultation with other United Nations agencies, has developed the *International Code of Conduct on the Distribution and Use of Pesticides* (FAO 1986), which was subsequently amended with the introduction of the principle of prior informed consent (PIC). This document benefits the international community by providing guidelines on the availability, regulatory activity, marketing, and safe use of pesticides. However, promoting the safe use of pesticides in developing countries is difficult, because of the lack of infrastructure needed for adequate implementation. Many international agencies and developed countries are assisting developing country governments to establish or strengthen such infrastructure. The Code of Conduct is in full agreement with the London guidelines (UNEP 1989).

One of the main prerequisites for efficient risk management is an appropriate registration procedure (FAO 1986). The FAO has developed specific guidelines for registration of pesticides and accompanying legislation (FAO 1985a, 1989). They describe designs for a regulatory scheme and list basic requirements for the various phases of the registration process. All pesticides, regardless of use, should undergo specific registration procedures. It is of utmost importance that, within such a registration scheme, restrictions on pesticide availability be elaborated, so that the more hazardous pesticides are available only to properly trained operators. Only substances unlikely to present a hazard under the recommended mode of use should be made available for general use, such as in households.

In the recommended registration procedure, special emphasis is placed on adequate labeling of every pesticidal product (FAO 1985b). The importance of the container label and its information cannot be overemphasized. Label instructions, particularly technical information, must be communicated in a clear, concise way, as the safe and effective use of a pesticide will depend on the user's understanding of statements on the label and his or her ability and willingness to read and follow the instructions. Even when warnings and other information about correct use appear on the labels, such information may be of no use to people who are illiterate. Colour coding and pictograms, recently developed by the pesticide industry and the FAO, may help alleviate this problem.

Conclusion

Developing countries already face enormous problems because they must conquer a large number of diseases and solve other long-term public-health problems affecting ever-growing populations. They will have great difficulty dealing in isolation with the health problems now posed by chemicals. If sustained development is to be ensured, national health-care policies can no longer ignore the problem of chemicals. In the long term, postponing preventive action could prove to be a costly error.

Several factors handicap developing countries:

- International trade in high-risk or inadequately tested chemicals, especially when the sale of such substances has been forbidden or restricted in the producing country;

- Lack of understanding of the issues on the part of decision-makers;

- Absence of domestic legislation and regulations;

- Inadequate institutional facilities;

- Lack of personnel trained in these fields;

- Absence of poison control centres;

- The power of private interests; and

- The pressure of more urgent needs.

These shortcomings have led some chemical companies to build new facilities or move existing ones to countries where legislation is either lax or nonexistent and has facilitated the export of toxic wastes to poor countries.

It has become clear that, in the absence of a concerted and cooperative effort, many governments, especially in developing countries, will fall increasingly behind in establishing the means and the expertise necessary to tackle the highly complex health and environmental problems resulting from the use of chemicals.

FAO (Food and Agriculture Organization of the United Nations). 1985a. Guidelines for the registration and control of pesticides. FAO, Rome, Italy. 42 pp.

_____ 1985b. Guidelines on good labelling practice for pesticides. FAO, Rome, Italy. 36 pp.

_____ 1986. International code of conduct on the distribution and use of pesticides. FAO, Rome, Italy. 31 pp.

_____ 1989. Guidelines for legislation on the control of pesticides. FAO, Rome, Italy. 15 pp.

UNEP (United Nations Environment Programme). 1989. London guidelines for the exchange of information on chemicals in international trade (amended). UNEP, Nairobi, Kenya. 22 pp.

WHO (World Health Organization). 1984. Guidelines for drinking water quality (vol. I, II, and III). WHO, Geneva, Switzerland.

_____ 1985. Safe use of pesticides: 9th report of the WHO Expert Committee on Vector Biology and Control. WHO, Geneva, Switzerland. Technical Report 720, 60 pp.

_____ 1990a. Principles for the toxicological assessment of pesticide residues in food. WHO, Geneva, Switzerland. Environmental Health Criteria 104, 117 pp.

_____ 1990b. The WHO recommended classification of pesticides by hazard and guidelines to classification, 1990–91. WHO, Geneva, Switzerland. 39 pp.

Regional information resources for pesticide research

R. Fernando

National Poisons Information Centre, General Hospital,
Colombo, Sri Lanka

Pesticide poisoning is a serious health problem in many developing countries. Making information available to health professionals and the general public is essential in its prevention. In view of the success of the National Poisons Information Centre established in Sri Lanka in 1988, a proposal for a regional information network for Asia and the Pacific is presented. With international support, such a network would collect and disseminate information on all aspects of pesticide poisoning to national centres.

Pesticides in Sri Lanka

Between 1980 and 1986, the domestic supply of formulated pesticides in Sri Lanka increased markedly: for insecticides, it rose 106%; for herbicides, 214%; and for fungicides, 128%. Pesticides, by definition, are harmful to living organisms. Therefore, it is not surprising that misuse or abuse of pesticides has been increasingly responsible for illness and death among humans.

Poisoning with chemical pesticides has been reported in Sri Lanka since the 1950s. In the late 1970s, the national morbidity rate from pesticide poisoning was 79 cases per 100 000 people. Suicides accounted for 73% of poisoning cases; occupational and accidental pesticide poisonings were responsible for 17% and 8%, respectively.

In 1980, Sri Lanka's suicide rate was one of the highest in the world: 29 per 100 000 people. Only Denmark and Hungary reported higher rates in that year (31.6 and 44.9 per 100 000, respectively). An investigation of 407 pesticide-related deaths in an agricultural area of Sri Lanka revealed that 373 cases or 92% were suicides. It is clear that the common mode of suicide in Sri Lanka is ingestion of liquid pesticides.

Table 1. Number of people admitted to hospital and deaths from poisoning in Sri Lanka, 1988.

Type of poison	Admissions	Deaths
Organochlorines	1 280	95
Cholinesterase inhibitors	9 522	1 190
Other pesticides	2 195	239
Other poisoning and toxic effects	9 677	836
Medicinal agents	3 331	47
Snake bites	6 843	156
Total	32 848	2 563

Over half of the full-time agricultural workers in Sri Lanka use pesticides. In an agricultural district of the North Central Province, nearly 32% of those admitted to hospital in 1983 were suffering from occupation-related pesticide poisoning, indicating that this is a major problem in the agricultural community.

The improper storage of pesticides can also be hazardous. In 1985, Colombo and its suburbs were affected when the malathion stock of the Ministry of Health caught fire. Over one million people inhaled malathion fumes for over 1 week; some people admitted to hospital complained of headache, nausea, vomiting, dizziness, chest pain, and drowsiness.

Accidental poisoning with pesticides is also well documented. Drinking liquid pesticides by accident, due to inadequate and unsatisfactory storage facilities in households, and death, due to ingestion of minute quantities of pesticide remaining in reused containers, have been reported. In 1988, pesticides caused 40% of poisoning cases admitted to hospital and 59% of deaths due to poison (Table 1). This amounts to a morbidity rate of 79 per 100 000 people and a mortality rate of 9 per 100 000. These numbers do not include out-patients treated at state and private hospitals, patients admitted to private and Ayurvedic hospitals, or deaths occurring outside state hospitals or before admission to hospital. As many as 50 000 people may be poisoned with pesticides annually, resulting in over 2 000 deaths. Because of this alarming situation, health professionals in Sri Lanka believed there was a need for a poisoning advisory institution.

Establishing a national poisons information centre

In 1988, the National Poisons Information Centre (NPIC) was established in the General Hospital, Colombo. Its main objective was to provide information on all aspects of poisons and poisoning to the medical profession and general public, 24 hours per day throughout the year. Between 1988 and 1989, NPIC

Table 2. Number of enquiries received at the National Poisons Information Centre, 1988–1989.

Type of chemical	No. of enquiries
Pesticides	235
Drugs and therapeutic agents	141
Industrial chemicals	89
Plant poisons	51
Household chemicals	51
Snake venom	20
Agrochemicals	3
Others	135
Total	725

Table 3. Pesticides responsible for poisonings, 1988–1989.

Pesticide	Number	%
Insecticides	120	51.1
Herbicides	54	23.0
Rodenticides	24	10.2
Fumigants	1	0.4
Snail baits	1	0.4
Others or unknown	35	14.9
Total	235	100.0

Table 4. Pesticides responsible for deaths, 1988–1989.

Pesticide	No. of deaths
Paraquat	8
Organophosphates	6
Propanil	3
Chlorophenoxy	3
Organophosphate and carbamate	1
Unknown	1
Total	22

received 725 enquiries; 235 were related to pesticides, which caused 22 deaths (Tables 2, 3, and 4).

In addition to providing telephone or written information, the Centre has helped to disseminate information through publications. *Management of Pesticide Poisoning,* sponsored by the Pesticides Association of Sri Lanka, is distributed free to all medical practitioners in Sri Lanka. Known as the "Green Book," it contains information on the management of victims of poisoning due to all common pesticides. In an emergency, doctors can refer to this book before contacting the Centre.

Other reference publications include *Pesticides in Sri Lanka* (Friedrich-Ebert-Stiftung), which contains a collection of articles on all aspects of pesticide use in Sri Lanka in the last few decades. It contains information on economic, agricultural, medical, occupational, and legal aspects of pesticide use and is a valuable source for researchers. A brochure — *First Aid for Pesticide Poisoning* — in English and two national languages was distributed mainly through trade unions on tea, rubber, and coconut plantations.

NPIC produced a poster in English and the national languages that includes information on first aid for poisoning. It is now a widely recognized institution in Sri Lanka and is represented on the recently appointed National Committee to Monitor Import and Use of Chemicals. The Centre receives requests to participate in scientific and public seminars and meetings where poisoning is discussed.

The Centre is one of four pilot organizations using the poisons information package for developing countries that was developed by the International Programme on Chemical Safety. The national Centre contributed to this project by preparing 12 monographs containing poison information, including one on the herbicide propanil.

Over the past 2 years, the Centre has faced difficulties in answering some queries about pesticides because of lack of information on their toxicity. An enquiry concerning the effects of skin exposure to phenyl mercury acetate required contacting the importer of this pesticide who contacted the suppliers in the West. The necessary information was finally telexed to Sri Lanka.

Another problem was the unavailability of antidotes. Intravenous methylene blue, necessary for the treatment of methemoglobinemia caused by propanil poisoning was not available in Sri Lanka. After several deaths from propanil poisoning, the Ministry of Health was asked to import methylene blue.

All pesticides imported into Sri Lanka must be approved by the registrar of pesticides. Nevertheless, some businesses continue to import and sell banned substances, such as compounds containing arsenic; the antidote is not available in Sri Lanka. The Centre has received several enquiries about these pesticides and has informed the registrar accordingly.

The need for a regional information network

The problems encountered in Sri Lanka are not unique. Many developing countries face similar situations in relation to pesticide poisoning. Although there is worldwide concern about pesticide residues in the food we consume, the lack of analytical and monitoring facilities limit our knowledge of the extent of this problem in developing countries. Similarly, the effects of pesticides on the environment have not been adequately studied in these countries.

Although developing countries use only one-quarter of all pesticides produced in the world, half the intoxications and three-quarters of deaths from pesticide poisoning occur there. The lack of scientific journals and publication facilities has resulted in a dearth of information on pesticide-related ill-health. Unusual and sometimes bizarre cases of poisoning in developing countries are occasionally reported in international scientific journals, although the number is insignificant. Intercountry projects, such as the Regional Network on Pesticides in Asia and Pacific and the Safety and Control of Toxic Chemicals and Pollutants deal with agricultural and environmental aspects of pesticides. However, their contribution to reducing morbidity and mortality resulting from pesticide poisoning has not been significant.

The lack of information on the extent of pesticide poisoning is an immediate concern of the World Health Organization (WHO), industry, and governments, especially those of the developing world. Information on poisoning in both developed and developing countries is not widely disseminated among medical personnel in developing countries. A regional information network could improve this situation. Such a network could also be a valuable resource for pesticide-related research in public health and safety in Asia and the Pacific.

Information on all aspects of pesticide poisoning must be collected, classified, catalogued, abstracted, indexed, and disseminated for the use of researchers in the region. It should include information on epidemiology and medical management of pesticide poisoning, environmental pollution, pesticide residues in food and water, application technology, public awareness, health-education activities on use and misuse of pesticides, adequacy of protective clothing, and legislation to prevent health hazards. If successfully and efficiently disseminated, information of this type could help to identify and set priorities for research and possibilities for collaborative research among countries in Asia and the Pacific.

The experience gained in planning and organizing NPIC in Sri Lanka would be useful in establishing a regional information network on pesticides poisoning. Through national coordinators or focal points in selected countries, the network would be able to collect and disseminate information.

National information networks are the foundation for building complete information systems at the subregional, regional, and global levels. Developing an information system can be like building a pyramid. If each country builds a well-developed system of its own to serve as the base, then the information systems at higher levels will be more effective and better maintained. The goal of networking is to have two-way communication. In this way, all members of a network have knowledge about the other levels and interruptions or gaps in the flow of information are avoided.

In addition to conventional methods such as bibliographies, journals, newsletters, directories, and seminars, advanced technologies such as telex, facsimiles, audiovisual material, and computer disks could be used for quick exchange of information. The regional information network could coordinate and monitor current research in various countries and experiment with mechanisms for information delivery. It should provide technical guidance, support, and advisory services to countries requiring poisons information centres. A regional bibliographic and statistical data base maintained by the information network should serve as a resource for clinicians, agricultural experts, policy planners, and researchers.

Networking at the international level has advanced in terms of capabilities, linkages, and technology. In addition to the cooperation of national coordinators or focal points, participation of the United Nations and its specialized agencies and of nongovernmental organizations in developed and developing countries should be actively sought for support.

Acknowledgment — This research was carried out with the aid of a grant from the International Development Research Centre, Ottawa, Canada.

Promoting the safe use of pesticides in Thailand

T. Boonlue

Faculty of Communication Arts, Chulalongkorn University,
Bangkok, Thailand

The benefits derived through the use of pesticides in Thailand have been accompanied by environmental and human-health costs because of improper usage. Efficient and effective communication programs are necessary to impart knowledge, change attitudes, and adapt appropriate technology for use by farmers. Communication objectives have four functions: message delivery; group communication; feedback and feedforward; and organizational intercommunication. The Thai communication system has served all four functions. However, problems still exist. Communicators lack persuasive skills and have limited access to audiovisual materials. The feedback and feedforward mechanisms necessary for effective communication and systematic program evaluation are rarely implemented. Communication experts, researchers, training specialists, curriculum-development experts, and program evaluators must join forces to develop effective communication programs for the safe use of pesticides.

As an agricultural country, Thailand is accelerating efforts to increase food production to meet both national consumption needs and to generate a surplus for sale abroad. In every 5-year National Economic and Social Development Plan, the government has affirmed its aim of increasing agricultural production; for the 1987–1991 plan, the target increment was increased to 2.9% per year. All sectors and agencies involved in agriculture are working toward realizing this goal. Various strategies include crop improvement, better management of agricultural zones, and extension of farm land (DOAE 1985).

Ideally, agricultural targets should be achieved by increasing crop yields per unit of land already under cultivation. New farm land is scarce, or otherwise limited, making a strategy of expansion untenable. Moreover, the expansion of cultivation is often detrimental to forested areas and watersheds.

Crop yields per unit of land can be increased in several ways. Output can be increased through improved agricultural technology and the use of fertilizers, herbicides, and pesticides. Using pesticides to reduce crop losses to pests is the most popular strategy used by farmers. The increased production and profits possible through the use of pesticides makes them very attractive. As an input that requires relatively little investment in terms of capital and labour, pesticides have become widely used in Thailand.

However, the overuse of pesticides has had many negative consequences: public health has been threatened through the consumption of food contaminated with chemicals; insects have built up resistance to normal doses of pesticides; the equilibrium of the field ecosystem has been upset through the unintentional killing of insects that prey on pests; and environmental pollution has resulted.

Importation, distribution, and uses

Insecticides are used mainly on rice, vegetables, and cotton; fungicides are used on fruit trees, vines, vegetables, rice, and orchids. The herbicide market focuses on sugarcane, pineapple, rubber plantations, and rice. Data compiled by interviewing 324 farmers who grew various pulse crops in 1983, revealed that 99.7% of farmers in six northern provinces (Uttaradit, Phichit, Phitsanulok, Nakhon Sawan, Kamphaeng Phet, and Sukhothai) used pesticides on their crops (Unjitwatana 1984). In 1985, Thailand imported 17 379 t of pesticides with a value of 1 572 million THB (Thai bhatt).

Pesticides are distributed locally by both the government and the private sector. The Department of Agricultural Extension (DOAE) distributes pesticides free of charge to the end users through its official regional network. The private sector sells to farmers through the Bank of Agriculture and Agricultural Cooperatives and depends, to a large extent, on local agrochemical dealers. The number of private-sector pesticide-distribution points in the country is estimated at more than 2 000. The marketing channels include dealers at provincial, district, and village levels.

If they can afford it, subsistence and near-subsistence farmers purchase pesticides at the nearest retailer as soon as a pest appears. According to a 1985 Ministry of Agriculture and Cooperatives survey of the eastern provinces, about 70% of the farmers relied on the retailer to advise which pesticides to use. Most of these retailers have no training other than that provided by the pesticide companies. This often results in a bias in favour of certain companies' products, which are not necessarily the best or least toxic for the job. The remaining 30% of farmers stated that they relied on chemicals used effectively by neighbours and recommendations from extension personnel.

If a pesticide is not effective, farmers usually return to the retailer for another substance. This cycle is repeated until an obvious positive effect (e.g., dead pests) is observed. Thus, the "quick knock-down" but generally very toxic, broad-spectrum organophosphates such as methyl parathion are popular. Application of systemic pesticides is less common.

Pesticide storage on the farm is almost always unsatisfactory. On larger farms, pesticides are often stored in separate buildings. However, on small farms, pesticides are commonly stored in the living area, within reach of small children. Storage sheds are often poorly ventilated (high midday temperatures promote the decomposition of the pesticides); containers are not closed properly (humidity causes degradation of the active ingredients); labeling and storage practices are poor (leading to mixups that can have severe consequences, especially when herbicides are confused with insecticides); expiry dates are not usually recorded; and unusable pesticides accumulate and in time their containers corrode and leak making the storage area dangerous.

When applying pesticides, about half of the farmers use higher concentrations than recommended in the belief that "if a little is good, then a lot must be better." This practice encourages pest resistance and resurgence problems, requiring even greater pesticide doses to achieve the desired results.

Reasons for inappropriate use

The problems arising from increased use of agricultural pesticides are critical, especially when one considers the impact of toxic chemical residues on humans, animals, and the environment. Difficulties arise from the complex interaction of many factors:

- Pressure to increase yields without the expansion of farmlands;

- Poor knowledge and understanding of the proper use and handling of agropesticides;

- Farmers pay no attention to protective measures when spraying agropesticides;

- Farmers do not follow the directions on labels that stipulate an interval between the last pesticide application and harvest, resulting in contaminated food crops;

- Measures controlling the use of agropesticides are not strict or effective;

- Information on pesticides from government agencies seldom reaches the farmers and the public;

- Coordination among government officials, producers, and users is limited; and

- The innate steadfastness of farmers in traditional beliefs makes it difficult for extension officials to disseminate new knowledge to them.

Communication of information on pesticides

Message delivery

Message delivery is transferring information, usually in a one-way mode, e.g., the education activities carried out through radio, television, and newspapers. The training of farmers by agricultural extension workers also frequently involves face-to-face message delivery.

Group communication

Communication in a group involves two-way discussion and can be used to create or reinforce group identity. Group communication has often been used to enhance the effectiveness of message delivery and to increase the likelihood of changes in behaviour. An example of group communication would be the organization of a number of local farmers into a group to discuss their problems, exchange solutions, and conduct a demonstration plot.

Feedforward and feedback

The feedback function of communication is the circulation of messages in a system for the purpose of adjusting the system. It is a means of gauging how well the delivery system is achieving its stated goals. Feedforward, on the other hand, is the collection of information from and about a target group that helps shape the system's objectives. An illustration of feedforward communication would be a needs-assessment survey conducted to ensure that the most useful kind of program is initiated. Feedback would be the subsequent monitoring of information from the field to see how well the program is able to meet those needs.

Communication within an organization

Organizations can be viewed as systems of communication. Organizational communication is often adopted to reach larger portions of a population, especially those in rural areas, and to overcome problems of transport, poorly trained personnel, and large size and complexity of an organization. The larger an organization and the more geographically dispersed it is, the greater the importance of good communication to manage and coordinate its activities. For example, poorly trained agricultural-extension workers at the rural-service level will rely heavily on communication links to provide consultation, administrative supervision, and continuing education.

At present, several agencies are trying to implement strategies to inform farmers on the hazards associated with pesticide use. The Department of Agricultural Extension within the Ministry of Agriculture and Cooperatives is one of principal agencies responsible for disseminating agricultural knowledge to farmers. Recognizing that many problems result from the use of agropesticides, the Department is cooperating with the Department of Agriculture and the Shell Company of Thailand Limited. In 1984, they launched a campaign called "Promoting safe use of agropesticides among farmers" in eight provinces in the western part of Thailand where large quantities of vegetables and cash crops are cultivated.

About 74% of the people of Thailand earn their living through agriculture. However, the number of agricultural extension officials is small and information on agropesticides and new technology does not reach most farmers.

Education programs

Thailand has developed many training programs to educate and motivate farmers and to prompt changes in attitudes and behaviour. The Plant Protection Service Division of DOAE established plant protection service units whose prime objective is to provide training. Personnel in its subunits receive training from the larger unit, then relay information to farmers.

Most training is based upon a traditional classroom situation. Trainers use texts and other printed materials. Electronic audiovisual aids are seldom used, because of the poor quality of such materials in Thailand and the scarcity of equipment and electricity in rural areas.

However, some programs have successfully used vans equipped with audiovisual materials. Training teams travel to rural areas to train a group of trainers who, in turn, train other local people using the inexpensive training materials left behind. The local trainer serves to reinforce the message and make it specific and understandable to the trainees, often through group discussion. This approach substantially changes the role of the trainer and requires different skills. Knowledge of communications theory and audience research can be helpful in acquiring such skills.

Using low-cost materials such as flip charts and simple visual identification cards can be effective with illiterate trainees. Research can be helpful in designing simple communication aids.

The problems of communication

Government policy has emphasized accelerated development of impoverished areas. This has already mobilized most of the agricultural officials in the approximately 6 000 subdistricts in 27 provinces. Uneven distribution of human resources in the government sector results in inequitable educational opportunities for some farmers. Moreover, farmers who live far from the local information sources or from the offices of the agricultural extension officials will not have the opportunity to attend training or to keep themselves abreast of new information.

Use of the mass media has not been satisfactory; they have little influence and are unable to motivate farmers to change unfavourable behaviour. Farmers still seek information from their peers. If a farmer uses innovative technology to improve production in economic terms, other farmers will become interested spontaneously. However, the current flow of information to farmers is scant in both quantity and quality. Even though there are some laws and regulations to limit chemical use or to provide application instructions (the *Poisonous Article Acts* of 1967 and 1973), some farmers continue to use chemicals carelessly. Farmers use their own judgement and are influenced by advertisements and salesmen's recommendations.

Lessons from the communication program

Drawing firm conclusions is difficult because only a few communication activities have been fully evaluated. However, an overview of 12 knowledge, attitudes, and practices (KAP) studies and experience offer some information on the impact of the programs.

- Both mass media and interpersonal communication disseminate information effectively to certain groups. Interpersonal communication depends partly on individual competence and ability to transfer knowledge. The effectiveness of mass-media programs depends on the messages.

- Some misuses of pesticides persist despite extensive exposure to correct information.

- Mass-media programs can motivate people to seek information from other sources.

- Group and organizational communication have encouraged discussion about the hazardous effects of pesticides among villagers and agricultural extension personnel.

- Young people are more likely to comply with new practices and to change their attitudes in the desired direction.

- Of the three types of beliefs — descriptive, inferential, and informative — descriptive beliefs are the most critical, but also the most difficult to change.

The lessons learned from the communications program include:

- Feedback and feedforward are crucial steps in the design of communication programs.

- Pretesting materials in training and mass-media programs must be done to secure reactions from members of the intended audience to the tone, content, and appearance of proposed messages and materials.

- Messages that arouse extreme fear attract attention, but how much they change behaviour is not clear. Economic incentives and information about hazardous consequences may attract interest leading to changes in unfavourable behaviour.

- Training is an important component of a communication system. The process of training is really a process of communication with the objectives of educating, motivating (causing attitude change and stimulating the desire to work on a given problem), and influencing changes in behaviour.

- Program evaluation is also crucial to the development and refinement of the programs.

Conclusion

The Thai communication system has been developed to prevent the misuse of pesticides in agriculture. However, problems still exist. Communicators lack communicative and persuasive skills and have limited access to audiovisual materials. The feedback and feedforward information chains, needed to develop communication strategies, as well as systematic program evaluation are rarely implemented. Communication experts, researchers, training specialists, curriculum or education development experts, and program evaluators must join forces to develop effective communication programs for the safe use of pesticides.

DOAE (Department of Agricultural Extension). 1985. Report of the first working committee conference on plant protection for developing academic techniques to prevent and eliminate pests, 16–17 June 1985. Department of Agriculture, Bangkok, Thailand. Mimeo.

Unjitwatana, T. 1984. Responding to questions. Toxic Matter Journal, 11 (Jan–Feb), 30–33.

Learning about pesticides in Thai medical schools: a community-oriented, problem-based approach

Robert Chase

Community Medicine Residency Program, Department of
Clinical Epidemiology and Biostatistics, McMaster University,
Hamilton, Ontario, Canada

*A package of teaching and reference materials concerned with the human-
health aspects of pesticide use is described in this paper. It was designed
for use in Thai medical schools, which employ tutorial-based learning
methods. The problem-oriented format of the package encourages medical
students to develop critical thinking and decision-making skills. The
package addresses the broader community context of this serious health
issue in Thailand and provides the opportunity to consider areas of public
health often neglected by traditional physician training and hospital-
based activities. The package uses nationally derived health data. Problem
scenarios depict the realities of pesticide use and poisoning in Thailand to
generate discussion and achieve learning objectives. The need to reorient
medical curriculums and training in developing countries to sensitize
physicians to national health priorities is discussed, because it is necessary
to educate physicians about the role they can serve in the multidisciplinary
activities of health and safety promotion in their communities.*

Community-oriented medical education and problem-based learning

Medical educators in developing countries, as elsewhere, are faced with the
challenge of how to direct their physicians-to-be to the common health problems facing society. Traditionally, training has been modeled on the textbooks,
technology, and programs used in industrial countries. In Thailand, expansion
of the health-care system over recent decades has favoured specialization (in
1988, about 50% of physicians were specialists) and has led to a proliferation

of costly medical technologies and services that may have little relevance to the most pressing health needs of the nation's population (Sitthi-amorn 1989). Increasing attention has been paid to how best to use fixed or shrinking resources to maximize their health impact.

Medical graduates in Thailand are often ill-prepared to deal with the health realities of their own country. This problem received wide attention at the latest Thai National Medical Education conferences (in 1979 and 1986), with particular focus on how to incorporate the primary-health goals set out in national strategies into Thai training institutions (Sitthi-amorn 1989). Graduates of medical schools were seen to lack a population perspective on health and were thought to have an insufficient appreciation for the social and behavioural determinants of health relevant to their own country. Many are unprepared for the realities of the communities and health-care facilities in which they are destined to work. Although providing curative care is their primary task, health promotion and disease prevention have been neglected. As a consequence there is little chance of effecting change in the overall status of health or the underlying causes of morbidity and mortality.

At question also is how the curriculum and training can promote the attributes necessary for health graduates to interact effectively with the community and to assume the diverse roles of educator, health-care manager, and researcher. For this, it is essential to develop the capability of physicians to formulate relevant research questions and use national data and resources to improve the effectiveness of their activities.

Recognizing these deficiencies, medical education reform aims to develop students' capacity for critical thinking; appraisal of basic health data; defining, measuring, and evaluating community-health needs; and acting as agents of change in their role as both physician and responsible community leader. In Thailand and elsewhere, these objectives have led to the development of community-targeted, problem-based (CTPB) learning methods (Neufeld 1989).

The CTPB initiative in medical education in Thailand is an example of the larger movement among health-science educational institutions taking place on an international scale. In 1979, the Network of Community-Oriented Educational Institutions for Health Sciences was established with the support of the World Health Organization (WHO) and the Pan-American Health Organization (PAHO). The network's objectives include strengthening community-oriented learning in institutions and the development of approaches and methods that contribute adequately to community-health needs (Neufeld 1989). The number of participating institutions has grown from the original group of fewer than 30–43 full-member and 75 associate-member institutions.

A network task force on priority health problems in medical education has been developing an approach to education planning based on the health problems of populations served by the institutions (Neufeld 1989). Out of this effort, institutions have adopted CTPB programs. The relevance of a given health problem is assessed using information on mortality, morbidity, and the efficacy and effectiveness of preventive or treatment measures (MacDonald et al. 1989). Problem-based learning methods are used to develop the students' problem-solving abilities.

Problem-based learning makes use of realistic clinical scenarios described and presented on paper. In this way, students acquire the background knowledge and concepts necessary to understand health issues as if they were actually the doctors and decision-makers depicted. This method contrasts with the lecture or textbook-oriented approach, which may cover similar material, but lacks the immediacy and clinical relevance that often determine the usefulness of the concepts and facts the student is learning.

Although learning by this method has been shown to be no more effective than conventional techniques, its strengths are twofold (Norman 1988). The students' level of enjoyment and engagement in the learning process is higher than with conventional approaches, probably as a result of the opportunity to put oneself "in the shoes" of the clinician and decision-maker. The second advantage, based on psychological theory, is that knowledge is much better remembered in the context in which it was originally learned. By associating the learning process with life-like scenarios, the assumption is that the relevant knowledge will be more easily retrieved and applied when an actual clinical problem arises. In the context of CTPB learning, the student also has the opportunity to apply and consolidate his or her ability to evaluate population health information; and to assist in determining the magnitude of a given health problem within a community and the effectiveness of a prevention program.

Goals in developing the CTPB pesticide module

As a predominately agricultural country, issues of population health and medical education in Thailand are particularly relevant in examining how the current system addresses health problems associated with the use of pesticides. Traditionally, medical training and the provision of services have been largely aimed at the treatment of poisoning cases at the hospital level. At present, it is unlikely that a conventionally trained physician has an appreciation of the scope of the problem or the skills and support to help implement effective programs, such as community-directed education and village surveys, that could lead to the prevention of most unintentional poisonings.

In 1989–1990, a problem-based learning package on pesticides and their health effects was developed for use in Thai medical schools at the Faculty of Medicine, Chulalongkorn University. This subject was selected because:

- The agricultural sector has by far the largest occupational labour force: 70% of Thailand's 23.2 million workers (Thailand, Department of Health 1988).

- Pesticide use is considered to be the greatest occupational health hazard to agricultural workers — this is the consensus of officials interviewed in the Ministry of Public Health's Occupational Health Division and Occupational Health faculties at two major Thai universities.

- The educational materials about pesticides used by Thai medical schools were believed to be inadequate by curriculum committee members.

To emphasize population health and promote critical thinking among medical students, several criteria were included in the conceptual design of the educational package:

- The model would incorporate national health statistics and surveys about pesticides. Students would be encouraged to assess the strengths and weaknesses of this data. For example, the students would be asked to assess the limitations of pesticide-related mortality data collected at district hospitals for estimating community morbidity or to contrast hospital data with figures derived from the Ministry of Labour's compensation statistics (estimates of pesticide-related mortality rates differ by a factor of 70).

- The module would depict a community-based management strategy, arising from a field investigation at a farm where a poisoning incident had occurred. This would be presented along with a conventional medical "work-up" and treatment of a poisoned individual. The aim was to encourage the students to think in a practical way about the feasibility of community surveys and investigations and about the opportunities they have to promote useful preventive strategies.

- The module would apply basic principles of clinical epidemiology to pesticide-related health issues, including: evaluating the effectiveness of a screening test for plasma cholinesterase inhibition by calculation of sensitivity and specificity; appreciating the problems inherent in establishing a case definition of pesticide poisoning, and how different diagnostic criteria lead to discrepancies in morbidity rates; and identifying factors that affect the selection of an appropriate denominator, particularly stressing the concept of an at-risk population.

Many aspects of the health effects resulting from pesticide use require careful evaluation: the variability in clinical presentation; the range of toxicities and routes of exposures; and the appropriateness and reliability of laboratory aids in diagnosis. Illustrating the complexities of the problem might challenge students to pursue learning in a self-directed manner rather than relying on an approach based on simplistic assumptions.

Throughout the module, the importance of sound clinical judgement is emphasized, particularly when it is based on good history-taking and a recognition of the prevalence and range of pesticide-related health effects. Pesticide poisoning often goes unrecognized because of the failure to consider it as a possible diagnosis. Promoting the skills of enquiry and critical appraisal of laboratory tests and the clinical grounds upon which a diagnosis of pesticide poisoning is based or excluded prepares medical students more realistically for their future roles as community physicians and decision-makers.

Description of the module

The pesticide module consists of four sections:

- An introductory section outlines the format, sequence, and the relevant learning concepts contained in the module (see Appendix to this paper). It includes a brief summary of a problem, along with questions, suggestions for discussion, and learning objectives.

- The students' section (10 pages) presents the clinical scenario and poses appropriate questions.

- A tutor's guide (17 pages) provides background information on the epidemiologic issues and pesticide toxicology, and suggests useful concepts and facts for discussion during the course. It is assumed that the tutor does not have expertise in toxicology or epidemiology.

- A list of 15 articles and texts that were sources for the module, including Thai Ministry of Public Health statistics, journal articles about pesticide poisoning incidents in Asia, and two useful textbooks (see Appendix to this paper).

A preventive role for physicians

The use of pesticides in Thailand's agricultural sector is widespread, intensive, and poorly regulated. Pesticide use is increasing and free-market competition has led to a proliferation of 150 pesticide formulations sold under 3 000 trade names, many of which are of substandard quality (Gaston 1988). The application of pesticides is often excessive and indiscriminate, and often leads to personal exposure as well as high residues on food and considerable environmental contamination (Foo Gaik Sim 1985). Basic protective gear is seldom used and adequate training in pesticide use, handling, and storage is rare. Recently instituted regulatory measures, aimed at compliance of the Food and Agriculture Organization's *International Code of Conduct on the Distribution and Use of Pesticides* (FAO 1986), are unlikely to have a great impact on use at the local level (Gaston 1988).

At the community level, the extent of morbidity occurring as a result of unintentional exposure of farm workers is largely unknown. Field investigations in 1981 found that all farmers interviewed (30 in the northeast and 30 in the south) reported symptoms suggestive of poisoning after using pesticides (Foo Gaik Sim 1985). In a community survey of 10 000 adults in the coastal province of Rayong, 24% of those who used pesticides indicated that they experienced symptoms, but 95% of these had never reported to a health-care worker (Wongphanich et al. 1984). In other Asian countries, incidence of poisoning among pesticide handlers is about 7% per year (Jeyaratnam 1987).

Hospital records of poisoning cases seldom contain such important details as the nature of the poison and the route of exposure (Wongphanich 1985). This is a tremendous loss of valuable information on which practical interventions and preventive community strategies could be based. Meaningful improvement can only come about with changes to the current attitude and performance of health-care workers, especially physicians because of their important leadership role.

Current physician training consists primarily of diagnosis and treatment of acute cases of pesticide poisoning presenting at hospitals. Little or no attention is given to the nature of the problem at the community level or the role a physician could play in promoting health and safety education programs, accepting responsibility for data collection, and helping to direct public-health resources toward preventive strategies.

A second more general concern arising from conventional physician training is the lack of critical evaluation of diagnostic tests and treatments. In Thailand, as elsewhere, there has been widespread proliferation of ineffective and costly laboratory tests, often used with unreserved faith in their accuracy and usefulness. For example, although the measurement of acetylcholinesterase activity may help confirm the biochemical effects of organophosphates and carbamates, sole reliance on this test can lead to gross underrecognition of cases, because the test currently used in screening has a sensitivity of less than 50%. Proper orientation to the clinical epidemiologic principles needed to evaluate such tests is essential. The wide range of normal enzyme activity levels and the inability of a single test to determine the actual decline in activity (in the absence of baseline measurements) are other factors that require consideration.

Conclusions

Two Thai medical schools are incorporating the CTPB pesticide module into their curriculum (Chulalongkorn and Thammasat universities). Both of these schools have taken on a major commitment to a community-oriented medical curriculum. If the module proves to be effective, it will be modified and updated.

Evaluation will be an important part of this process. Key questions to consider include:

- At what level or stage of medical training should this material be introduced for best results?

- How compatible are the concepts and approaches with the rest of the curriculum the students use?

- Will the module contribute to changes in behaviour and activities of the future graduates, i.e., involvement in health-promotion activities, in collection of essential data and record-keeping in rural hospitals, and in approaches to diagnosis and management of pesticide-related poisoning incidents?

The trend toward community-oriented education in institutions teaching health sciences throughout the developing world presents an important opportunity to reform the attitudes and behaviours of physicians, particularly as they relate to the health problems associated with pesticide use. Pesticide use, and other occupational or environmental health issues, are especially suited to this learning approach because of their multidisciplinary nature.

Acknowledgments — This project was made possible by the Young Canadian Researchers' Award program of the International Development Research Centre, Ottawa, Canada. I am deeply grateful to Dr Chitr Sitthi-amorn, my supervisor at Chulalongkorn University. His dedication and vision of a healthier world with dignity for all has been an inspiration.

FAO (Food and Agriculture Organization of the United Nations). 1986. International code of conduct on the distribution and use of pesticides. FAO, Rome, Italy. 31 pp.

Foo Gaik Sim. 1985. The pesticide poisoning report: a survey of some Asian countries. International Organization of Consumers' Unions, Penang, Malayasia.

Gaston, C. 1988. Promoting safe and efficient use of pesticides: a report on trends in pesticide use in Southeast Asia and the level of compliance with the FAO international code of conduct on the distribution and use of pesticides. Kasetsert University, Bangkok, Thailand.

Jayaratnam, J.; Lun, K.C.; Phoon, W.O. 1987. Survey of acute pesticide poisoning among agricultural workers in four Asian countries. Bulletin of the World Health Organization, 65, 521–527.

MacDonald, P.J.; Chong, J.P.; Chongtrakul, P.; Neufeld, V.R.; Tugwell, P.; Chambers, L.W.; Pickering, R.J.; Oates, M. 1989. Setting educational priorities for learning the concepts of population health. Medical Education, 23, 429–439.

Neufeld, V. 1989. Community-based medical education: some recent initiatives toward making medical education more responsive to national health priorities. Annals of Community-Oriented Education, 2, 65–84.

Norman, G.R. 1988. Problem-solving skills, solving problems and problem-based learning. Medical Education, 22.

Sitthi-amorn, C. 1989. Clinical epidemiology: a population targetted approach to health reform. Chulalongkorn University Printing House, Bangkok, Thailand.

Thailand, Department of Health. 1988. Occupational health organization in Thailand. Occupational Health Division, Department of Health, Ministry of Public Health, Bangkok, Thailand.

Wongphanich, M.; Prasertsud, P.; Samathiwat, A.; Kongprasart, S.; Kochavej, L.; Bupachanok, T.; Samarnsin, S. 1985. Pesticide poisoning among agricultural workers. Report to the International Development Research Centre, Ottawa, ON, Canada. 186 pp.

Appendix: Community-targeted, problem-based (CTPB) learning module in occupational and community health — Pesticide poisoning

Problem summary

This CTPB learning module deals with pesticide poisoning, a prevalent occupational-health issue in Thailand.

The clinical scenario presented describes a woman reporting to an out-patient clinic with symptoms of mild–moderate poisoning by organophosphate insecticide. As is often true in occupational exposure to pesticides, the diagnosis is made mainly by a thorough history-taking rather than clinical examination and diagnostic testing. Severe poisoning does, of course, occur (more commonly in Thailand by intentional ingestion), and there are opportunities in this module to learn proper emergency management.

The module encourages students to think critically, formulate important questions, and evaluate the effectiveness (and limitations) of current diagnostic tests; epidemiologic data available in Thailand and; community health aspects of prevention.

The subjects covered in this module include:

■ History-taking and clinical skills;

■ Signs and symptoms (acute and chronic) associated with organophosphate pesticides;

- Medical and public-health management;

- Burden of illness appraisal and critical appraisal of epidemiologic data;

- Critical appraisal of a diagnostic screening test;

- Toxicology and pharmacology of organophosphate pesticides and their antidotes;

- Basic knowledge and classification of pesticides; and

- Health and safety aspects of agricultural work with pesticides.

If all topics are covered, five tutorial sessions will be required. A tentative schedule is attached. Study of the case scenario is divided into four sections:

- Clinical presentation;

- Management and diagnostic testing;

- Community screening and agricultural health and safety; and

- Epidemiology of pesticide poisoning in Thailand.

Sections of this module may be omitted or covered briefly if there is a shortage of time or the topics are not relevant to the tutorial group.

Copies of the tutor's notes can be handed out to the students if that is useful.

Schedule for pesticide poisoning learning module.

Day	Tutorial session	Homework
1	Begin Part I	I.3 Signs and symptoms I.4 Pesticide classification I.5 Case definition of organophosphate poisoning
2	Complete Part I, begin Part II	Treatment Cholinesterase physiology and biochemistry
3	Complete Part II, begin Part III	Sensitivity and specificity of screening test Question delayed neuropathy
4	Complete Part III, begin Part IV	Burden of illness Health statistics Epidemiology Health and safety
5	Complete Part IV	

Outline for learning module using organophosphate poisoning as an example.

	Tutor's guide (discussion and questions)	Learning concepts	Resources[a]
Part I: Scenario A woman presents at an out-patient department with symptoms of mild–moderate organophosphate poisoning for 2 weeks. Students must enquire further to arrive a proper diagnosis; use tutor's notes for supplying information when requested.	1. Signs and symptoms enquiry	- History-taking - Physical exam - Neurology - Physiology	5 5, 7, 12
	2. Develop differential diagnosis		
	3. Clinical picture of organo-phosphate poisoning: acute/chronic, mild/severe	- Classification of pesticides	5, 7
	4. What kinds of pesticides are there?	- Case definition	1, 5, 12, 14
	5. Can there be a standard definition of pesticide poisoning?		
Part II: Case management This part calls for the student to make clinical decisions in an out-patient department situation, as well as consider a response to a community problem.	1. List five priority steps in managing this patient; and a severe poisoning case	- Case management - Physiology	5, 6, 7, 12
	2. Is there a lab test to confim diagnosis? Availability? Usefulness?	- Clinical biochemistry - Clinical epidemiology - Physiology	3, 4, 5, 12
Part III: Community survey/screening The other probable cases are considered and there is a site visit to the farm in question for a survey and screening with a plasma cholinesterase test. The first patient has continuing symptoms. The causes of the poisoning are investigated.	1. What strategy would be most appropriate?		Tutor's notes
	2. What are the objectives of a site visit?	- Site visit	Tutor's notes
	3. Compare the two pesticides	- Toxicities of pesticides	1, 7, 13
	4. Chronic symptoms. Question delayed neuropathy	- Neurology	5, 12, 13
	5. Routes of exposure to pesticides	- Toxicology - Health and safety	5, 7, 14
	6. Human factors contributing to exposure		See III.5
	7. Reactive paper test	- Clinical biochemistry	11, Epidemiology textbook
	8. Reactive paper test	- Clinical epidemiology	See III.7
	9. What can be concluded from the test results?	- Clinical epidemiology	Tutor's notes
Part IV: Epidemiology of pesticide poisoning in Thailand: burden of illness Different information sources of poisoning data are reviewed and the student is asked to evaluate them critically. The concept of "the population at risk" is discussed. Information from a Thai community survey is presented. The results and limitations of the study are discussed. Alternative methods to collect better hospital and community data are discussed.	1. Source and validity of statistics	- Epidemiology	Tutor's notes, etc., 2, 8, 9
	2. Source and validity of statistics		See IV.1
	3. Incidence rates; the effect of the denominator	- Epidemiology	15
	4. Compare annual incidence rate of poisoning to other health problems		Various
	5. Alternatives to collecting community information	- Research methods	Group discussion, 5

[a] Numbers refer to sources listed below.

226

Resources used in preparing the module

1. Baker, E.L.; Zack, M.; Miles, J.W.; Alderman, L.; Warren, M.; Dobbin, R.D.; Miller, S.; Teeter, W.R. 1978. Epidemic malathion poisoning in Pakistan malaria workers. Lancet, 1978 (1), 31–34.

2. Boon-Long, J.; Glinsakon, T.; Potisiri, P.; Srianjuta, S.; Suphakarn, V.; Wongphanich, M. 1986. Toxicological problems in Thailand. In Ruchirawat, M.; Shank, R.C., ed., Environmental toxicity and carcinogenesis: proceedings of the regional workshop on health in the workplace. Mahidol University, Bangkok, Thailand.

3. Coye, M.J.; Barnett, P.G.; Midtling, J.R.; Velasco, A.R.; Romero, P.; Clements, C.L.; Rose, T.G. 1987. Clinical confirmation of organophosphate poisoning by serial cholinesterase analyses. Archives of Internal Medicine, 147, 438–442.

4. Crane, C.R.; Sanders, D.C.; Abbott, J.K. 1975. Cholinesterase: use and interpretation of cholinesterase measurements. In Sushin, I., ed., Methodology for analytical toxicology. CRC Press, Boca Raton, FL, USA. Pp. 86–87.

5. Davies, J.E.; Freed, V.H.; Whitlemore, F.W. 1983. An agromedical approach to pesticide management: some health and environmental considerations. Paper prepared in cooperation with the Agency for International Development Consortium for International Crop Protection, University of Miami, School of Medicine. [Contact: J.E. Davies, Department of Epidemiology, Rm. 669, University of Miami, School of Medicine, PO Box 016069, Miami, FL 33101, USA.]

6. Hayes, W.J., Jr. 1975. Toxicology of pesticides. Williams and Wilkins, Baltimore, MD, USA. 336 pp.

7. ILO (International Labour Organization). 1979. Guide to health and hygiene in agricultural work (Chapter 3). ILO, Geneva, Switzerland.

8. Jeyaratnam, J.; Lun, K.C.; Phoon, W.O. 1987. Survey of acute pesticide poisoning among agricultural workers in four Asian countries. Bulletin of the World Health Organization, 65, 521–527.

9. Thailand, Ministry of Public Health. 1988. The situation of occupational health, 2528–2531 [1985–1988]. Division of Occupational Health, Ministry of Public Health, Bangkok, Thailand. P. 3.

10. _____ n.d. Occupational health organization in Thailand. Division of Occupational Health, Ministry of Public Health, Bangkok, Thailand. P. 7.

11. Husbumner, C.; Rochanachim, M.; Punpong, T.; Bumpenyu, C.; Eamsobhana, P.; Sinlapanapaporn, S. 1985. The development of reactive paper for the screening test of early organophosphorous and some carbamate insecticides poisoning in fieldwork. Division of Occupational Health, Department of Health, Ministry of Public Health, Thailand. Internal document [in Thai].

12. Namba T.; Nolte, C.T.; Jackrel, J.; Grab, D. 1971. Poisoning due to organophosphate insecticides. American Journal of Medicine, 50, 475–492.

13. Senanayake, N.; Johnson, M.K. 1982. Acute polyneuropathy after poisoning by a new organophosphate insecticide. New England Journal of Medicine, 306, 155–157.

14. WHO (World Health Organization). 1982. Recommended health-based limits in occupational exposure to pesticides. WHO, Geneva, Switzerland. Technical Report 677, 7–38.

15. Wongphanich M.; Prasertsud, P.; Samathiwat, A.; Kongprasart, S.; Kochavej, L.; Bupachanok, T.; Samarnsin, S. 1985. A research report: pesticide poisoning among agricultural workers. Report to the International Development Research Centre, Ottawa, ON, Canada. 186 pp.

The role of health-service units in monitoring and prevention of pesticide intoxication

Shou-Zheng Xue,[1] Song-Ling Zhong,[2] Shi-Xin Yang,[2] Jin-Ling Li,[3] and Cheng-Nong Xie[4]

[1]Department of Occupational Health, School of Public Health, Shanghai Medical University, Shanghai, People's Republic of China; [2]Section of Labour Hygiene, Municipal Centre of Health and Anti-Epidemics, Shanghai, People's Republic of China; [3]County Health and Anti-Epidemic Station, Song-Jiang, Shanghai, People's Republic of China; [4]Section of Labour Hygiene, Provincial Centre of Health and Anti-Epidemics, Nan-Jing, People's Republic of China

The occurrence of acute pesticide poisoning in a province and rural area around Shanghai is presented and analyzed. The incidence of occupational cases tripled at the beginning of the rural economic reformation, but declined by half in 5 years. The death rate from nonoccupational poisonings was reduced to one-third in the 1980s. Preventive measures responsible for this reduction, especially the important role of peripheral health-service units and an active primary health-care network, are described.

Between 1 and 2.9 million cases of acute pesticide poisoning occur annually in developing countries, resulting in 20 to 220 thousand deaths (Jeyaratnam 1985). Addressing this problem is one of the highest priorities in occupational health and medicine, but it is often neglected in developing countries (Christiani et al. 1990). Action is urgently required to reduce the hazards posed by pesticide use (Shih 1983, 1985; Xue 1987).

In China, the prevalence of pesticide poisoning increased after the reformation of rural production and economic systems, which began in the 1980s. With the dissociation of the rural collective production system, highly toxic pesticides began to be handled by unqualified people and to be more widely used. Safety regulations that had previously been effective could no longer be enforced. As an inevitable result of the loosening of restrictions for handling such high-risk chemicals, the number of acute poisoning cases increased dramatically.

In the past decade, tremendous efforts have been made to cope with this new challenge. As acute pesticide poisoning is widespread among the rural population, and because both curative and preventive medicines are concerned, its control must depend on the general health-care system. Fortunately, the rural primary health-care system remained active and was strengthened during this time in areas around Shanghai. In particular, the peripheral health-service units played an important role in the prevention of pesticide intoxication.

The peripheral health-care units, which operate at the subnational level, consist of Health and Anti-Epidemic stations in township hospitals and infirmaries in the villages. The latter are run by two or three rural health workers (the "barefoot" doctors). They provide medical and health care to about 2 000 residents living within a 30-min walking distance (or a 10-min bicycle ride). Most of the "barefoot" doctors have received 3 years of training in general medicine and public health, including an internship in a county hospital. They have passed county, city, and provincial examinations. Most have more that 20 years' experience practicing in rural areas.

The Health and Anti-Epidemic Station in the township hospital usually consists of three to five staff with intermediate-level training in medicine and public health. They serve the countryside mainly through instructing and guiding rural-health workers. In addition, they are involved with infectious-disease control, vaccination programs, sanitary inspection of food and the environment, health education, and industrial-health services, especially prevention of pesticide intoxication.

There is also a Health and Anti-Epidemic Station in every county hospital, which is responsible for public health and disease control. Industrial health is usually the responsibility of 5–10 staff including several hygienists. This group is in charge of implementing intoxication-prevention programs and promoting the improvement of working conditions. The three levels cooperate with relevant sectors and the community to form a medical-health network. They communicate and integrate their activities through monthly meetings, common programs, projects, and reports.

Usually the program for prevention of pesticide poisoning is led by the municipal or provincial public-health authorities and implemented with the participation of all three levels of the peripheral health-service units. Annual meetings are held to summarize and evaluate the progress of the programs. Strategies and plans for the next year, including points needing to be reinforced and certain special research projects, are discussed. County-level meetings are held once a year in May or June, before the time for pesticide application. The chief physician of the county hospital frequently delivers short lectures on the diagnosis and treatment of pesticide poisoning, particularly with regard to new pesticides to be used in the area. Educational posters and booklets are distributed to rural paradoctors free of charge.

Methods used to monitor pesticide poisoning

Through this public-health network, pesticide poisoning is monitored at all levels (village, township, county, and province). In 1957, a reporting system for occupational diseases was established. Recording of pesticide poisonings began in 1961 in Shanghai and 1973 in Jiang-Su using the following procedures.

- Pesticide poisoning cases were first handled by the village "barefoot" doctors. After diagnosis and treatment, they were registered and reported directly to the county Health and Anti-Epidemic Station using a special index-card form. These cards were gathered and analyzed by the county station. Quarterly, they were sent with the occupational disease reports to the higher-level organization.

- The municipal or provincial station summarized the reports and calculated the incidence of poisoning cases and the resultant death rate. Etiologic and relevant factors were evaluated to modify programs and initiate special projects for further study.

- Research projects were undertaken to: survey hygienic conditions surrounding the use of pesticides in the field; assess exposure by measuring air concentration, skin contamination, blood cholinesterase activity, and urinary changes; and observe any other adverse effects. Comparisons were made among sprayers in teams and among individual farmers.

- Organophosphates (e.g., dichlorvos, trichlorophon, dimethoate, parathion, methyl parathion, etc.) were measured in air samples collected in tubes containing silica gel. Cotton swab samples were used to assess dermal contamination by gas chromatography using flame photodetector (Wang and Du 1985). The activity of blood cholinesterase was determined by the modified Hestrin method. Chlordimeform in the air samples, cotton swabs, and urine samples was measured colorimetrically with chromogen naphthylene-ethyldiamine.

- Training courses to renew knowledge on prevention, diagnosis, and treatment of poisoning were delivered to personnel at various levels: physicians in county hospitals, hygienists in health stations, staff in township hospitals, and the rural paradoctors. These groups, in turn, educated rural people in preventive measures to be taken when using pesticides. As part of their training, "barefoot" doctors are required to complete a curriculum of courses and an internship of about 3 years. Diagnosis, treatment, and prevention of pesticide poisoning is one of the most important parts of their study.

Results and discussion

Jiang-Su province

By the late 1980s, the number of cases of work-related acute pesticide poisoning in Jiang-Su province had decreased gradually to half the number in the 1970s (Fig. 1). In 1981, the number of occupational poisonings tripled, but, because the number of people applying pesticides had increased five times, the overall incidence of poisoning cases was not elevated. However, after 1982, the number of nonoccupational poisonings remained steady at triple the 1970's rate. Fortunately, over the same period, the death rate due to nonoccupational poisoning decreased (Fig. 2). Thus, the overall mortality rate from pesticide poisoning did not increase at all.

Of the 80 million people in the province of Jiang-Su, 70% live in rural areas; 70% of these people are engaged in agriculture as their main occupation. On average, 1.2–1.3 people/family are involved in the application of pesticides. The estimated rate of occupational poisoning among those exposed to pesticides is 0.4 per 1 000. The use of pesticides is an occupational hazard that has resulted in the loss of life, health, and economic productivity. Most of these incidents are easily preventable compared with other occupational hazards.

In contrast, nonoccupational intoxication is a complex social and mental-health issue as most of the victims result from attempted suicide. Although Jiang-Su has a huge population, the incidence of nonoccupational intoxication was low. The crude mortality rate from nonoccupational pesticide poisoning

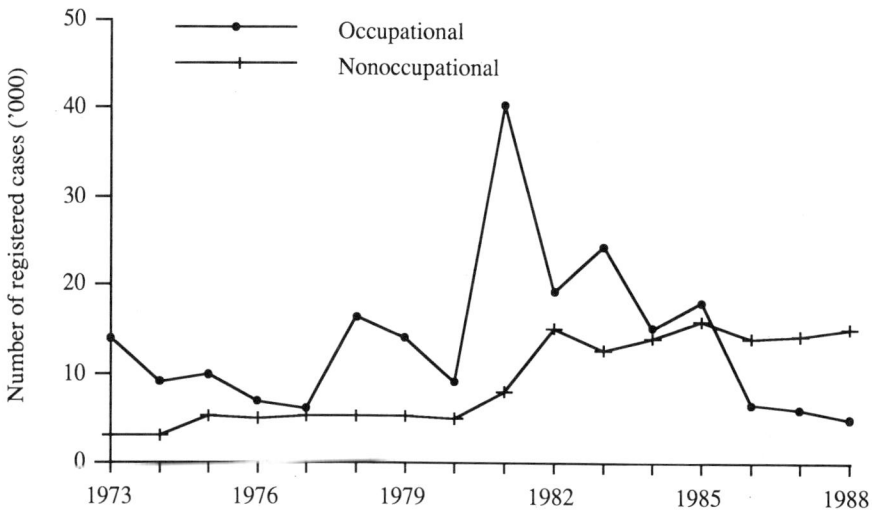

Fig. 1. Recorded cases of acute occupational and nonoccupational poisoning in Jiang-Su province, 1973–1988.

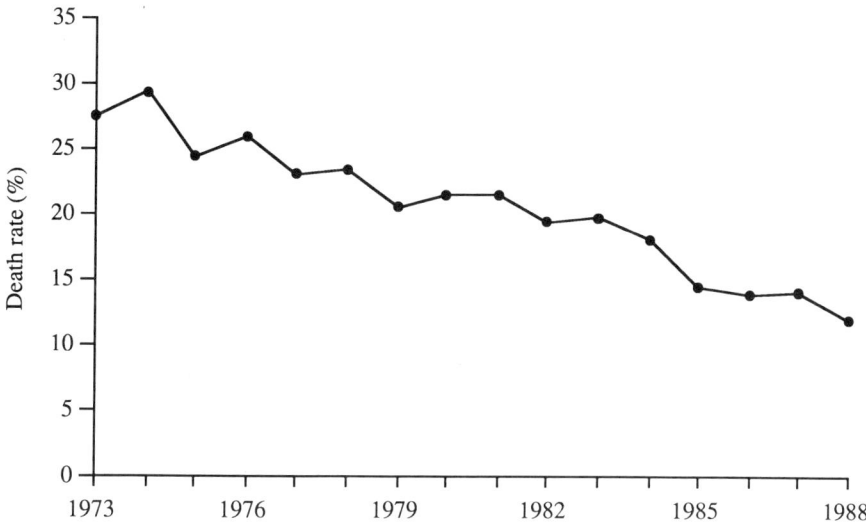

Fig. 2. Death rate (%) among cases of nonoccupational poisoning in Jiang-Su province, 1973–1988.

was 0.4 per 100 000 annually. Moreover, effective medical intervention (including prompt first aid, rapid transfer of the patient to hospital, improvement of emergency treatment, appropriate therapy, and intensive medical care) has successfully reduced the death rate of nonoccupational poisoning from 30% to 11%.

The township and village-level health-service units take care of more than 98% of the occupational poisoning cases and 50% of the nonoccupational cases. When severe, life-threatening cases are transferred to better-equipped facilities, the rural doctor usually accompanies the patient to provide artificial respiration and administer atropine if needed.

The rural area of Shanghai

The rural area of Shanghai consists of 10 counties with a population of 4.5 million; most are engaged in agriculture. Over a similar time period, the incidence of pesticide poisoning was lower than in the Jiang-Su area. Recently, the number of cases has increased after remaining at a low level for nearly two decades (1965–1982). Moreover, the types of pesticides causing most of the poisoning cases have changed. Since the late 1970s, demeton has not been used; parathion has not been used since 1983. The chemicals currently responsible for occupational intoxication are methamidophos, phosphamidon, carbofuran, and chlordimeform, although these are much less toxic than parathion and demeton.

As a result of deregulation, the number of cases of pesticide poisoning has increased because more people are handling and using pesticides. Before the rural economic reformation, handling of highly toxic pesticides was limited to specially organized and trained teams of five to eight people. Now, anyone can purchase and use such pesticides. Pesticides are used by an average of 1.3 people/family. Although the absolute number of poisoning cases has increased, the incidence rate is the same or has even declined (below 1 per 1 000).

The severity of poisonings has also been reduced. The proportion of severe cases (involving coma, pulmonary edema, and respiratory failure) declined from 10% in the 1960s to 1.1% in the 1970s and 0.5% by the 1980s. Although more people are handling pesticides, the amount used and the area covered has remained relatively constant, i.e., the amount of pesticide each sprayer handles has been reduced.

Song-Jiang

Song-Jiang is one of the ten counties under the Shanghai municipal authorities. During the 1980s, only 150 cases of acute pesticide poisoning were recorded in this county. Although Song-Jiang is a large county with 0.5 million people, the incidence of pesticide poisoning was very low because comprehensive, multisectoral efforts were adopted to prevent both occupational and nonoccupational acute pesticide poisoning. This required an intensive effort at every stage of pesticide production, distribution, storage, and use, as well as careful training of the spray operators, instruction in the selection of pesticides, equipment maintenance, field inspection, medical surveillance, early diagnosis, and adequate treatment.

The key to success in this county has been the collaboration among plant-protection divisions (under the agricultural bureau), distribution agencies (commerce bureau), county hospitals and health stations, and health staff at the township and village levels.

Pesticides, in dilute solution, are applied collectively using high-powered sprayers that project a rain-like spray as far as 30 m. This equipment is much safer than knapsack sprayers. The practice survived the period of agrarian reform, because the equipment could not be dismantled and distributed to individual families. As a result, farmers have had to work together, and the spraying equipment continues to be operated by qualified people.

In Song-Jiang, nonoccupational poisonings are also far fewer than in other counties due to strict regulations and proper storage of highly toxic pesticides, which prohibit access to these substances by the general population. In the 1980s, there were an average of 39 poisoning cases resulting in 5 deaths (12.8%) per year.

Preventive measures

Substitution with less toxic pesticides

Substituting less toxic pesticides for highly toxic ones is an effective and simple preventive measure. Certain highly toxic pesticides, such as parathion, are responsible for most poisonings (Table 1). Their replacement with less dangerous pesticides has been successful in many respects. However, the hazards of the replacement pesticide should not be underestimated; low toxicity pyrethroids have caused poisonings as well. Changing to a less hazardous formulation can also be an effective safety measure. For example, although carbofuran ranks high in toxicity, a 3–5% granular preparation can be used safely.

Selection of the appropriate devices for application

To minimize, as much as possible, dermal contamination and uptake of the toxicant, careful attention must be paid to the selection of appropriate and safe devices for the application of pesticides (Table 2).

Table 1. Proportion of acute poisonings involving parathion.

Year	% parathion among pesticides used	% of poisonings caused by parathion
1981	4.73	85.5
1982	3.78	81.5
1983	5.01	87.7
1984	4.02	89.8

Source: Huang 1985.

Table 2. Degree of dermal contamination after spraying parathion using various devices in a rice field.

Device and method of application	Dermal contamination after 20 min (g/cm^2)				Concentration in air at breathing zone (mg/m^3)
	Hand	Front of leg	Foot	Trunk	
Fine mist spraying	15.40	10.40	7.55	2.44	0.25
Long distance projection	20.85	0.05	1.46	—	0.01
High-volume spraying	1.27	—	0.94	0.43	—
Tractor driven equipment	0.08	—	—	0.47	0.01
Hand pressure	12.55	6.28	6.45	—	0.04

Source: Chen 1985.

Table 3. Proportion of workers using preventive measures when handling pesticides according to amount of training.

Sex	Training	Proportion using protective measures (%)		
		Full	Incomplete	None
Male	Yes ($n = 86$)	43.0	27.9	29.1
	No ($n = 206$)	20.9	37.9	41.3
Female	Yes ($n = 11$)	91.0	9.1	0.0
	No ($n = 91$)	7.7	51.6	40.7

Training and education

The effectiveness of training is obvious (Table 3). In the village, the advice of "barefoot" doctors is easily accepted. Farmers dealing with toxic pesticides are frequently reminded by health workers of the fundamental principle that health cannot be given by others, but can only be attained through one's own efforts. This is equally true in the prevention of acute pesticide poisoning.

Early detection and treatment of pesticide intoxication

Rural doctors in the village infirmaries play an important role in supervising the safe use of pesticides by providing pre- and postexposure check-ups, giving instructions on precautions for avoiding poisoning, etc. When a case of pesticide intoxication does occur, rural doctors are on the front line with the patient. During their evening rounds, they are in the village seeking possible patients and providing early detection and diagnosis as part of their routine work. They provide first aid and treatment. Mild cases are usually kept at home and visited by the rural doctors. Patients suffering intoxication of intermediate severity are kept at the observation room in the infirmary for therapy. More severe cases are referred and accompanied to the township hospital. Reduction in the death rate is mainly due to the quick and efficient emergency treatment by "barefoot" doctors.

Other measures

Other preventive measures occur at the national level. They include legislation, registration and regulation of pesticides, inspection of pesticide factories, and monitoring special and long-term effects.

Conclusion

The experiences in China over the last decade may be helpful to pesticide workers facing similar conditions in developing countries and elsewhere. To summarize:

- Although pesticide use is necessary in agricultural production and helps control vectors of infectious diseases effectively, the potential for acute poisoning can result in health and economic losses. Acute pesticide poisoning remains an important health problem in many developing countries.

- A well-organized and active primary health-care system can provide a solid foundation for implementing all public-health activities and addressing new challenges, such as the prevention and therapy of acute pesticide poisoning. Although the principles of prevention are clear, the policy and main strategies are decided by higher authorities. The successful implementation of these plans depends upon a collaborative network of all levels of the health-care system. Feedback is necessary to evaluate and modify policy decisions.

- Monitoring of pesticide intoxication is an important component of prevention and control strategies.

- People's participation must be based upon their understanding of the principle that health can only be achieved through one's own effort. Avoiding overexposure to pesticides and preventing acute intoxication episodes is everyone's responsibility, but will ultimately be of benefit to one's own health and welfare. Prevention is better than cure.

Chen, G.X. 1985. Measures for controlling adverse effects of pesticides in rural areas. *In* Xie, C.N., ed., Proceedings of the symposium of prevention of pesticide poisoning, May 1985, Zheng-Jiang, Jiang-Su Provincial Station of Health and Anti-Epidemics, People's Republic of China. Pp. 107–112.

Christiani, D.C.; Durvasula, R.; Myers, J. 1990. Occupational health in developing countries: review of research needs. American Journal of Industrial Medicine, 17, 393–401.

Huang, Y. 1985. Substitution of highly toxic pesticides with safer ones as the main measures in preventing pesticide poisoning. Labour Medicine, 2(1), 53–54.

Jeyaratnam, J. 1985. Health problems of pesticide usage in the Third World. British Journal of Industrial Medicine, 42, 505–506.

Shih, J.H.; Zhang, Y.X.; Wu, Z.Q.; Wang, Y.L.; Xue, S.Z.; Gu, X.Q. 1983. The control of pesticide poisoning. Chinese Journal of Industrial Hygiene and Occupational Disease, 1(1), 17–20.

Shih, J.H.; Wu, Z.Q.; Zhang, Y.X.; Xue, S.Z.; Gu, X.Q. 1985. Prevention of the acute Parathion and Demeton poisoning in the farmers around Shanghai. Scandinavian Journal of Work and Environmental Health, 11 (suppl. 4), 49–54.

Wang, X.C.; Du, L.B. 1985. Study on the collection and analysis organophosphorus insecticides with gas chromatography. In Xie, C.N., ed., Proceedings of the symposium of prevention of pesticide poisoning, May 1985, Zheng-Jiang, Jiang-Su Provincial Station of Health and Anti-Epidemics, People's Republic of China. Pp. 55–61.

Xue, S.Z. 1987. Health effects of pesticides: a review of epidemiologic research from the perspective of developing nations. American Journal of Industrial Medicine, 12, 269–279.

PART IV

ENVIRONMENTAL CONCERNS

AND ALTERNATIVES TO

MANUFACTURED PESTICIDES

IN AGRICULTURE AND PUBLIC HEALTH

Pesticides and wildlife: a short guide to detecting and reducing impact

P. Mineau and J.A. Keith

National Wildlife Research Centre, Canadian Wildlife Service,
Environment Canada, Hull, Quebec, Canada

In this review, the hazards posed to wildlife, especially birds, by the use of pesticides and the complexities of predicting those hazards are described. Examples of problem pesticides and situations and suggestions for detecting and reducing the damage done to wildlife are presented. Although the material is offered from a Canadian perspective, the specific problems of developing countries are also addressed.

The definition of wildlife differs among cultures, countries, and jurisdictions. In this paper, we focus on birds. Birds are an important and visible part of our environment. Traditionally, they have been used as sentinels of general environmental quality (Diamond and Filion 1987) and a large body of literature exists on avian toxicology (Tucker and Leitzke 1979; Hill and Hoffman 1984). In North America, most bird species are federally protected from unlicensed capture or kill. However, given their mobility, it is difficult to protect or exclude them from areas that are being treated with pesticides. Birds are particularly sensitive to some of the more toxic classes of pesticides such as organophosphate and carbamate insecticides (Walker 1983; Peakall and Tucker 1985) and their reproduction has been found to be affected by organochlorines (Burgat-Sacaze et al. 1990).

Although the effects of pesticides on birds are well known, it is clear that most pesticides were not developed with birds in mind. More disturbing yet is that the focus on bird studies has resulted in virtual ignorance of the effects of pesticides on most other groups of nontarget organisms, with the possible exception of fish. Biologists around the world are sounding an alarm about the state of reptile and amphibian diversity and population numbers (Rabb 1990). These wildlife groups have been largely ignored by most developed countries.

Scope of the problem

Although the effects of pesticide use are varied, direct mortality of wildlife, in and around treated fields and forests, is one of the most visible signs of a pesticide's impact. More than 30 pesticides registered in North America and Europe have been known to result in wild bird or mammal kills, even when used according to the relatively stringent regulations in those regions. Such cases of direct mortality usually have two characteristics in common: acutely toxic insecticides are often, although not solely, responsible; and the pesticides reach wildlife through a ready pathway.

Wildlife species are exposed to pesticides through many routes. They may ingest toxic substances in food or through preening and grooming; they can absorb pesticides through their skin, encountering droplets directly or by rubbing against foliage and other contaminated surfaces; and they can inhale vapour and fine droplets. The degree to which each of these routes of exposure contributes to the total dose depends on the crop being sprayed, the chemical, the animal species exposed, and other factors.

Cases of secondary poisoning can occur when predators consume prey contaminated by pesticides. Such predators are few because of their position at the end of a food chain. The death of a predator may constitute a significant reduction in the local population of that species. Avian predators are especially important agents of control for a number of species considered to be pests, e.g., rodents. Poisoning of this kind has been closely associated with organochlorine insecticides and other substances that are not readily metabolized and, therefore, accumulate in tissue. Other currently registered pesticides, even if readily metabolized, can also cause secondary poisoning under the right conditions, e.g., when the predator encounters a high concentration of the pesticide in its prey.

Generally, insecticides are more toxic to animals than herbicides, fungicides, and most other pesticides, except vertebrate control agents. Two groups of insecticides, the organophosphates and the carbamates, cause most deaths; they affect the nervous systems of insects and vertebrates alike. Because wild animals and birds know nothing of "safe reentry intervals" for treated areas, they run the risk of being exposed to very high doses simply by being in the wrong place at the wrong time. The use of a toxic insecticides affects not only the wildlife present in a treated area, but also species with habits (hunting, foraging, or migratory) that may cause them to wander into areas where pesticides are being applied.

The impact of pesticides on plants and invertebrates, whether target or non-target, may indirectly affect birds, mammals, and other vertebrate wildlife, particularly when a "pest" (which may be food or shelter for another species) is eradicated or drastically controlled. To make pest-control programs compatible with the preservation of wildlife, the most target-specific pesticides must be chosen and, ideally, unsprayed areas must be provided as refuges.

Using target-specific products minimizes the effect on desirable nontarget species in the crop as well as in areas adjacent to it. As agriculture in developing areas becomes increasingly intensive, wildlife species become relegated to remnants of natural habitat next to farm land, such as small wetlands or woodlots in the middle of cultivated fields, drainage ditches, fence lines, hedgerows, or even small rock piles. Inadvertently, even these habitats can be affected through the direct application of pesticides, spray drift, or runoff.

An example of this situation in Canada can be found in the Prairies where small wetlands used by nesting ducks are surrounded by agricultural fields. Most of the insecticides in use have the potential to reduce the number of aquatic invertebrates in these wetlands, thereby reducing the food supply for ducklings. The effect is greatest when insecticides are applied aerially because the small wetland areas are sprayed directly. In addition, the use of herbicides in the Prairies is also likely to reduce the quantity and quality of cover available for nesting ducks (Mineau et al. 1987; Sheehan et al. 1987).

Long-term contamination of the wildlife environment is the most lasting and widespread problem associated with pesticide use. Bioaccumulation of substances that are poorly metabolized and excreted by wildlife species and that accumulate in some tissue, such as fat in the case of lipophilic substances, causes particular concern. The most notorious and best documented examples are a number of organochlorine insecticides such as dichlorodiphenyltrichloroethane (DDT) and dieldrin.

Although death is the surest sign of pesticide impact, sublethal and delayed mortality effects are also significant. Many pesticides can affect the normal functioning of exposed individuals at doses insufficient to cause death. At high doses, the organophosphates and carbamates cause respiratory failure and death. However, wild birds exposed to these agents in lesser amounts have experienced impaired coordination and loss of appetite. Other consequences of sublethal exposure to these agents include birds spending less time at the nest, providing less food for their brood, being less able to escape predation, and being more aggressive with their mates. Notable effects of sublethal pesticide ingestion also include reproductive failure through reduced hatching or fledging success. Several pesticide products can cause embryonic mortality when sprayed directly onto eggs.

It is difficult to assess the extent to which environmental factors combine with sublethal pesticide exposure to cause delayed mortality. Exposure to some pesticides makes wildlife more vulnerable to predation. It also causes weight loss and an inability to maintain body temperature, both of which increase the chances of a small bird dying in inclement weather. Finally, exposure to pesticides may be linked with reduced resistance to disease.

Although intuitively evident to the ecologist, it is most difficult to demonstrate the indirect effect of pesticides on wildlife. In one of the best examples available to date (Potts 1980), long-term data on population numbers and pesticide use, as well as a great deal of fundamental research, were required to demonstrate the causal relation between herbicide use and the decline in populations of Partridge (*Perdix perdix* and *Alectoris rufa*). Much of the research into the long-term ecological effects of modern (i.e., nonorganochlorine) pesticides is in its infancy. For this reason, much of the following account deals only with the direct effect of pesticide use on wildlife.

Theoretical consideration of the hazards posed by pesticides

In the absence of comprehensive field-testing of a pesticide submitted for registration in Canada or other developed countries, regulations require that the potential environmental hazard be estimated on the basis of toxicity to a few test species and on the projected use of the product. This hazard assessment often follows a "quotient" method, whereby levels that are toxic or cause mortality in test species are compared to predicted levels of exposure. In theory, safety factors are introduced into this calculation to allow for errors in estimation or extrapolation. In practice, the level of uncertainty is so high that most of the acutely toxic pesticides, such as organophosphates and carbamates, cannot adequately be assessed without field-testing.

Because of the importance of the pesticide market in the USA and because US registration is highly desirable to pesticide manufacturers, environmental toxicology data required by the US Environmental Protection Agency (EPA) are usually available for most of the pesticides registered worldwide (Table 1). Canada currently accepts all environmental toxicology data generated to US

Table 1. Environmental toxicology information, at the tier 1 and 2 levels, required for most pesticides used in the field.

Vertebrate tests
Avian single-dose oral LD_{50}[a]
Avian dietary LC_{50}
Wild-mammal toxicity test (seldom available, data on laboratory rodents used instead)
Acute toxicity test for freshwater fish
Avian reproduction test
Fish early life-stage test
Invertebrate tests
Acute toxicity test for freshwater invertebrates
Honey bee: contact LD_{50} and toxicity of foliage residues
Aquatic invertebrate life-cycle test

Note: Supplementary tests on estuarine and marine fish and invertebrates are also requested where relevant.

[a] LD_{50} = lethal dose to 50% of animals tested; LC_{50} = lethal concentration to 50% of organisms tested.

specifications. Other data are assessed on a case-by-case basis. Canada has its own guidelines for environmental chemistry and fate (Agriculture Canada et al. 1987) and is currently developing guidelines for testing nontarget plants.

Urban and Cooke (1986) describe how these data are used by the EPA to generate hazard scenarios. Canada uses a similar procedure. The goal is to assess the likely risk that the pesticide in question will cause mortality to exposed wildlife and hence the need for fieldwork that would confirm or negate this assessment.

Some of the difficulties in accurately predicting the impact of a pesticide on wildlife on the basis of laboratory-derived results are described here. Emphasis is placed on factors that might cause the hazards to be underestimated.

The community at risk

The first step in a pesticide evaluation is to gain knowledge of the biological communities that are potentially at risk in the area of pesticide use. In a large country with distinct physiographic regions or for countries with incomplete faunal surveys, this can be a formidable challenge. Species may change their food habits in response to an overabundant supply (such as during an insect outbreak) and, therefore, even an intimate knowledge of the "normal" ecology of a species may not be sufficient.

The extent to which pesticide use modifies the propensity of wildlife to feed in treated fields or adjoining areas is another question that has not received much attention. A given array of species in the general area of pesticide use does not necessarily mean that those species will be exposed.

Fletcher and Greig-Smith (1988) have proposed methods to quantify the use of fields by birds. Given the complex nature of ecosystems, evaluation must be focused on a few indicator species. These must be chosen, not so much for their inherent physiological susceptibility to pesticide use (this is not usually known), but for the likelihood that their life habits will lead to maximum exposure. Unfortunately, the choice of indicator species has often been made on the grounds of cost, logistics, and overall feasibility rather than on the grounds of more scientifically desirable criteria. This can give rise to misleading signals and a false sense of security.

Representativeness of the test species

Those evaluating the safety of pesticides to humans benefit by being able to study several surrogate species to extrapolate to the single species of interest. In contrast, wildlife evaluators can look at only a few species to predict effects on a diverse fauna. In some cases, they can work directly with the species of interest, but often interspecies extrapolation is necessary. There are 1–3 million species of vertebrates living in the world today. Testing all nontarget species for likely pesticide impact is impractical as well as unethical.

On the basis of data compiled for a variety of pesticides, Tucker and Leitzke (1979) suggested that, in 95% of the cases, if the acute toxic dose of any given compound is known for one species of bird, then the acute toxic dose of the same compound will be within a factor of 10 (plus or minus) for a second species. However, one must be careful of such generalizations. Reanalysis of part of the same data base (bird toxicity data for 44 organophosphate insecticides) determined that the range in acute toxicity values between species was highly dependant on, and therefore an artefact of, the number of species tested. The average number of species tested was 3.9 for compounds showing a range of toxicity less than one order of magnitude; those showing larger variation were tested on an average of 6.8 species.

It has been argued (Schafer and Brunton 1979) that some species appear to show an inherent susceptibility or resistance to a wide range of environmental toxicants. Thus, current US regulatory testing using the Mallard (*Anas platyrhynchos*) and Bobwhite Quail (*Colinus virginianus*) should be revised, because these species do not give a realistic assessment of the hazards of pesticides to the largest group of bird species, the passerines or perching birds.

A reanalysis of some of the large data bases on acute toxicity of pesticides to birds (Schafer and Brunton 1979; Tucker and Leitzke 1979) has shown that phylogenetically related birds are not necessarily similarly sensitive to any given pesticide (Mineau 1991). This casts doubt on our ability to predict the hazard to a species on the basis of that in a closely related species. This type of extrapolation has been especially common in dealing with rare or endangered species or with species that are not amenable to direct investigation.

Assessing dietary exposure

Knowing which nontarget species are likely to be exposed during pesticide applications requires an understanding of exposure routes. Ingestion of contaminated food has been identified as the most likely route of intoxication for wild birds or mammals and is still the only route that is commonly assessed in standardized hazard-assessment procedures. However, the dietary route may not be the only or even the main route in all cases of wildlife exposure (Mineau et al. 1990).

The food intake of a small organism is greater than that of a larger one when expressed as a ratio of body weight. Thus, other things being constant, smaller species tend to be more vulnerable to acute pesticide intoxication. Also, at certain times of the year, wildlife species may have higher energy requirements, and hence higher food intakes, than at other times. For example, a bird feeding young at the nest may have much higher energy requirements than normal. Climatic conditions and physical factors, such as nutritional status, disease, and parasite load also influence the toxicity of pesticides to the organism directly, whereas, indirectly, they may alter food consumption patterns causing the organism to ingest more or less of the pesticide.

Formulation-specific concerns

The extent to which wildlife species are exposed to the active ingredient in a pesticide formulation is usually unclear. Similarly, it is generally not known how long the various other elements of a pesticide formulation remain with the active ingredient.

Predictably, granular formulations of organophosphate and carbamate insecticides are equal or less toxic than the equivalent technical grade materials (Hill and Camardese 1984). Unfortunately, granular formulations tend to enhance the availability of a pesticide to birds, which negates any advantage of reduced toxicity. Liquid formulations, on the other hand, are typically more toxic than the pesticide parent material (Hill 1986).

Application rates and expected residue levels

Under operational conditions, considerable variation in application rates of pesticides can be expected. This is generally not recognized as a problem by pesticide users, because of the relatively wide margin of safety of pesticides in terms of both efficacy and safety to the crop. Exact application of a pesticide according to label instructions would require accurate measuring of the various components, perfect calibration of the equipment being used, faultless technique on the part of the applicator, and ideal weather conditions and terrain. Even under the highly regulated and mechanized conditions of developed countries, these requirements cannot be met. It is more reasonable to expect that the rate of delivery of the pesticide follows a broad distribution about the desired application rate. This is especially true in the developing world where most spraying is done using by knapsack sprayers and is more vulnerable to human error.

Other situations also give rise to a higher-than-intended rate of application of a pesticide. Drift is a problem during both ground and aerial applications. For hazard assessment, an important aspect is the additive nature of droplet drift associated with multiple-swath applications. This can give rise to high pesticide application levels in downwind fields and beyond (Maybank et al. 1978).

If the route of exposure of wildlife species is primarily through the consumption of contaminated food, the degree of exposure is only approximately related to the amount of pesticide delivered to the crop and nearby noncrop areas. In currently accepted procedures for risk assessment (Urban and Cooke 1986), residue levels on foodstuffs are estimated on the basis of standard factors that assume that the rate of application and surface area are the only factors having a bearing on residue levels. However, widespread mortality of waterfowl caused by the use of the insecticide diazinon (Stone and Gradoni 1985; Frank et al. 1991) illustrates the difficulties associated with accurately predicting residue levels on plant surfaces (Mineau 1991). Residue on grass blades after a mechanized application of diazinon at the rate of 1.1 kg active ingredient/ha ranged from 17 to 181 ppm. Yet, well-tended turf is as uniform and structurally simple a "crop" as one is likely to encounter in a hazard evaluation.

The relation between the intended rate of application and the likely exposure of wildlife species is, therefore, complex and difficult to predict accurately even when the route of exposure is known. This is an especially serious problem for acutely toxic products with low margins of safety.

Persistence of pesticides

Most nonorganochlorine pesticides are relatively short lived, at least in plant and animal tissue. They are more likely to persist in abiotic components of the environment, such as soil, aquatic sediments, or groundwater. The rapid disappearance and lack of bioaccumulation characteristic of organophosphate and carbamate insecticides made these classes an attractive alternative when the persistence problems associated with organochlorine insecticides became widely known.

A recent compilation of residue dissipation rates from plant surfaces (Willis and McDowell 1987) cites average half-lives of 3.0 days and 2.4 days for organophosphates and carbamates, respectively. However, there are site-specific examples of long environmental half-lives about which generalizations must be made with caution. For example, the relatively lipophilic organophosphate, fenitrothion, has been found to persist at low levels bound to the waxy epicuticle of conifer needles (Sundaram 1987) and these residues may persist from one spray season to the next.

A proper evaluation of the persistence of a particular pesticide requires exact knowledge of the conditions under which it is to be used. For example, Getzin (1973) reported that the persistence of carbofuran was 7- to 10-fold greater in acidic soils (pH 4.3–6.8) than in alkaline soil (pH 7.9). The stability of carbofuran granules under acidic conditions has been identified as a factor in large kills of waterfowl in Canada. Carbofuran granules remain toxic to waterfowl throughout the autumn and winter when fields are often flooded and attractive to birds.

Examples of problem pesticides or problem situations

Organochlorine insecticides

It is beyond the scope of this paper to review the evidence of ecological damage resulting from the use of persistent organochlorines. The use of most of these products has been cancelled or significantly curtailed in most developed countries; DDT, aldrin, dieldrin, endrin, chlordane, heptachlor, toxaphene, mirex, and chlordecone have been banned or severely restricted in North America. However, some, like DDT, continue to be a concern because of local contamination (Fleming et al. 1983) and because of the high residue levels in birds that migrate to the near tropics (DeWeese et al. 1986). These birds in turn pass the residues along the food chain and continue to hamper the full recovery of some species such as the Peregrine Falcon (*Falco peregrinus*) in North America.

Organochlorine insecticides are inexpensive to manufacture and, as long as pest resistance has not resulted, relatively efficacious. They have been credited with numerous successes in reducing the extent of some notable vector-borne diseases, at least in the short term. Also, the human-health effects, at least the short-term effects, of some of these products may be less severe than those of the more toxic alternatives, especially under conditions in which pesticides are used in developing countries. For these reasons, some members of the international aid community advocate the continued use of these products in the developing world. As our collective memory of the ecological ravages of organochlorines fades, some would have us believe (Coulston 1985) that thoroughly documented effects, such as the eggshell thinning induced by DDE [1,1-dichloro-2,2-bis (4-dichlorophenyl) ethylene] — a metabolite of DDT — did not take place and that the developed world overreacted by banning organochlorine insecticides. Unfortunately, such revisionist rhetoric, however ill-informed, is likely to be popular with debt-burdened developing countries faced with an impending pest crisis.

Persistent organochlorines may need to be used in small quantities under conditions that minimize release to the environment, e.g., underground injections for termite control, until safer alternatives can be made available or at least until aid budgets are increased to allow for the purchase of alternative products that are more expensive but cause less ecological damage. Given the evidence on the environmental effects of persistent organochlorines, anyone advocating widespread use of these products, such as is now being proposed for malaria control in Brazil, clearly does not have the long-term ecological interests of that country or of the rest of the biosphere in mind.

Arguments of species diversity or ecological integrity may be insufficient to convince governments to limit the use of persistent organochlorine pesticides. However, there are other reasons to avoid these hazardous products. The use of organochlorines has fostered secondary pest outbreaks after the selective removal of insect and bird predators. Some of nature's most efficient rodent control agents (predatory birds) are ultimately the hardest hit by the use of persistent organochlorines.

Organochlorines that are still widely used in North America (Table 2) continue to be assessed. For example, dicofol has recently been shown to cause egg-shell thinning, similar to DDE (Schwarzbach et al. 1988), and its future use is, therefore, uncertain.

Table 2. Organochlorine insecticides still in use in North America.

Compound	Issues of concern
Dicofol	Egg-shell thinning, bioaccumulation, and effects on behaviour and reproduction
Methoxychlor	Aquatic impact and accumulation in fish
Endosulfan	Aquatic impact
Lindane	Widespread contamination by other benzenehexachlorides (BHCs)

Toxic organophosphate and carbamate insecticides

These two classes of cholinesterase-inhibiting insecticides currently account for the majority of insecticides registered world wide. As mentioned previously, birds are ill-equipped to deal with these products. Grue et al. (1983) concluded that kills were largely predictable on the basis of toxicity and extent of use. Because these products are short-lived, problems have generally arisen from situations where there was a clear route of exposure.

Foliage eaters

Consumers of freshly sprayed foliage obviously represent a high-risk category. Grazing or browsing species must ingest large quantities of non-nutritious foodstuffs. Some, such as waterfowl, have inefficient digestive systems and compensate by bulk feeding. In addition to the previously described diazinon example, other problems with anticholinesterase insecticides include die-offs of ducks and geese in alfalfa fields treated with carbofuran (Hill and Fleming 1982) and of Sage Grouse (*Centrocercus urophasianus*) feeding on alfalfa crops treated with dimethoate or on potato foliage and weeds sprayed with methamidophos (Blus et al. 1989).

As a result of widespread waterfowl mortality, the use of diazinon on turf grass (golf courses and sod farms) was banned in the USA. With increased awareness of this problem, other cases began to be reported, for example, in Canada (Frank et al. 1991). Carbofuran is currently under review in Canada, in part because of its use in waterfowl-breeding habitats and the demonstrated problem with waterfowl mortality in alfalfa fields.

Granular insecticides

Granular formulations of insecticides are notorious for promoting hazardous exposure. A number of products have been linked to massive bird kills and to some cases of mammalian mortality as well. These products are used world-wide on crops, such as corn (maize), rice, and rapeseed. The granules themselves are attractive to birds and possibly other species. Depending on the specific granule base, birds mistake them for grit or seeds or both (Balcomb

1984). The base for pesticide granules is usually clay, gypsum, silica, crushed corn cobs, or coal. All are attractive to birds, although coal possibly less so. When they are readily available, birds will consume large quantities of granules (Kenaga 1974).

The problem of ingestion is primarily a result of the limitation of agricultural engineering. At planting time, pesticide granules are frequently incorporated into the soil over the seed or actually in the seed furrow. Unfortunately, even the best modern machinery cannot adequately cover the granules so that none remains on the surface (Erbach and Tollefson 1983; Hummel 1983; Maze et al. 1991). In rice cultivation, the granules are simply applied to shallow water or to fields about to be flooded, giving rise to severe effects on waterfowl (Flickinger et al. 1980; Littrel 1988). As a result of an EPA proposal to prohibit all uses of granular carbofuran, the manufacturer voluntarily withdrew the substance for all but a few minor uses in the USA. This product is also under regulatory review in Canada. Manufacturer-conducted field tests have shown that the material cannot be used safely, at least for corn, and kills are being recorded world wide (Mineau 1988).

As little as one granule of carbofuran is usually sufficient to kill a small songbird. Eagles and other raptors have been killed as a result of consuming birds killed by granular carbofuran. However, carbofuran is not the only hazardous granular product (Table 3). The rating scheme outlined in Table 3 was derived from concentration and toxicity values of the pesticide as well as on the basis of available field evidence. The ranking is tentative as evidence varies greatly from one product to the next. The manner in which these products are used in various countries may differ; therefore, a use pattern cannot easily be incorporated into the rating. The physical integrity of these products has also not been considered. It is reasonable to expect that, all else being equal, granules that are more friable and break down quickly (e.g., clay

Table 3. Tentative ranking of the hazard to wildlife of common granular insecticides.

High hazard	Parathion	Fensulfothion
	Carbofuran	EPN[a]
	Fenamiphos	Aldicarb
	Phorate	
Moderate hazard	Diazinon	Oxamil
	Ethoprop	Disulfoton
Low hazard	Fonophos	Ethion
	Chlorpyrifos	Terbufos
Very low hazard	Tefluthrin	

[a] o-ethyl o-4-nitrophenyl phenyl phosphonothionate.

and gypsum) are preferable to more resilient particles (e.g., silica and corn cob). Also, not all available granular products have been included in this table.

Based on this information, the safest product is not an organophosphate or carbamate, but rather a pyrethroid. Provided that pyrethroid insecticides can be kept out of aquatic systems, they should be used in preference to organophosphates or carbamates. They usually provide a large margin of safety for terrestrial vertebrates and are, therefore, ideal for granular formulations.

Seed dressings

Seed dressings or seedcoats, whether factory applied or added to the seed just before planting, constitute another well-known wildlife hazard. Many species are naturally attracted to seeds. Awareness of the hazard first came to light following massive kills of birds by organochlorine and organomercurial seed dressings used in North America and Europe. The offending organochlorines were mainly aldrin, dieldrin, and heptachlor. In contrast, one organochlorine, lindane (γ-benzene hexachloride, BHC), appears to be safe for wildlife and is currently the material of choice for cereal treatments in Canada. Lindane does not seem to accumulate in wildlife tissue to any degree, although it does give rise to some contamination of the environment by assorted BHCs that are either present as impurities or metabolized from the gamma isomer. This product is currently thought to be preferable to the more toxic organophosphate or carbamate alternatives, which have been linked to wildlife kills (Greig-Smith 1988) in such products as carbophenothion and chlorfenvinphos.

Contaminated insects or other prey

Spraying of an insecticide often results in a local abundance of food in the form of dead or moribund insects, either of the pest species or of assorted nontarget species. A number of bird species make use of this temporary bounty and will gorge themselves, often to their detriment. For example, California Gulls (*Larus californicus*) died from ingesting lethal quantities of grasshoppers sprayed with a liquid formulation of carbofuran (Leighton 1988). Grackles (*Quiscalus quiscula*) have been killed under the same conditions. Carbofuran has also been shown to have a dramatic impact on a rare ground-nesting owl species in Canada, the Burrowing Owl (*Athene cunicularia*) (Fox et al. 1989). The owls are probably poisoned as they feed on contaminated grasshoppers, although they may also be exposed through ingesting surface-coated small mammals. The liquid formulation of carbofuran is under regulatory reevaluation in Canada.

Forestry insecticides

In contrast to cultivated field situations, pesticides are applied directly to complex wildlife habitats in forested areas, where a large number of individuals of many species can be exposed. For this reason, forestry products must be more stringently reviewed. In Canada, the forestry insecticide fenitrothion is currently under regulatory review. Although fenitrothion is not as acutely toxic as a number of other anticholinesterase insecticides used in agriculture, its impact on birds may make the product unacceptable for broad-scale forest spraying (Busby et al. 1989). Evidence against this product is primarily in the form of data indicating severe and widespread inhibition of brain acetylcholinesterase in several songbird species. Similar levels of inhibition have been associated with a number of sublethal reproductive effects as well as some mortality. Fenitrothion is readily absorbed dermally and birds may be at risk merely through exposure to the spray cloud (Mineau et al. 1990).

Avicides

The use of avicides in North America is limited and most bird control in agricultural areas is nonlethal. However, one product, fenthion, is widely used in Africa for the control of the Red-Billed Quelea (*Quelea quelea*). Fenthion (queletox) is an organophosphate with exceptional dermal penetrating properties. The common method of applying this avicide is by spraying roosts at night using aircraft. In Canada, it is registered to control Rock Doves (*Columba livia*), House Sparrows (*Passer domesticus*), and a variety of icterid species (primarily *Agelaius phoeniceus*). However, fenthion spraying is not permitted, and exposure must be achieved by means of a fenthion-impregnated perch. Target birds have only to land briefly on such perches to contract a lethal dose, but may fly for several hours before dying.

Even under these conditions, which minimize contact, raptors are extremely vulnerable to secondary poisoning if they should prey on a contaminated bird (Hunt et al. 1991). Furthermore, target birds that have come into contact with treated perches are much more vulnerable to predation than unexposed individuals, thus increasing the risk of secondary poisoning (Hunt et al. 1992). The use of fenthion as an avicide is likely to have serious consequences on the local raptor community as well as on any scavenger.

Rodenticides

Rodent control is a relatively minor agricultural practice in the developed world. In developing countries, however, people are often in direct competition with rodents for both stored and standing crops. Rodenticides are not specific to their intended targets and can have a significant impact on nontarget species. Only detailed knowledge of the habits of the target species and use of specific baiting locations or specialized bait holders can reduce the taking of "innocent" species.

More problematic is secondary poisoning. The newer, more efficacious anticoagulants present an even greater hazard to predators than the older products (Wenz 1984; Hegdal and Colvin 1988). Compounds such as difenacoum, brodifacoum, flocoumafen, and similar "super" coumarin-type products should not be used in situations where the target species is likely to be preyed upon or scavenged. Generally, the use of these products should be restricted to human dwellings and grain-storage areas. Other strategies should be employed for field-control of rodent pests.

The use of endrin to control rodents in orchards in North America has severe consequences for raptors (Blus et al. 1983) and is clearly unacceptable. The use of DDT to control bats is also ill-advised and resulted from the mistaken notion that bats were more sensitive to DDT than other mammals. This is true only when bats are awakened from their hibernacula and have low levels of body fat. Moreover, DDT does not always provide lethal control and can actually increase the risk that humans will be bitten by debilitated bats.

Some products that potentially give rise to secondary poisoning can be used safely under conditions that restrict the availability of the affected target to predators or where the predator, through its normal feeding practice, discards the dangerous body part(s). Strychnine, for example, is likely to be dangerous only when the predator consumes the gastrointestinal tract of the affected target. This characteristic limits exposure of some raptor species even though they may consume large numbers of contaminated prey.

To limit secondary poisoning, zinc phosphide is probably the best product available. Its use should be encouraged where it is of sufficient efficacy. Again, on ecological grounds, care should be taken to preserve the raptor population of any given area because they are an integral part of a long-term rodent-control strategy.

Studying and monitoring the impact of pesticides on wildlife in the field

Compound-specific enquiries

Because of the requirements for registration, compound-specific enquiries are the most common type of wildlife investigation used to demonstrate the safety of a particular pesticide. However, no single strategy is adequate for all situations. It is often best to start with testable hypotheses, then devise ways to prove or disprove these hypotheses.

In probabilistic terms, type I errors (i.e., concluding that there has been an effect from a pesticide when there have been no effects) are uncommon. For example, when wildlife mortality is encountered in a treated field, the probability that this mortality resulted from reasons unrelated to the application are small for the simple reason that it is uncommon to witness random wildlife mortality under normal circumstances.

On the other hand, the probability of type II errors (i.e., failure to detect an occurring problem) is much higher. Often the implicit assumption is that all effects are observable. In practice, however, this belief is not always tenable.

For example, searching for carcasses on treated fields is often thought to be a reliable method of detecting impact. Yet, it has not been ascertained whether affected wildlife die locally or, if they are able, choose to leave the area and die further afield (Mineau and Collins 1988).

Compound-specific studies are conducted in several ways. Some involve the controlled application of a pesticide by the investigators; others may take place during the normal operational use of the product, with or without the knowledge of the user. Again, the research question being asked determines which strategy is most appropriate. As well, considerations of cost and logistics often weigh heavily in this decision. The interpretation of results will have different emphasis, depending on whether the application was carefully controlled or conducted under variable conditions typical of operational use.

Pesticides known to have high acute toxicity and identified routes of exposure for several species usually prompt these types of investigations. Reports of wildlife kills connected with the use of a particular pesticide product may also trigger an enquiry. Common field-testing protocols are listed in Table 4, along with examples and critiques. An overview by Fite et al. (1988) provides guidance on most of the study types mentioned. Other recent assessments of wildlife field-testing can be found in Greaves et al. (1988) and Conservation Foundation (1989).

Use of reference compounds

It is often impractical or not economically feasible to test every potentially hazardous pesticide. Extensive studies of one or several products with a similar pattern of use can prove helpful as benchmarks. Untested related

Table 4. Common field-testing strategies for compound-specific enquiries.

Strategy	Protocols or critique	Examples
Carcass searches	Mineau and Collins 1988	Brewer et al. 1988
Surveys	Ralph and Scott 1980 Edwards et al. 1979 Milliken 1988	Peakall and Bart 1983
Use of biomarkers	Hill and Fleming 1982 Mineau and Peakall 1987 Fairbrother et al. 1991	Busby et al. 1989
Use of sentinel species	—	Robinson et al. 1988
Monitoring reproduction in naturally nesting species	—	Powell 1984 Busby et al. 1990 Fox et al. 1989
Multiple approach	—	Bunyan et al. 1981

products can then be compared and confirmed with laboratory toxicity studies or limited, but highly focused, field investigations. Sheehan et al. (1987), for example, used field data on the effect of carbaryl (specifically the ability of ducklings to obtain enough invertebrate prey for adequate growth and maintenance after an overspray of their pond habitat) to rank a number of insecticides with regard to their hazard to young waterfowl. This required modeling duckling growth as well as the persistence and partitioning of insecticides in a pond environment. The toxicity of the insecticides to a few indicator species in the laboratory was also factored into the final assessment.

In Canada, fenitrothion is used as a forestry benchmark compound against which other cholinesterase-inhibiting insecticides are assessed (Busby et al. 1989).

Crop and use-pattern enquiries

It is not always possible to focus a wildlife investigation on a single pesticide. In a number of cropping situations, a variety of different pesticides is used in quick succession making the identification of compound-specific impact difficult. This is especially true in crops where the cosmetic appearance of the product is deemed to be important and no amount of damage is considered permissible.

Where treatment is so complex that it is difficult to assess exposure to any one pesticide, two approaches are possible: treated sites or landscapes can be compared to untreated areas; and the "severity" of treatment (the a priori expectation of toxicity) for any given site can be used as a variable against which a number of different parameters (such as reproductive success) can be assessed through regression analysis. Caution must be exercised in comparing treated and untreated areas because they are likely to differ in several ways making interpretation difficult. In Canada, these approaches have recently been used to investigate the impact of multiple sprays on the bird fauna of fruit orchards and to compare the number and diversity of wildlife in conventional farms and "organic" farms (where synthetic pesticides or fertilizers are not used).

Population enquiries and baseline wildlife surveys

Regional or national survey data of wildlife population levels are rarely adequate to demonstrate pesticide impacts. However, in the case of some particularly damaging compounds (e.g., DDT and dieldrin), it has been possible to document regional and near-global population effects. Granular formulations of carbofuran may also have the potential to affect population levels of a number of bird species (EPA 1988).

Although systematic survey data are not usually available in the developing world, local lore and indigenous knowledge of the local fauna should not be overlooked. In developing countries, the bulk of pesticide applications are by hand; thus it is reasonable to expect that farmers and fieldworkers develop a rather intimate knowledge of the field and surrounding areas. Fieldworkers should be encouraged to report any instances of mortality, abnormal behaviour, or disappearance of wildlife species from the field area. In addition, the establishment of standardized surveys to estimate wildlife abundance and diversity in intensively farmed landscapes should be considered. Wherever relevant, rare, vulnerable, or ecological keystone species can play an important role as indicator species.

To carry out wildlife monitoring in treated areas, it is necessary to have a sound knowledge of the normal complement of species and the likelihood that these species will be exposed during pesticide treatments. Investigators should keep in mind that the diversity or abundance of species may have already been affected by past pesticide use. Fletcher and Greig-Smith (1988) have suggested several methods for quantifying the use of cropland by wildlife as part of an impact-assessment program.

Baseline agricultural surveys

The ability to define wildlife impacts is often hampered by a lack of knowledge about prevailing agricultural practices. Such information can be difficult to obtain when growers are not willing to cooperate with investigators. In some cases, lack of support from the local community may indicate a failure to comply with existing pesticide regulations. Realistically, however, no amount of policing can replace a good grower-education system and the full participation of landowners and fieldworkers in promoting sound environmental protection.

It is often necessary to conduct pure engineering or residue monitoring studies to define the extent of hazard to a local wildlife population, e.g., measurement of the proportion of granular insecticides or treated seed remaining on the soil surface after agricultural machinery has been used. Another study of this type might be measuring the level of residue remaining on foliar surfaces after application of a liquid spray and at intervals before harvest.

Practical considerations for users of pesticides and regulatory authorities

The following recommendations are made as the starting point toward more environmentally friendly pesticide use to minimize damage to wildlife and wildlife habitat.

- Select the least toxic and the least persistent product available for the use required. Consider whether spraying is essential. Assume that the product is at least as toxic to birds and other wildlife as it is to humans.

- Avoid using products that are known to move away from the area of application through vapour drift or runoff. Avoid applying pesticides under conditions that are conducive to droplet drift.

- If possible, spray at a time that does not coincide with the breeding season for wildlife species. Avoid spraying near nests, dens, or burrows.

- Follow label instructions scrupulously. Take heed of any special warnings concerning fish or wildlife and abide by specified buffer zones.

- Avoid the use of granular formulations of acutely toxic insecticides. They cannot be used safely because complete incorporation into the soil is never achieved. If these products must be used and are applied with mechanized equipment, shut off delivery long before the end of crop rows and avoid spillage over bumps, in turn areas, and at loading sites. Cover any visible spills. If applied by hand, try to ensure complete soil incorporation.

- Protect valuable wildlife areas by staying well away from field edges, woodlots, wetlands (even if temporarily dry), ditches, hedges, fence-lines, and rock piles. Do not allow pesticide sprays to drift onto these habitats.

- Inspect fields carefully. Avoid the repeat use of any product that causes any wildlife mortality. Experience has shown that what is observed is only the "tip of the iceberg."

- Treat and dispose of empty containers as directed. Where programs are available, recycle containers.

- Avoid contamination of any body of water, whether permanent or temporary.

- Never wash spray equipment in lakes, ponds, or rivers. Avoid drawing water directly from these areas. Rather, take water to the spray equipment, keeping chemicals and spray gear well away from the body of water in case of spills. If necessary use backflow devices.

- If carrying out a vertebrate-poisoning program, ensure that bait placement minimizes exposure to nontarget species. Locate and remove all carcasses to avoid scavenging. Avoid using products of high secondary toxicity if the target species is subject to predation or scavenging. Protect all species of raptors. These often fall prey to farmers who hold the misguided view that they represent a threat to livestock. Yet their benefits to agriculture almost always outweigh their occasional taking of small farm animals.

- Report any incident of wildlife mortality to competent authorities. Only through such feedback will it be possible to minimize wildlife impacts in the future.

- Establish monitoring programs to detect wildlife problems. If expertise is not available locally, seek outside help to design field-monitoring programs to suit specific situations.

Agriculture Canada; Environment Canada; Fisheries and Oceans Canada. 1987. Environmental chemistry and fate: guidelines for registration of pesticides in Canada. Supply and Services Canada, Ottawa, ON, Canada. 56 pp.

Balcomb, R.; Bowen, C.A., II; Wright, C.A.; Law, M. 1984. Effects on wildlife of at-planting corn applications of granular carbofuran. Journal of Wildlife Management, 48, 1 353–1 359.

Blus, L.J.; Henny, C.J.; Kaiser, T.E.; Grove, R.A. 1983. Effects on wildlife from use of endrin in Washington State orchards. *In* Transcripts of the 48th North American wildlife and natural resources conference. Wildlife Management Institute, Washington, DC, USA. Pp. 159–174.

Blus, L.J.; Stanley, C.S.; Henny, C.J.; Pendleton, G.W.; Craig, T.H.; Craig, E.H.; Halford, D.K. 1989. Effects of organophosphorus insecticides on Sage Grouse in southeastern Idaho. Journal of Wildlife Management, 53(4), 1 139–1 146.

Brewer, L.W.; Driver, C.J.; Kendall, R.J.; Lacher, T.E., Jr; Galindo, J.C.; Dickson, G.W. 1988. Avian response to a turf application of Triumph 4E. Environmental Toxicology and Chemistry, 7(5), 391–401.

Bunyan, P.J.; van den Heuvel, M.J.; Stanley, P.I.; Wright, E.N. 1981. An intensive field trial and a multi-site surveillance exercise on the use of aldicarb to investigate methods for the assessment of possible environmental hazards presented by new pesticides. Agro-Ecosystems, 7, 239–262.

Burgat-Sacaze, V.; Rico, A.G.; Petit, C. 1990. Effets des pesticides sur la reproduction des oiseaux. Annales de l'Association Nationale de la Protection des Plantes, 2(1/1), 187–204.

Busby, D.G.; White, L.M.; Pearce, P.A.; Mineau, P. 1989. Fenitrothion effects on forest songbirds: a critical new look. *In* Ernst, W.R.; Pearce, P.A.; Pollock, T.L., ed., Environmental effects of fenitrothion use in forestry: impacts on insect pollinators, songbirds and aquatic organisms. Environment Canada, Atlantic Region, Canada. Pp. 43–108.

Busby, D.G.; White, L.M.; Pearce, P.A. 1990. Effects of aerial spraying of fenitrothion on breeding White-Throated Sparrows. Journal of Applied Ecology, 27, 743–755.

Conservation Foundation. 1989. Pesticides and birds: improving impact assessment. Conservation Foundation, Washington, DC, USA. 68 pp.

Coulston, F. 1985. The dilemma of DDT. Regulatory Toxicology and Pharmacology, 5, 329–331.

DeWeese, L.R.; McEwen, L.C.; Hensler, G.L.; Petersen, B.E. 1986. Organochlorine contaminants in passeriformes and other avian prey of the Peregrine Falcon in the western United States. Environmental Toxicology and Chemistry, 5, 675–693.

Diamond, A.W.; Filion, F.L., ed. 1987. The value of birds. International Council for Bird Preservation, Cambridge, UK. Technical Publication 6, 267 pp.

Edwards, P.J.; Brown, S.M.; Fletcher, M.R.; Stanley, P.I. 1979. The use of a bird territory mapping method for detecting mortality following pesticide application. Agro-Ecosystems, 5, 271–282.

EPA (US Environmental Protection Agency). 1988. Carbofuran special review support document. EPA, Washington, DC, USA.

Erbach, D.C.; Tollefson, J.J. 1983. Granular insecticide application for corn rootworm control. Transactions of the American Society of Agricultural Engineers, 26, 696–699.

Fairbrother, A.; Marden, B.T.; Bennett, J.K.; Hooper, M.J. 1991. Methods used in determination of cholinesterase activity. *In* Mineau, P., ed., Cholinesterase-inhibiting insecticides: their impact on wildlife and the environment. Elsevier, Amsterdam, Netherlands. Pp. 35–72.

Fite, E.C.; Turner, L.W.; Cook, N.J.; Stunkard, C. 1988. Guidance document for conducting terrestrial field studies. Environmental Protection Agency, Washington, DC, USA. EPA 540/09-88-109, 67 pp.

Fleming, W.J.; Clark, D.R., Jr; Henny, C.J. 1983. Organochlorine pesticides and PCBs: a continuing problem for the 1980s. *In* Transcripts of the 48th North American wildlife and natural resources conference. Wildlife Management Institute, Washington, DC, USA. Pp. 186–199.

Fletcher, M.R.; Greig-Smith, P.W. 1988. The use of direct observations in assessing pesticide hazards to birds. *In* Greaves, M.P.; Smith, B.D.; Greig-Smith, P.W., ed., Field methods for the study of environmental effects of pesticides. British Crop Protection Council, Cambridge, UK. Monograph 40, pp. 47–56.

Flickenger, E.L.; King, K.A.; Stout, W.F.; Mohn, M.M. 1980. Wildlife hazards from Furadan 3G applications to rice in Texas. Journal of Wildlife Management, 44(1), 190–197.

Fox, G.A.; Mineau, P.; Collins, B.T.; James, P.C. 1989. The impact of the insecticide carbofuran (Furadan 480F) on the Burrowing Owl in Canada. Canadian Wildlife Service, Ottawa, ON, Canada. Technical Report Series 72.

Frank, R.; Mineau, P.; Braun, H.E.; Barker, I.K.; Kennedy, S.W.; Trudeau, S. 1991. Deaths of Canada Geese following spraying of turf with diazinon. Bulletin of Environmental Contamination and Toxicology, 46, 852–858.

Getzin, L.W. 1973. Persistence and degradation of carbofuran in soil. Environmental Entomology, 2(3), 461–467.

Greaves, M.P.; Smith, B.D.; Greig-Smith, P.W., ed. 1988. Field methods for the study of environmental effects of pesticides. British Crop Protection Council, Cambridge, UK. Monograph 40, 370 pp.

Greig-Smith, P.W. 1988. Hazards to wildlife from pesticide seed treatments. *In* Martin, T., ed., Application to seeds and soil. British Crop Protection Council, Cambridge, UK. Monograph 39, pp. 127–133.

Grue, C.E.; Fleming, W.J.; Busby, D.G.; Hill, E.F. 1983. Assessing hazards of organophosphate pesticides to wildlife. *In* Transcripts of the 48th North American wildlife and natural resources conference. Wildlife Management Institute, Washington, DC, USA. Pp. 200–220.

Hegdal, P.L.; Colvin, B.A. 1988. Potential hazard to Eastern Screech-Owls and other raptors of brodifacoum bait used for vole control in orchards. Environmental Toxicology and Chemistry, 7, 245–260.

Hill, E.F. 1986. Caution: standardized acute toxicity data may mislead. US Fish and Wildlife Service, Washington, DC, USA. Research Information Bulletin 6-86, 2 pp.

Hill, E.F.; Camardese, M.B. 1984. Toxicity of anticholinesterase insecticides to birds: technical grade versus granular formulations. Ecotoxicology and Environmental Safety, 8, 551–563.

Hill, E.F.; Fleming, W.J. 1982. Anticholinesterase poisoning of birds: field monitoring and diagnosis of acute poisoning. Environmental Toxicology and Chemistry, 1, 27–38.

Hill, E.F.; Hoffman, D.J. 1984. Avian models for toxicity testing. Journal of the American College of Toxicology, 3(6), 357–376.

Hummel, J.W. 1983. Incorporation of granular soil insecticides by corn planters. American Society of Agricultural Engineers, St Joseph, MI, USA. Publication 83-1017, 14 pp.

Hunt, K.A.; Bird, D.M.; Mineau, P.; Shutt, L. 1991. Secondary poisoning hazard of fenthion to American Kestrels. Archives of Environmental Contamination and Toxicology, 21, 84–90.

_____ 1992. Selective predation of organophosphate-exposed prey by American Kestrels. Animal Behaviour, 43(6), 971–976.

Kenaga, E.E. 1974. Evaluation of the safety of chlorpyrifos to birds in areas treated for insect control. Residue Reviews, 50, 1–41.

Leighton, F.A. 1988. Some observations of diseases occurring in Saskatchewan wildlife. Blue Jay, 46(3), 121–125.

Littrell, E.E. 1988. Waterfowl mortality in rice fields treated with the carbamate, carbofuran. California Fish and Game, 74(4), 226–231.

Maybank, J.; Yoshida, K.; Grover, R. 1978. Spray drift from agricultural pesticide application. Journal of the Air Pollution Control Association, 28(10), 1 009–1 014.

Maze, R.C.; Atkins, R.P.; Mineau, P.; Collins, B.T. 1991. Measurement of surface pesticide residue in seeding operations. Transactions of the American Society of Agricultural Engineers, 34(3), 795–799.

Milliken, R.L. 1988. A comparison of spot, transect and plot methods for measuring the impact of forest pest control strategies on forest songbirds. Forest Pest Management Institute, Canadian Forestry Service, Sault Ste Marie, ON, Canada. Information Report FPM-X-83.

Mineau, P. 1988. Avian mortality in agro-ecosystems: 1 — The case against granular insecticides in Canada. *In* Greaves, M.P.; Smith, B.D.; Greig-Smith, P.W., ed., Field methods for the study of environmental effects of pesticides. British Crop Protection Council, Cambridge, UK. Monograph 40, pp. 3–12.

_____ 1991. Difficulties in the regulatory assessment of cholinesterase-inhibiting insecticides, *In* Mineau, P., ed., Cholinesterase inhibiting insecticides: their impact on wildlife and the environment. Elsevier, Amsterdam, Netherlands. Pp. 277–300.

Mineau, P.; Collins, B.T. 1988. Avian mortality in agro-ecosystems: 2 — Methods of detection. *In* Greaves, M.P.; Smith, B.D.; Greig-Smith, P.W., ed., Field methods for the study of environmental effects of pesticides. British Crop Protection Council, Cambridge, UK. Monograph 40, pp. 13-27.

Mineau, P.; Peakall, D.P. 1987. An evaluation of avian impact assessment techniques following broad-scale forest insecticide sprays. Environmental Toxicology and Chemistry, 6, 781–791.

Mineau, P.; Sheehan, P.J.; Baril, A. 1987. Pesticides and waterfowl in the Canadian prairies: a pressing need for research and monitoring. *In* Diamond, A.W.; Filion, F.L., ed., The value of birds. International Council for Bird Preservation, Cambridge, UK. Technical Publication 6, 267 pp.

Mineau, P.; Sundaram, K.M.S.; Sundaram, A.; Feng, C.; Busby, D.; Pearce, P.A. 1990. An improved method to study the impact of pesticide sprays on small song birds. Journal of Environmental Science and Health, B25(1), 105–135.

Peakall, D.B.; Bart, J.R. 1983. Impacts of aerial application of insecticides on forest birds. CRC Press, Boca Raton, FL, USA. Critical Reviews in Environmental Control, 13(2), 117–165.

Peakall, D.B.; Tucker, R.K. 1985. Extrapolation from single species studies to populations, communities and ecosystems. *In* Vouk, V.B.; Butler, G.C.; Hoel, D.G.; Peakall, D.B., ed., Methods for estimating risk of chemical injury: human and non-human biota and ecosystems. Scientific Committee on Problems of the Environment, Paris, France.

Potts, G.R. 1980. The effects of modern agriculture, nest predation and game management on the population ecology of partridges (*Perdrix perdrix* and *Alectoris rufa*). Advances in Ecological Research, 8, 2–79.

Powell, G.V.N. 1984. Reproduction by an altricial songbird, the Red-Winged Blackbird, in fields treated with the organophosphate insecticide fenthion. Journal of Applied Ecology, 21, 83–95.

Rabb, G.B. 1990. Declining amphibian populations. Species, 13/14, 33–34.

Ralph, C.J.; Scott, J.M., ed. 1980. Estimating numbers of terrestrial birds: proceedings of a conference, Asilomar, CA, USA. Studies in Avian Biology, 6.

Robinson, S.C.; Kendall, R.J.; Robinson, R.; Driver, C.J.; Lacher, T.E., Jr. 1988. Effects of agricultural spraying of methyl parathion on cholinesterase activity and reproductive success in wild starlings (*Sturnus vulgaris*). Environmental Toxicology and Chemistry, 7, 343–349.

Schafer, E.W., Jr; Brunton, R.B. 1979. Indicator bird species for toxicity determinations: is the technique usable in test method development? *In* Beck, J.R., ed., Vertebrate pest control and management materials. American Society for Testing and Materials, Philadelphia, PA, USA. STP 680, pp. 157–168.

Schwarzbach, S.E.; Shull, L.; Grau, C.R. 1988. Eggshell thinning in Ring Doves exposed to *p,p'*-dicofol. Archives of Environmental Contamination and Toxicology, 17, 219–227.

Sheehan, P.J.; Baril, A.; Mineau, P.; Smith, D.K.; Harfenist, A.; Marshall, W.K. 1987. The impact of pesticides on the ecology of prairie nesting ducks. Canadian Wildlife Service, Ottawa, ON, Canada. Technical Report 19.

Stone, W.B.; Gradoni, P.B. 1985. Wildlife mortality related to the use of the pesticide diazinon. New York State Department of Environmental Conservation, Albany, NY, USA. 27 pp.

Sundaram, K.M.S. 1987. A comparative evaluation of dislodgable and penetrated residues, and persistence characteristics of aminocarb and fenitrothion, following applications of several formulations onto conifer trees. Journal of Environmental Science and Health, B21(6), 539–560.

Tucker, R.K.; Leitzke, J.S. 1979. Comparative toxicology of insecticides for vertebrate wildlife and fish. Pharmacology Therapeutics, 6, 167–220.

Urban, D.J.; Cooke, N.J. 1986. Ecological risk assessment. Hazard Evaluation Division, Environmental Protection Agency, Washington, DC, USA. EPA 540/9-85-001.

Walker, C.H. 1983. Pesticides and birds: mechanisms of selective toxicity. Agriculture, Ecosystems and Environment, 9, 211–226.

Wenz, C. 1984. New chemicals under fire. Nature, 309(28), 741.

Willis, G.H.; McDowell, L.L. 1987. Pesticide persistence on foliage. Reviews of Environmental Contamination and Toxicology, 100, 23–73.

Control of disease vectors:
a current perspective

P. Wijeyaratne

Health Sciences Division, International Development Research Centre,
Ottawa, Ontario, Canada

A brief history of the use of pesticides to control vector-borne diseases is presented. Integrated vector control is defined and some promising new approaches are described, with emphasis on environmental management and methods that lend themselves to greater community participation. The role of the community in vector control is discussed, and a brief account is given of some community-based vector control projects in Africa and South America.

The use of pesticides in public health has been mainly to control vector-borne diseases. Vectors are small animals, mostly insects, that transmit pathogens from one host to another. Insects were incriminated as vectors of human disease as early as 1878, when Manson connected mosquitoes with filariasis. Soon afterwards, Ross and others showed that mosquitoes transmitted malaria (Harrison 1978). Some of the tropical diseases that most affect human well-being are vector borne (Table 1).

Most of these diseases are associated with bloodsucking insects, and many are associated with water, because the vectors spend all or part of their life cycles in or near it. For example, culicine mosquito vectors of filariasis and Japanese encephalitis breed in water, ranging from organically polluted drains to rice fields; anopheline vectors of malaria in seepage puddles, ponds, and streams; *Cyclops*, the intermediate host in guineaworm disease, in wells and ponds; *Bulinus*, *Biomphalaria*, and *Oncomelania* snails, the intermediate hosts in schistosomiasis, in streams, low earth-dam reservoirs, and rice-irrigation systems; and *Simulium* vectors of onchocerciasis (river blindness) in fast-flowing waters including spillways of dams.

Other vector-borne diseases that have caused disastrous epidemics are typhus (*Rickettsia prowazeki*), transmitted by body lice (*Pediculus humanus*), and bubonic plague (*Yersinia pestis*), transmitted from rodents to humans by fleas (*Xenopsylla cheopis* and *Pulex irritans*). These diseases are not major problems

Table 1. Prevalence and distribution of some vector-borne diseases of humans.

Disease	Vectors or intermediate hosts	Estimated world prevalence[a]	Main areas of distribution[b]
Malaria	Mosquitoes (*Anopheles* spp.)	100×10^6/yr	Pantropical
Filariasis	Mosquitoes (*Anopheles, Culex, Aedes, Mansonia*)	90×10^6	Africa, Asia
Yellow fever	Mosquitoes (*Aedes* spp., *Haemagogus*)	—	Africa, South America
Dengue	Mosquitoes (*Aedes* spp.)	—	Pantropical
Japanese encephalitis	Mosquitoes (*Culex* spp.)	—	Asia
Onchocerciasis	Blackflies (*Simulium* spp.)	18×10^6	Africa, Central and South America
Leishmaniasis	Sandflies (*Lutzomyia, Phlebotomus*)	12×10^6	Africa, mid-East, Asia, South America
Sleeping sickness	Tsetse flies (*Glossina* spp.)	20×10^{3} [c]	Africa
Chagas' disease	Triatomine bugs (*Triatoma, Panstrongylus, Rhodnius*)	$16\text{-}18 \times 10^6$	Central and South America
Schistosomiasis	Snails (*Biomphalaria, Bulinus, Oncomelania*)	200×10^6	Africa, Southeast Asia, Middle East
Guineaworm	Crustacean (*Cyclops*)	$5\text{-}15 \times 10^6$ [d]	Africa, mid-East, India, Pakistan

[a] WHO (1989a,' unless otherwise stated.
[b] WHO (1989b).
[c] Number of new cases reported each year, but "...the real incidence is certainly much higher" (WHO 1989a).
[d] Hopkins (1988).

now, but they continue to threaten; typhus, especially, soon becomes epidemic in times of war, famine, and overcrowding.

Once their role was established, it was soon realized that the most effective way to control the diseases was to control their vectors. Before 1914, yellow fever in Cuba and the Panama Canal Zone was controlled by draining, destroying, oiling, covering, or burying all water containers in which *Aedes aegypti* could breed (Harrison 1978). By 1940, *Ae. aegypti* had been eradicated from Brazil by similar methods, rigorously enforced by almost militaristic campaigns (Soper et al. 1943). Malaria in Italy and other countries was controlled by draining the marshes that the *Anopheles* larvae inhabited, applying the larvicide Paris Green (copper acetoarsenite), releasing larvivorous fish (*Gambusia*), and screening windows (Harrison 1978).

The insecticide era

Although dichlorodiphenyltrichloroethane (DDT) was only discovered to kill insects in 1939, by 1944 this insecticide was being used to control typhus and malaria in Italy, and its use soon became worldwide to control pests and vectors of all kinds. For the control of malaria, houses and cattle sheds were sprayed twice a year with DDT wettable powder (WP) to kill resting *Anopheles* adults.

In 1955, following the spectacular success of this technique in Guyana, Italy, Taiwan, and other countries, the World Health Organization (WHO) inaugurated a global malaria-eradication program. By 1958, 75 countries had joined and, in 1961–1962, at the peak of the campaign, 69 500 t of pesticides, mainly DDT, were applied each year by 130 000 spraymen to 100 million dwellings occupied by 575 million people. The number of people contracting malaria fell from about 300 million/year before 1946 to about 120 million/year by the late 1960s; in a population that had doubled in size, malaria was eradicated from 10 countries.

After 1965, however, malaria made spectacular "comebacks" in many countries, especially India, Pakistan, and Sri Lanka. Between 1972 and 1976, the number of malaria cases increased 2.3 times worldwide, although the late 1970s saw some improvement as control campaigns were revived (Bull 1982). The resurgence of malaria was due to a variety of factors, including:

- Premature slowing of the eradication campaigns;

- Poor management and unsustainable approaches;

- Inadequate understanding of the habits of the mosquitoes (many spent too little time indoors to be vulnerable to the spray deposits); and

- Insect resistance.

By 1980, resistance to at least one insecticide had been reported in over 150 species of arthropods of public-health importance, including 93 species of mosquitoes (WHO 1984). This led to failure of some control programs and a switch to alternative insecticides, although "in some instances, resistance has become a convenient scapegoat for failures due to other factors" (Davidson 1989). DDT has been banned in some countries, but is still used for vector control in others. Where resistance has become a problem, DDT has been replaced by organophosphates (e.g., malathion and fenitrothion), carbamates (e.g., propoxur and bendiocarb), or synthetic pyrethroids (e.g., permethrin and deltamethrin). These insecticides are, for the most part, less persistent and much more expensive than DDT. Antimalaria campaigns continue in many countries, but the goal is no longer eradication, just control.

Other vector control campaigns have also been based mainly on insecticides. Many Caribbean and South American countries have had programs to control *Ae. aegypti* since 1960 or earlier, attacking larvae with the organophosphate temephos and perifocal spraying (for emerging adults) with another organophosphate, fenthion, supplemented during dengue epidemics by fogging or aerial spraying with malathion. Although eradication was achieved in some countries, by 1977 most campaigns in the region were "ineffectual, poorly supported and lacking adequate professional supervision and leadership" (Giglioli 1979).

Many countries in South America have promoted house-spraying with dieldrin and other residual insecticides to control triatomine bugs. Large tracts of land in west and central Africa have been cleared of tsetse flies by aerial spraying, especially with endosulfan.

The most successful campaign using insecticides in recent years was the onchocerciasis control program in West Africa, which now covers 14 000 km of rivers in 11 countries. Launched in 1974, the program has involved aerial spraying of blackfly-breeding sites with the organophosphate insecticide, temephos, now replaced in some areas by *Bacillus thuringiensis* because of resistance to temephos. Onchocerciasis control has been good in the central area, but reinvasion by blackflies in the periphery is still a problem (Remme and Zongo 1989).

An international campaign against guineaworm involved treating drinking-water sources with temephos to kill *Cyclops*, the intermediate host.

Side effects of pesticides used in public health

In spite of the millions of houses sprayed with DDT for malaria control, no accidental deaths of spraymen or householders due to DDT poisoning have been reported. However, some domestic animals were killed, especially cats, with the result that rat populations increased in some sprayed areas (Bull 1982). Bedbugs, cockroaches, and other household pests soon developed resistance to DDT, and became more abundant because the DDT had killed many of their predators. This led some householders to oppose spraying.

In the past, it was claimed that environmental contamination by DDT used in malaria control was relatively minor because it was sprayed inside houses, and the quantities were much smaller than those used in agriculture. However, washing of equipment, containers, and overalls, and the unauthorized use of DDT for other purposes (e.g., fish poisoning) spread the substance outside the houses. DDT used for malaria control accounted for an estimated 8% of global DDT contamination (Bull 1982).

Where pesticides are stockpiled for use in epidemics (e.g., dengue), environmental contamination due to leaking containers (some insecticides are highly corrosive), fire, theft, war, or natural disasters is always a danger.

Extensive studies in Africa of the effects on nontarget organisms of aerial spraying of endosulfan against tsetse flies and temephos used in rivers against blackfly larvae have revealed no permanent damage to treated ecosystems. In the case of tsetse fly control, it has been argued that any changes to the ecosystems caused by spraying are insignificant compared to the changes that will follow human settlement of tsetse-cleared land. However, there is no need for complacency in these matters, and further studies are required.

Agriculture and vector-borne diseases

The links between agricultural development and human health are poorly documented and "the health sector has up to now had minimal involvement in agricultural policies or projects" (Lipton and de Kadt 1988). However, people who clear forests for cultivation are at a high risk of infection with diseases such as malaria, leishmaniasis, African trypanosomiasis, and other diseases, because of increased exposure to the vectors. Moreover, insecticide resistance in mosquito vectors has been induced by the use of the insecticides in agriculture (Bull 1982). Selection acts upon the mosquito larvae as they develop in insecticide-contaminated waters. Particularly implicated in this context have been dieldrin, DDT, malathion, parathion, and propoxur (used mainly on rice and cotton crops). Among the affected anopheline species have been *Anopheles acconitus* (Indonesia), *A. albimanus* (Central America), *A. gambiae* s. 1 (West Africa), *A. maculipennis* (Turkey), *A. culcifacies* (India), and *A. sinensis* (China).

Irrigation presents a special problem, because many vectors live or develop in water. Some vectors develop directly in irrigated fields, others in canals, seepages, artificial lakes, or dam spillways. Increasing vector populations and more human contact with them have increased the transmission of malaria and schistosomiasis in the China, Kenya, Sri Lanka, Sudan, and other countries. Japanese encephalitis has been associated with irrigation practices in Sri Lanka and elsewhere in Asia. Poorly planned resettlements of people displaced by artificial lakes have also increased the transmission of vector-borne diseases (Service 1989).

Integrated vector control

From 1945 on, DDT, lindane, and other residual insecticides provided a single, highly effective method of vector control that eclipsed all other methods until about 1970. Since then, the development of resistance, concerns about environmental contamination and human safety, and the high cost of alternative insecticides have led to a revival of interest in other methods of vector control. Some of these methods were already known, but were neglected during the DDT era: personal protection (e.g., screens and repellents), source reduction (e.g., draining or removing artificial breeding sites), the use of fish to prey on mosquito larvae, and community-based health education. Other methods, such as genetic control, synthetic attractants, insect growth regulators, and the use of remote sensing by satellite to detect vector habitats have also been attempted experimentally and in endemic situations.

Integrated vector control has been defined as "the utilization of all appropriate technological and management techniques to bring about an effective degree of vector suppression in a cost-effective manner" (WHO 1983). It demands an adequate knowledge of the biology, ecology, and behaviour of the vector, nontarget organisms, and the human population to ensure not only effective control of the vector, but also human safety and prevention of other unacceptable side effects, including environmental damage.

Although there are many integrated approaches to vector control, only some of the more promising ones are considered here, with emphasis on those that could be used in community programs to achieve greater sustainability. (See Axtell (1972) for a comprehensive discussion of this issue.)

Pesticides

Chemicals

Although DDT and other organochlorines have been banned in many countries and replaced in others because of vector resistance or adverse effects, many other synthetic insecticides are still available. WHO (1984) listed 37 compounds in common use for the control of vectors and pests of public-health importance (for a partial list see Table 2). In some places, such as California (USA), vast amounts of insecticides are still used each year for mosquito control. DDT was replaced initially by organophosphates, such as malathion, and more recently by synthetic pyrethroids, such as permethrin, which are costly but effective at very low doses. Unfortunately, there are

Table 2. Some pesticides used in public health.

Pesticide class and name	Vectors and modes of application	Remarks
Insecticides		
Organochlorines		
DDT	*Anopheles* spp.; house spray	Banned in many countries
BHC (lindane)	*Anopheles* spp.; house spray	Banned in many countries
Endosulfan	*Glossina* spp.; aerial spray	
Organophosphates		
Malathion	*Anopheles* spp.; house spray	
	Aedes spp.; fog, aerial spray	
Fenitrothion	*Anopheles* spp.; house spray	
	Aedes spp.; fog, aerial spray	
Chlorpyrifos	*Culex quinquefasciatus* larvae	Used in polluted waters
Temephos	*Aedes aegypti* larvae, *Cyclops*	Used in drinking water
	Simulium larvae; aerial spray	Used in OCP,[a] W. Africa
Carbamates		
Bendiocarb	*Anopheles* spp.; house spray	
	Triatomines; house spray	
Propoxur	*Anopheles* spp.; house spray	
Pyrethroids		
Pyrethrum extract	Mosquitoes; coils	
Permethrin	Mosquitoes; nets, clothing, soap	
Deltamethrin	*Anopheles* spp.; house spray, nets	
Microbials		
Bacillus thuringiensis israelensis	*Aedes* spp. larvae	
	Simulium larvae; aerial spray	Used in OCP, W. Africa
B. sphaericus	*Culex* spp. larvae	
Molluscicides		
Niclosamide	Snail hosts of *Schistosoma*	
Endod	Snail hosts of *Schistosoma*	Extract of *Phytolacca* seeds

[a] OCP = onchocerciasis control program.

already many reports of mosquito and biting-fly resistance to pyrethroids, some of them also because of cross-resistance to DDT (Miller 1988). Reports of such proven resistance to pyrethroids have come from Israel, Saudi Arabia, Sri Lanka, Tanzania, and the USA among others.

Pathogens

Some effective microbial pesticides are now available for vector control, especially spore/crystal preparations of *B. thuringiensis* serotype H-14 var. *israelensis* (*Bti*) and *B. sphaericus*. These microbials are highly toxic and specific to the targeted larvae of mosquitoes and blackflies. However, they are relatively expensive and difficult to formulate because the toxic crystals sink and become inaccessible to most larvae, although floating, slow-release formulations of *Bti* are now available (Hudson 1985). *Bti* is widely used in the onchocerciasis control program in West Africa and, increasingly, for mosquito control in California and elsewhere. Because these microbial pesticides are virtually nontoxic to mammals, they can be applied by community volunteers.

Plant extracts

Endod, an extract of seeds of the Ethiopian plant *Phytolacca dodecandra*, and damsissa, a product of *Ambrosia maritima* in Egypt, are effective against the snail intermediate hosts of *Schistosoma*. Other local natural products could be developed for vector control. For example, fruit pods of the tree *Swartzia madagascarensis*, widely used in Africa as a fish poison, were also found to be toxic to *Anopheles* larvae and *Bulinus* snails (Minjas and Sarda 1986); Alpha T from the marigold flower (*Tagetes*) is toxic to mosquito larvae (Arnason et al. 1987).

Personal protection

Personal protection includes all measures taken at the individual or the household level to prevent biting by vectors. Anklets impregnated with repellents significantly reduced biting rates of mosquitoes on volunteers in Tanzania (Curtis et al. 1987). Washing with soap containing a repellent (diethyl toluamide, DEET) or an insecticide (permethrin) reduced mosquito biting rates in the Gambia (Lindsay and Janneh 1989). Bed netting has been used for centuries to give personal protection against biting insects (Lindsay and Gibson 1988). When impregnated with insecticides, the netting provides community protection as well; mosquitoes rest on the treated fabric and are killed. In numerous large-scale trials in various parts of the world, malaria transmission appears to have been reduced by the systematic use of nets impregnated with permethrin or deltamethrin (Rozendaal and Curtis 1989).

House improvements such as screening, insecticidal paints (Lacey et al. 1989), and filling in cracks in the walls could provide definitive measures against triatomine bugs (Schofield and White 1984).

Predators

Larvivorous fish such as *Gambusia affinis* have been used for controlling mosquito larvae for many years. Among the more promising recent developments is the use of young Chinese catfish (*Clarias fuscus*) to control *Ae. aegypti* in household water containers in China (Wu et al. 1987) and a community-based malaria-control scheme in India, which paid for itself by selling carp and prawns that were reared in the same group of ponds as the guppy fish used for controlling mosquitoes (Gupta et al. 1989).

Many other organisms have been tested for the biological control of vectors. Candidates for controlling larvae of *Ae. aegypti* and other mosquitoes that develop in small containers are dragonfly larvae in Myanmar, the copepod crustacean *Mesocyclops aspericornis* in French Polynesia (Rivière et al. 1989), and the predatory mosquito species, *Toxorhynchitis* (WHO 1988).

Trapping

Mechanical and other types of traps have been used to reduce populations of tsetse flies. Several designs have been developed, some of them incorporating chemical attractants and insecticides. In Uganda, an effective tsetse trap has been made from old tires and locally available plant materials (Okoth 1986). Light traps, installed in pig sties, have been tested for the control of *Culex tritaeniorhynchus* in Japan (Wada 1988).

Environmental management

Changing the environment to prevent vector breeding or to minimize contact between vectors and people can be an effective control mechanism (WHO 1982). Environmental management methods (Birley 1989) include:

- Environmental modification, i.e., any permanent or long-lasting change in land, water, or vegetation, such as filling, draining, or forest clearance;

- Environmental manipulation, e.g., flushing streams, changing water salinity, and removing shade plants; and

- Modifying human habitation or behaviour, e.g., locating new settlements away from vector populations, modifying house design, and changing water supply and waste disposal.

Intermittent irrigation was used to prevent the development of mosquito larvae in rice fields (Lacey and Lacey 1990) and layers of expanded polystyrene beads prevented *Culex quinquefasciatus* from laying their eggs in wet pit latrines in Tanzania (Maxwell et al. 1990). Much environmental management work can be done by community volunteers with guidance in the initial stages from vector-control specialists.

Training and education

Integrated control strategies require more people trained in vector biology. In addition to the usual sources of health education, such as schools and clinics, information can reach the public through billboards, newspapers, radio, and television. A dengue-control theme has even been incorporated into a "soap opera" in Puerto Rico (Gubler 1989).

The role of the community

Many problems and failures in vector control have been due, not only to technical difficulties, poor management, and lack of continuity, but also to the fact that not enough attention had been paid to the beliefs and attitudes of the affected communities. For example, many *Ae. aegypti* control campaigns in the past 20 years have relied too heavily on ultra-low-volume spraying, which is not always effective. The use of this method has given people a false sense of security, reinforced their belief that *Ae. aegypti* control is the government's responsibility, and taken away the pressure to get rid of larval habitats in their own backyards (Gubler 1989).

A recent WHO report (1987) explores the ways in which more responsibility for vector control can be transferred from the national to the district level and ways of getting people more involved in protecting themselves against vector-borne diseases because "community participation makes people more aware of their ill-health and general underdevelopment and of how they can overcome these problems." Vector control at the community level has to compete with more basic needs, such as food, shelter, and employment, and the need for it may not be appreciated during periods of little or no disease.

Nevertheless, examples of successful community participation in 15 countries include: setting tsetse traps; draining, filling, or clearing weeds from mosquito breeding sites; rearing larvivorous fish; source reduction of *Ae. aegypti*; and distribution of nylon filters to keep *Cyclops* out of drinking water. Vector-control campaigns should work closely with primary health-care programs to achieve greater effectiveness and sustainable results (WHO 1987).

Research on community strategies for vector control

Building research capacity, producing new knowledge, and creating linkages among researchers are perceived as essential components of development by the International Development Research Centre (IDRC). However, IDRC-supported projects must contribute to improving the welfare and standard of living, particularly of the poor and disadvantaged who are to be the ultimate beneficiaries of the research. IDRC tries to ensure that the activities it supports meet the long-term goals of development as viewed from the perspective of these beneficiaries: sustainable growth, equity, and participation.

Environmental and community control of dengue in China

This IDRC-supported project (91-0032) of China's Hainan Island Bureau of Public Health focuses on feasible and sustainable intervention at the community level to prevent the occurrence of dengue epidemics on the island. The research team will work closely with affected communities to establish long-term, integrated, preventive measures through health education aimed at controlling vector breeding in the home environment and surrounding areas.

Village volunteers will participate, through committees, in the periodic reporting of suspected dengue cases and in the active surveillance by boats in the local harbours for *Aedes* breeding sites. Larvivorous fish will also be used in drinking-water storage vats.

It is hoped that this project will not only lead to a sustainable strategy to prevent dengue epidemics, but will also pave the way for preventive health in general in the island.

Malaria-vector control

Nepal

The Nepal Malaria Eradication Organization (NMEO) is examining sustainable strategies to minimize problems due to malaria in rural communities. Under the *Environmental control of malaria* project (88-0212), NMEO is exploring environmental management methods of malaria-vector control that will be suitable under local conditions and that can be implemented with community involvement. The study will be carried out through the existing community-based political system by the rural units of the NMEO. The results will be examined closely by the national health authorities with a view to rapid establishment of a sustainable malaria-control system in the endemic areas of Nepal.

Peru

Malaria control is a major challenge in Peru, where the disease assumes serious proportions in the poorest jungle and Amazonian areas. Rising costs of insecticides, insecticidal resistance of the insect vectors, and environmental contamination have now forced exploration of alternatives to insecticide use.

The use of safe and efficient biological control agents such as *Bti* depends largely on costs and feasibility of production. The *Biological control of malaria* project (88-0213) aims to develop and field-test a simple technique for local production of *Bti* using coconut milk as the medium. Whole coconuts will also be used as a convenient medium for inoculation of *Bti* by the communities themselves and to facilitate transportation and application in different parts of the endemic region.

Tanzania

Community trials of insecticide-treated bed netting result in mass protection from malaria. However, in most malaria regions, the cost of bed nets is beyond the resources of the population. The *Community prevention of malaria* project (89-0216) will test the original and practical idea of using polypropylene fibres from locally available sacking material, used for agricultural products, to make "grass skirt" style bed curtains. It is known that bed nets with holes perform as well as intact bed nets when they are impregnated with insecticide.

The project will compare interventions in three communities — one with impregnated bed nets, one with impregnated bed curtains, and one with conventional control — in terms of reduction of clinical malaria episodes, malariometric indices, vector transmission indices, relative durability of materials, persistence of their effects, acceptability of the measures by the population, and the relative costs.

Leishmaniasis vector control in Peru

A primary health-care strategy was proposed for the control of Andean leishmaniasis (project 90-0081), based upon the epidemiology and ecology of the disease identified during the first phase of the project. In addition to homes and their immediate surroundings other rural sites contribute significantly to leishmaniasis transmission and to the maintenance of *Leishmania* in the environment. Research in the second phase involves early diagnosis and treatment of the disease and insecticide spraying of disease-transmission sites in homes and immediate surroundings. Most significantly, the strategy will also involve insecticide spraying and reforestation of rural areas to eliminate these previously neglected sites of *Leishmania* transmission. The impact of this approach, which will be carried out by members of the community and through the health post with the support of project staff, will be compared with that of classic measures alone in a valley with similar characteristics.

Chagas' disease vector control in Paraguay

Chagas' disease is one of the most serious tropical diseases found in Latin America, both in terms of its prevalence throughout the region and its impact on morbidity and potential employability. About 65 million people are directly exposed to the risk of *T. cruzi* infection, a further 15–20 million are actually infected and, of these, about 10% develop chronic Chagas' disease.

In the absence of drugs and vaccines suitable for large-scale treatment, the only effective measure lies in a preventive approach that focuses on the control of triatomine insects, which transmit the causative agent to man, within the domestic and peridomestic environment. Triatomine-control strategies include insecticide spraying and housing improvements resulting from increased community awareness of the problem.

The *Chagas' disease prevention* project (87-0342) will compare three types of interventions in three similar communities: insecticide application, housing improvement, and a combined insecticide and housing intervention. Pre- and postintervention assessments will be made of changes in human *T. cruzi* infection measured by serology, changes in triatomine infestation levels in houses, and changes in awareness and knowledge about the disease. The generated information will be invaluable, not only to Paraguay's health authorities, but also to the rest of the endemic region.

Schistosomiasis

Egypt

The control of schistosomiasis in the Nile delta, through the use of the plant molluscicide *Ambrosia maritima* (damsissa), has been the subject of IDRC-supported research for the past 10 years. Phase IV of the project (87-0204) will expand the study protocol to a new area, conduct a knowledge, attitudes, and practices (KAP) study, and study the toxicology of *A. maritima*. Specifically, the objectives are:

- To confirm the application dose of *A. maritima* in various water situations in established and newly reclaimed areas;

- To assess an integrated approach to schistosomiasis control using *A. maritima* for snail control, combined with praziquantel case treatment, in an established farming area of high prevalence and in a reclaimed resettlement area at risk of schistosomiasis spread;

- To conduct a KAP study, in both the established and resettlement areas, to determine sociobehavioural aspects of acceptance of *A. maritima* by the community and to estimate the requirements for sustained self-help in *A. maritima* control strategies; and

- To perform toxicologic studies on *A. maritima*, including determination of its effect on aquatic nontarget organisms.

If successful, the investigators will be able to provide convincing evidence that *A. maritima* is an effective and long-lasting plant molluscicide, suitable for use in the Nile Delta and without major toxic effects on humans or nontarget aquatic organisms.

Zimbabwe

Effective synthetic molluscicides have proved to be too costly for most developing countries where schistosomiasis is endemic. Plant molluscicides offer an alternative. Of these, the best known is the soapberry *Phytolacca dodecandra* (endod), a plant indigenous to Zimbabwe. During the past 5 years, the most potent varieties have been identified and field trials have shown that the plant is effective in reducing snail intermediate hosts. The present study (90-0278) is designed to evaluate the efficacy, acceptability, sustainability, and cost-effectiveness of different approaches to reducing prevalence, morbidity, and transmission of schistosomiasis following mass chemotherapy. Approaches include: application of *P. dodecandra* through community effort; through health services effort; and conventional chemotherapy through health services. The project also includes studies of local ecotoxicity, community participation, and health economics of the alternative interventions.

Despite promising results, long-term community participation in vector control must be secured, because operations, such as the control of container-breeding mosquitoes, may have to be continued indefinitely. Little is known about extending pilot projects into permanent national programs. Community volunteers may become victims of political struggles or professional rivalries if their work is not given proper recognition. The best chance of maintaining community support seems to lie in integrating vector control into the primary health-care system, which is now established in many countries (MacCormack 1990). More research is also needed on how to coordinate vector control with work in agriculture, forest and water management, and on the role of migrant workers in disease ecology and control.

Community-based vector control is not a way to reduce government spending. Although local initiatives should be encouraged, each country will still need teams of professional vector-control workers, using well-established methods, to meet its obligations under international health regulations (WHO 1972).

Conclusions

Disease vectors will probably remain with us indefinitely. Optimizing use, doses, and safety of control measures and balancing vector control with consideration for the environment is a challenge we must face.

Future considerations must include such questions as whether an insecticide-free environment is possible, or desirable. All development projects should include ecological planning to prevent increases in vector-borne diseases. Environmental management must be considered on both the large and small scale. The role of the primary health-care system must be defined and intersectoral cooperation obtained.

The goal is to ensure that vector-control strategies satisfy ecological requirements as well as local needs and priorities, that they are shaped around people's lifestyles and living patterns, and that they promote community self-reliance with respect to ongoing development.

Arnason, J.T.; Philogène, B.J.R.; Morand, P.; Scaiano, J.C.; Werstiuk, N.; Lam, J. 1987. Thiophene and acetylenes: phototoxic agents to herbivorous and blood-feeding insects. *In* Heitz, J.R.; Downum, K.R., ed., Light-activated pesticides. American Chemical Society, Washington, DC, USA. Symposium, 339, 255–264.

Axtell, R.C. 1972. Principles of integrated pest management in relation to mosquito control. Mosquito News, 39, 706–716.

Birley, M.H. 1989. Guidelines for forecasting the vector-borne disease implications of water resources development. World Health Organization, Geneva, Switzerland. PEEM Guidelines Series 2, unpublished report, WHO/VBC/89.6.

Bull, D. 1982. A growing problem: pesticides and the Third World poor. Oxfam, Oxford, UK. 192 pp.

Curtis, C.F.; Lines, J.D.; Ijumba, J.; Callaghan, A.; Hill, N.; Karimzad, M.A. 1987. The relative efficacy of repellents against mosquito vectors of disease. Medical and Veterinary Entomology, 1(2), 109–119.

Davidson, G. 1989. Insecticide use: an anti-alarmist view of its advantages and disadvantages. Tropical Diseases Bulletin, 86(6), R1–R6.

Giglioli, M.E.C. 1979. *Aedes aegypti* programs in the Caribbean and emergency measures against the dengue pandemic of 1977–1978: a critical review. *In* Dengue in the Caribbean, 1977. Pan-American Health Organization, Washington, DC, USA. Scientific Publication 375, pp. 133–152.

Gubler, D.J. 1989. *Aedes aegypti* and *Aedes aegypti*-borne disease control in the 1990s: top down or bottom up. American Journal of Tropical Medicine and Hygiene, 40(6), 571–578.

Gupta, D.K.; Sharma, R.C.; Sharma, V.P. 1989. Bioenvironmental control of malaria linked with edible fish production in Gujarat. Indian Journal of Malariology, 26(1), 55–59.

Harrison, G. 1978. Mosquitoes, malaria and man. John Murray, London, UK. 314 pp.

Hopkins, D.R. 1988. Dracunculiasis eradication: the tide has turned. Lancet, 1988, 148–150.

Hudson, J.E. 1985. The development of *Bacillus thuringiensis* H-14 for vector control. Tropical Diseases Bulletin, 82(8), R1–R10.

Lacey, L.A.; d'Alessandro, A.; Barreto, M. 1989. Evaluation of a chlorpyrifos-based paint for the control of Triatominae. Bulletin of the Society of Vector Ecologists, 14(1), 81–86.

Lacey, L.A.; Lacey, C.M. 1990. The medical importance of riceland mosquitoes and their control using alternatives to chemical insecticides. Journal of the American Mosquito Control Association, 6 (suppl. 2), 1–93.

Lindsay, S.W.; Gibson, M.E. 1988. Bednets revisited — old idea, new angle. Parasitology Today, 4(10), 270–272.

Lindsay, S.W.; Janneh, L.M. 1989. Preliminary field trials of personal protection against mosquitoes in The Gambia using DEET or permethrin in soap, compared with other methods. Medical and Veterinary Entomology, 3(1), 97–100.

Lipton, M.; de Kadt, E. 1988. Agriculture–health linkages. World Health Organization, Geneva, Switzerland. Offset Publication 104, 111 pp.

MacCormack, C.P. 1990. Appropriate vector control in primary health care. *In* Curtis, C.F., ed., Appropriate technology for vector control. CRC Press, Boca Raton, FL, USA. Pp. 221–227.

Maxwell, C.A.; Curtis, C.F.; Haji, H.; Kisumku, S.; Thalib, A.I.; Yahya, S.A. 1990. Control of Bancroftian filariasis by integrating therapy with vector control using polystyrene beads in wet pit latrines. Transactions of the Royal Society of Tropical Medicine and Hygiene, 84(5), 709–714.

Miller, T.A. 1988. Mechanisms of resistance to pyrethroid insecticides. Parasitology Today, 4(7), S8–S9, S12.

Minjas, J.N.; Sarda, R.K. 1986. Observations on the toxicity of *Swartzia madagascarensis* (Leguminosae) extract to mosquito larvae. Transactions of the Royal Society of Tropical Medicine and Hygiene, 80(3), 460–461.

Okoth, J.O. 1986. Community participation in tsetse control. Parasitology Today, 2(3), 88.

Remme, J.; Zongo, J.B. 1989. Demographic aspects of the epidemiology and control of onchocerciasis in West Africa. *In* Service, M.W., ed., Demography and vector-borne disease. CRC Press, Boca Raton, FL, USA. Pp. 367–386.

Rivière, F.; Kay, B.; Klein, J.M.; Sechan, Y. 1989. *Mesocyclops aspericornis* (Copepoda) and *Bacillus thuringiensis* var. *israelensis* for the biological control of *Aedes* and *Culex* vectors (Diptera: Culicidae) breeding in crab holes, tree holes and artificial containers. Journal of Medical Entomology, 24(4), 425–430.

Rozendaal, J.; Curtis, C.F. 1989. Recent research on impregnated mosquito nets. Journal of the American Mosquito Control Association, 5(4), 500–507.

Schofield, C.J.; White, G.B. 1984. House design and domestic vectors of disease. Transactions of the Royal Society of Tropical Medicine and Hygiene, 78, 285–292.

Service, M.W. 1989. Irrigation: boon or bane? *In* Service, M.W., ed., Demography and vector-borne disease. CRC Press, Boca Raton, FL, USA. Pp. 237–254.

Soper, F.L.; Wilson, D.B.; Lima, S.; Antunes, W.S. 1943. The organization of permanent, nation-wide anti-*Aedes aegypti* measures in Brazil. Rockefeller Foundation, New York, NY, USA.

Wada, Y. 1988. Strategies for control of Japanese encephalitis in rice production systems in developing countries. *In* Smith, W.H., ed., Vector-borne disease control in humans through rice agroecosystem management. International Rice Research Institute, Los Baños, Philippines. Pp. 153–160.

WHO (World Health Organization). 1972. Vector control in international health. WHO, Geneva, Switzerland. 144 pp.

_____ 1982. Manual on environmental management for mosquito control, with special emphasis on malaria vectors. WHO, Geneva, Switzerland. Offset Publication 66, 283 pp.

_____ 1983. Integrated vector control. WHO, Geneva, Switzerland. Technical Report Series 688, 72 pp.

_____ 1984. Chemical methods for the control of arthropod vectors and pests of public health importance. WHO, Geneva, Switzerland. 108 pp.

_____ 1987. Vector control in primary health care. WHO, Geneva, Switzerland. Technical Report Series 755, 61 pp.

_____ 1988. Urban vector and pest control: 11th report of the WHO Expert Committee on vector biology and control. WHO, Geneva, Switzerland. Technical Report Series 767.

_____ 1989a. Tropical diseases: progress in international research, 1987–1988 — 9th programme report of the UNDP/World Bank/WHO Special Programme for Research and Training in Tropical Diseases (TDR). WHO, Geneva, Switzerland. 128 pp.

_____ 1989b. Geographic distribution of arthropod-borne diseases and their principal vectors. WHO, Geneva, Switzerland. Unpublished report, WHO/VBC/89.967, 134 pp.

Wu, N.; Wang, S.; Han, G.X.; Xu, R.; Tang, G.; Qian, C. 1987. Control of *Aedes aegypti* larvae in household water containers by Chinese catfish. Bulletin of the World Health Organization, 65(4), 503–506.

Botanical pesticides: optimizing pest control and minimizing environmental impacts

B.J.R. Philogène[1] and J.D.H. Lambert[2]

[1]University of Ottawa and [2]Department of Biology, Carleton University, Ottawa, Ontario, Canada

Various types of botanical pesticides and their mode of action are examined: photosensitizers, antifeedants, and lignans. The role of naturally occurring synergists is also discussed. For any new botanical substance to be accepted as a safe biocide, it must be subjected to the same rigorous toxicity tests as synthetic products. The steps in this process are identified and outlined, using the molluscicide endod as an example.

The use of chemicals of plant origin for the control of destructive insects or vectors of diseases is not new. Rotenone, nicotine, and pyrethrins have been available for a long time (Matsumura 1975). Only pyrethrins remain in use as efficient, low-cost insecticides.

With increasing problems of toxicity and resistance of target pests to the synthetic compounds currently in use, interest in "natural" pesticides has revived. Pest-management specialists have been looking elsewhere for pest control as a result of the environmental and health problems brought about by the intensive and extensive use of organochlorines, organophosphates, and carbamates. Moreover, third-generation insecticides (insect hormones and their analogues) have not produced the results expected.

Botanical pest control attempts to provide environmentally acceptable methods of insect control through the naturally occurring substances in the crop plant or application of compounds derived from other plants. Such compounds and their derivatives should minimize losses at all the stages of agricultural and forest production and provide new molecules for the control of disease vectors. The search for new chemicals to be used as insecticides may also provide compounds with a mode of action different from the classic neurotoxicant, against which so many arthropod species have become resistant. This important development in pest-management strategies has been made possible by close collaboration among plant scientists, entomologists, chemists, and toxicologists (Miller and Miller 1986; Morgan and Mandava 1990).

Developments in plant–insect studies

The selective pressures brought about by phytophagous insects and other plant-eating organisms have resulted in the development, in plants, of a multitude of compounds that are either toxic or modifiers of physiology or behaviour. Current research has led to the identification and characterization of some particularly interesting molecules (Morgan and Mandava 1990). It has also permitted a more meaningful characterization of biochemically based resistance in plants.

Natural photosensitizers

Light is often forgotten or misunderstood as a factor in the study of insects (Philogène 1982) and, until recently, little attention has been paid to its role in plant–insect relations. Our research team looked specifically at the activation of secondary plant substances by light and their subsequent photosensitizing effects on insects (Arnason et al. 1983).

Secondary plant compounds with phototoxic action have been identified in several plant families, but particularly in the Asteraceae. The secondary metabolites of plants that are capable of photosensitizing insects include furanocoumarins, furanoquinoline alkaloids, beta-carboline alkaloids, poly-acetylenes and their thiophene derivatives, and extended quinones. These compounds may have direct lethal effects, retard larval development, or be ovicidal.

Polyacetylenes and their thiophene derivatives are toxic to a broad range of organisms, especially some insect species. One compound in this group particularly attracted our attention: alpha-terthienyl (α-T), which occurs naturally in marigold (*Tagetes*). We investigated its phototoxic effect on mosquito larvae and plant-feeding Lepidoptera (Philogène et al. 1985; Champagne et al. 1986; Iyengar et al. 1987). The results have been so encouraging that we are contemplating the commercialization of α-T.

Phototoxic acetylenes and thiophenes presumably provide enhanced protection to the plant by virtue of their involvement in high-energy photochemical processes and the catalytic nature of singlet-oxygen generation, which they mediate (Arnason et al. 1987a). Without photosensitizing radiation, these compounds still possess many of the insect-deterrent effects observed with other nonphotosensitizing secondary plant metabolites: feeding deterrence, growth reduction, and reduced nutrient utilization. Because α-T is not a neurotoxin but a cytotoxin, cross-resistance to organophosphates does not seem to occur.

The isoquinoline alkaloid, berberine, which is present in over 60 plants and 9 botanical families, was found to possess antifeeding activity on *Euxoa messoria* (Devitt et al. 1980) and insecticidal action against larvae of *Aedes atropalpus* (Philogène et al. 1984). A slow singlet-oxygen generator capable of producing cytogenetic damage, berberine had an LC_{50} (lethal concentration to 50% of organisms tested) of 8.8 ppm under near-ultraviolet light for 24 h. Larval, pupal, and adult survival were all significantly affected after a brief initial exposure of the larval stage, suggesting long-term carry-over effects. Lamentably, because berberine can intercalate with the DNA molecule, it is an unlikely candidate as a commercially useful insecticide.

Antifeedants

Many compounds have been known for years to act as feeding deterrents or antifeedants (Philogène 1974). Plants contain quite an array of such molecules, which could be exploited either as components of plant-resistance mechanisms or as exogenous compounds.

Most known antifeedants are complex molecules that are difficult to synthesize. They are active at very low concentrations and, consequently, are not found in large quantities in plants. Even with the best extraction procedures, it is impossible to obtain quantities that could be used on a large scale and the process is costly. In spite of this, the search goes on and new molecules are reported frequently. Over 350 compounds have been identified since 1976 (Wharten and Morgan 1990); only a small number of these molecules are effective antifeedants.

Our research team has investigated one group — the sesquiterpene lactones. These secondary plant substances from the Asteraceae have a number of biological effects on phytophagous insects, suggesting that they have evolved as deterrents to insect herbivory. These compounds also deter oviposition, inhibit growth, lengthen development, and are toxic.

The effects of tenulin, a sesquiterpene lactone from bitterweed (*Helenium amarum*) on the European corn borer (*Ostrinia nubilalis*), the variegated cutworm (*Peridroma saucia*), and a polyphagous grasshopper (*Melanoplus sanguinipes*) have been examined (Arnason et al. 1987b). The influence of the compound on the biological activity of these insects was observed at levels that were much lower than its concentration in leaves of *H. amarum* (15–25 mol/g fresh weight). When applied to leaf disks of corn, tenulin reduced feeding by *O. nubilalis* at a concentration of 0.3 mol/g fresh weight and above. In spite of the potent antifeedant effects, larvae could be induced to feed on artificial diets containing tenulin at low concentrations. At 5 mol/g in artificial diets, tenulin reduced growth and delayed larval development of *O. nubilalis* and *P. saucia*. *Melanoplus sanguinipes* suffered 50% larval mortality when injected with 0.88 mol of the sesquiterpene lactone.

Azadirachtine, a tetramortriterpernoid from the neem tree *(Azadirachta indica)* is the most effective and the most promising of all characterized antifeedants (Jacobson 1983; Koul et al. 1990; Schmutterer 1990). Neem and its byproducts have been used for centuries in India for protection of stored cereals and various standing crops and as drugs for all kinds of aliments (Koul et al. 1990). The greatest obstacle to its use is the limited shelf-life of neem seeds, which contain the highest titre of oils. There are, nevertheless, several formulations available in the subcontinent (Repelin, Wellgro, Nimbasol, Biosol, and Neemark) and one formulation has been approved by the Environmental Protection Agency in the USA (Margosan-O).

Lignans

The toxicity of insecticidal compounds can be significantly increased by the addition of synergists, i.e., chemicals that increase the lethality of insecticides given in sublethal amounts (Brindley and Selim 1984). The identification of sesamin in chrysanthemum flowers where it synergizes the activity of pyrethrum (Donskotch and El Feraly 1969) triggered interest in the potential interactions between the numerous secondary metabolites synthesized by plants.

Lignans are one group of synergists that reduce the efficacy of detoxification enzymes produced by insects that have been subjected to toxic compounds: more than 200 lignans have been found in about 50 plant families and 146 species (MacRae and Towers 1984). They are most abundant in the Pinaceae, Podophyllaceae, Rutaceae, and Lauraceae, with a few occurrences in the Asteraceae. Lignans alone have limited activity on insects; their effectiveness is felt in association with synthetic pesticides or with other plant compounds. Sesamin, sesamolin, sesangolin, asarinin, myristicin, savinin, and hinokinin are effective in enhancing the activity of a wide variety of insecticides, particularly pyrethrins and carbamates. They have, therefore, been used to develop numerous and economically important synergists.

Some lignans of the Asteraceae and Piperaceae (diasesartamin, sesamolin, cubebin, epiyangambin, and dill apiole) have been examined as potential inhibitors of the mixed-function oxydases of *O. nubilalis* (Bernard et al. 1988). At concentrations ranging from 10^{-7} to 10^{-3} M, diasesartamin, dill apiole, sesamolin, and, to a lesser extent, cubebin were found to be inhibitors of mixed-function oxydase activity. Epiyangambin had little effect.

Most lignans occur in plants with other toxins. It is, therefore, reasonable to hypothesize that these compounds have a role in synergizing the biological activity of these toxic compounds. More research is required to establish how this chemical capability of plants can be successfully exploited in pest-management strategies.

283

Molluscicide- and pesticide-development strategy

Molluscicides are crucial for the control and prevention of schistosomiasis. Of the commercial molluscicides available, only bayluscid (niclosamide) is recommended by the World Health Organization (WHO). However, bayluscid is costly: in 1974, 1 t cost 26 000 USD. Most countries where the disease is endemic cannot afford to spend such sums to attack transmission sites.

The possibility of using a naturally and locally produced plant molluscicide to remove the intermediary vector of schistosomiasis has been recognized since the early 1930s. The most promising plant, *Phytolacca dodecandra* (endod), was identified in 1964 (Lemma 1965) and tested in field trials at Adwa, Ethiopia, between 1969 and 1973 (Lemma et al. 1978). Several researchers have suggested that a plant molluscicide grown by farmers and integrated with chemotherapy could provide a cheap and efficient means to control schistosomiasis in village and farm communities in infected areas (Webbe and Lambert 1983; Lambert et al. 1985).

One reason why endod had not received support in the past was because the toxicologic tests required to determine its suitability as a molluscicide were too expensive for the Ethiopian authorities and university laboratories. Few developing countries have the resources (technical or financial) to conduct a full-scale analysis of the potential risks of pesticides before allowing their use in local environments. At the same time, there is little incentive for industry to complete the tests without obtaining patent rights.

No industrialized country will advocate the use of a botanical pesticide that has not first been exposed to well-established testing procedures. Although endod has been used for many years as a detergent for washing clothes, it has not been subjected to safety evaluation tests. If it is to be used as a molluscicide, such tests are mandatory.

In 1985, the International Development Research Centre (IDRC) and the United Nations Financing System for Science and Technology for Development (UNFSSTD) agreed to examine the basic requirements for evaluation of the toxicology of endod and make recommendations for research and development. Before 1985, numerous toxicity tests had been carried out using extracts of endod (Lemma et al. 1984). However, none of these tests used a standard endod preparation.

At a 1986 meeting sponsored by IDRC, it was concluded that, before endod could be considered for general field use, an endod standard had to be developed and tested. The standard would meet established composition parameters and exhibit consistency of efficacy and safety. Second, tier 1 tests for registration patterned after premarketing guidelines of the Organisation for Economic Co-operation and Development (OECD) would meet the minimal data requirements. The following tests were to be carried out:

- Acute toxicity studies;

- Mutagenicity studies; and

- Environmental studies.

After the 1986 meeting and removal of all outstanding patent rights, IDRC agreed to fund the standard preparation and basic toxicology tests essential for providing internationally acceptable data describing the human acute toxicity and ecological toxicity potentials of endod. Tests for product chemistry, acute toxicity, and environmental fate of premarket chemicals followed tier 1 criteria for registration of pesticides. Tests for acute toxicology effects included:

- Approximate lethal dose;

- Primary dermal and eye irritation;

- Dermal sensitization;

- Inhalation; and

- 28-day repeat oral.

To determine the ecotoxicologic effects on common aquatic nontarget organisms, tests were carried out to determine:

- Algal growth inhibition;

- Fish acute toxicity; and

- *Daphnia* species 14-day reproduction.

Mutagenicity was determined using the Ames and sister chromatid exchange tests.

The cost to complete this preliminary phase was about 150 000 USD. If any test for acute toxicity indicates a potential hazard, carrying out tests for chronic toxicity should be considered. Such tests are included under tier 2 and are considerably more expensive. We believe that if a botanical compound proves hazardous under tier 1 criteria, the chances of it being acceptable are limited; costs would inhibit further tests, and research on the product would be discontinued.

Development strategy

The criteria for development of botanical pesticides are extremely complex. They involve the collaboration of a number of disciplines and functions including: plant taxonomy, agroecology, organic chemistry, entomology, toxicology, regulatory agencies, processing, and marketing. All these activities involve time and money. Producing a synthetic pesticide may cost more than 20 million USD before it can be marketed.

A development plan has been identified for synthetic products (Braunholtz 1977), but none is available for natural plant products. Our experience with endod provided insight into problem areas, the linkages between steps in the process, and likely costs for the different stages of development. Through trial and error, we learned that the process must be followed carefully if such a product is to be internationally approved and recommended as a molluscicide.

The initial phase of the development strategy begins with the discovery of the new plant biocide and identification of the active chemical compound(s) (Fig. 1). A simple laboratory screening test to determine efficacy will indicate its possible value. This can be followed by a controlled preliminary field trial.

Once the potential role of the botanical compound has been identified, the process that leads to its successful approval is a complex one involving three interactive phases.

Chemistry and toxicology

- The physicochemical characterization of the active principle;
- Identification of a standard extraction protocol; and
- Production of a bulk sample for acute toxicology, ecotoxicology, and mutagenicity tests.

Agrobotanical studies

- Identification of plant type with highest yield (HYV, high-yielding variety), potency, and pesticidal activity;
- Propagation;
- Insect/plant pathogen resistance;
- Cropping, harvesting, and processing; and
- Cost–benefit analysis

Field trials

- Vector ecology;
- Identification of transmission sites and peak infection periods; and
- Formulations and biodegradation.

Identification of the plant type with the highest biocidal potential, insect resistance, and potential for cultivation must be determined very early in the process. The standard product must be reproducible, not only for the toxicity tests but for long-term production, distribution, and use. The identification of a standard extraction protocol under "good laboratory practices" is a mandatory requirement for registration. Without such a protocol, the toxicity tests will not be accepted by any registration agency. The potential toxicologic

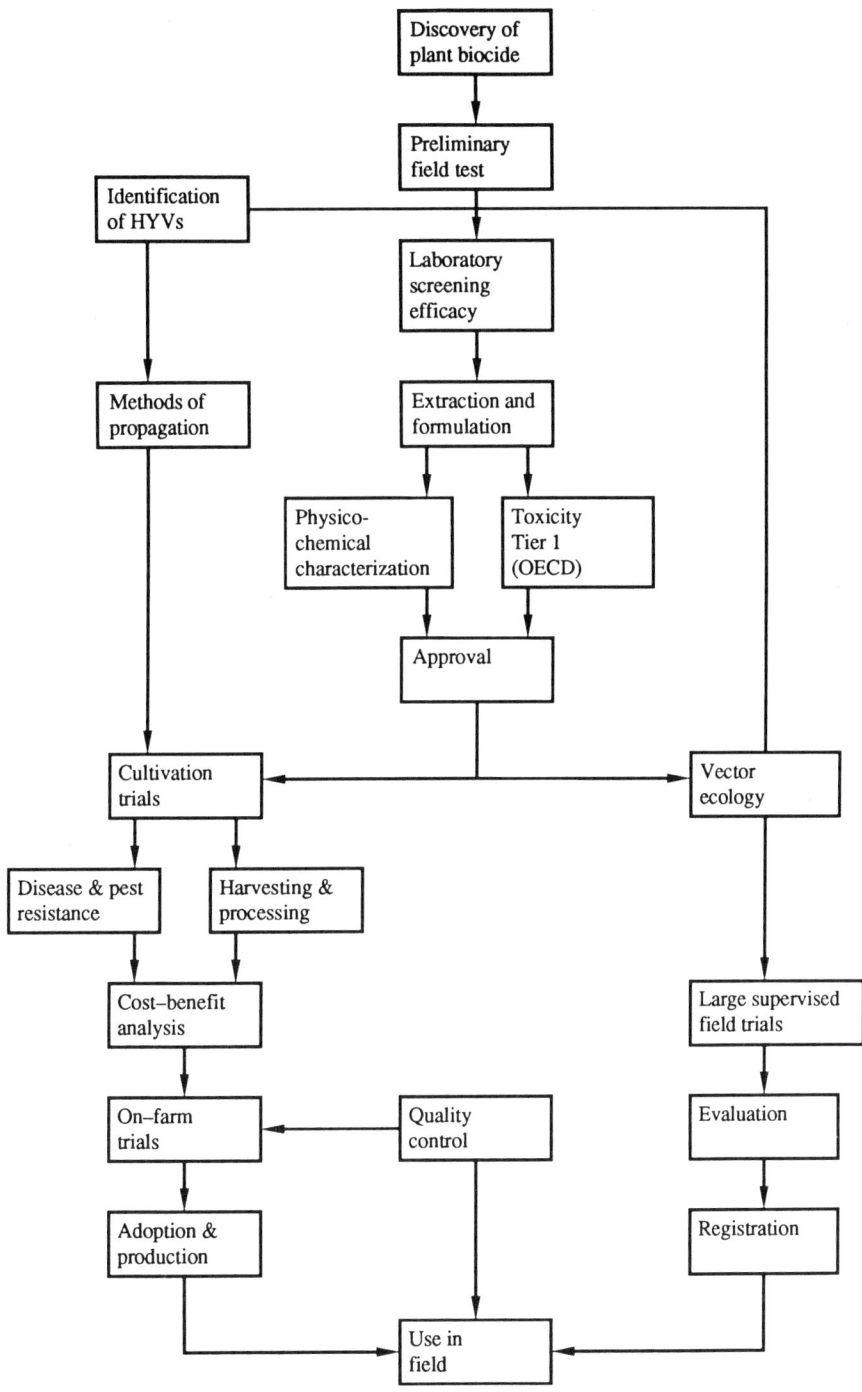

Fig. 1. Steps and interactions between stages from discovery of a biocide to approved field application.

and environmental effects must be determined using prescribed OECD guidelines.

If the botanical compound is acceptable under tier I criteria, all other phases can be carried out. Preliminary cropping trials and collection of information on effects on the target pest and related environmental data must also be carried out. During toxicity testing, data on vector ecology, transmission sites, peak periods of infection, and human use can be collected. Additionally, agrobotanical studies to identify methods of propagation, insect and disease resistance, and cultivation trials can be carried out.

If toxicity tests indicate that the compound is safe to use, the final set of tests will be the field trials. These must follow WHO protocols to determine time and mode of application, concentration, environmental fate, rate of degradation, and efficacy toward the target organism. The trials should be similar to those used for synthetic products. The approval process for botanicals will not take any less time than that needed for synthetic pesticides.

We are seeking pesticides that are rapidly biodegradable and attack sites at which pests cannot readily develop resistance through mutation. Although botanical pesticides have been present for millennia, few pest species have developed resistance, unlike the situation with synthetic chemicals.

If, as we expect, endod proves to be safe for the control of schistosomiasis-bearing snails, then it could replace niclosamide. Using a cheaply produced, easily extracted molluscicide, integrated with chemotherapy, the health of people in endemic regions could be improved.

Conclusion

The potential for further and meaningful developments in botanical pest control is enormous. We are aware that the cost of developing botanical pesticides can be as great as those for synthetic pesticides. Nevertheless, we believe it is an avenue that should be pursued because plant and animal chemical defense mechanisms have proven highly effective in inhibiting predator or pest activity.

One should also consider other avenues. For example, more work is required on the behaviourial, physical, and biochemical adaptations of insects exposed to phototoxins, antifeedants, and the other types of compounds produced by plants. Insects are able to metabolize the defense compounds of their natural hosts, whereas those not usually feeding on those hosts cannot. This is an important consideration in the identifying selective botanical insecticides and the potential for resistance.

At the molecular level, more information is required on adaptation, such as the quenching of molecules through compounds such as ascorbic acid, carotenes, tocopherols, or enzymes (e.g., superoxide dismutase, catalase, and glutathione reductase). The importance of proteases to hydrolyse ingested proteins in the presence of protein-digestion reducers must also be studied further.

The role of botanical substances in plant–insect interactions must be investigated with emphasis on the basic mechanisms that underlie the insects' response to and avoidance of xenobiotics. Such studies will contribute to our understanding of the mechanisms involved in the development of insect resistance to chemicals and will facilitate biotechnological developments and improve pest-management strategies.

The search for new, naturally occurring insecticides and synergists represents a challenge to pest-management specialists and toxicologists. The need to control an increasing number of resistant species requires the identification of compounds with novel modes of action. No significant improvement in vector control can be envisaged without a drastic change in the nature of the biochemical tools available. They must have long-term economic benefits and be environmentally acceptable.

Arnason, J.T.; Isman, M.B.; Philogène, B.J.R.; Waddell, T.G. 1987b. Mode of action of the sesquiterpene lactone tenulin from *Helenium amarum* against herbivorous insects. Journal of Natural Products, 50, 690–695.

Arnason, J.T.; Philogène, B.J.R.; Morand, P.; Scaiano, J.C.; Werstiuk, N.; Lam, J. 1987a. Thiophene and acetylenes: phototoxic agents to herbivorous and blood-feeding insects. *In* Heitz, J.R.; Downum, K.R., ed., Light-activated pesticides. American Chemical Society, Washington, DC, USA. Symposium, 339, 255–264.

Arnason, J.T.; Towers, G.H.N.; Philogène, B.J.R.; Lambert, J.D.H. 1983. The role of natural photosensitizers in plant resistance to insects. *In* Hedin, P., ed., Plant resistance to insects. American Chemical Society, Washington, DC, USA. Symposium, 208, 139–151.

Bernard, C.B.; Arnason, J.T.; Philogène, B.J.R.; Lam, J.; Waddell, T. 1988. Effects of lignans and other representative secondary metabolites of the Asteraceae on the MFO activity of the European corn borer, *Ostrinia nubilalis* Hubner (Lepidoptera: Pyralidae). Phytochemistry, 28, 1373–1378.

Braunholtz, J.T. 1977. Pesticide development and the chemical manufacture. *In* Proceedings of the 15th international congress of entomologists. Entomological Society of America, College Park, MD, USA. Pp. 747–755.

Brindley, W.A.; Selim, A.A. 1984. Synergism and antagonism in the analysis of insecticide resistance. Environmental Entomology, 13, 348–353.

Champagne, D.E.; Arnason, J.T.; Philogène, B.J.R.; Morand, P.; Lam, J. 1986. Light-mediated allelochemical effects of naturally occurring polyacetylenes and thiophenes from Asteraceae on herbivorous insects. Journal of Chemical Ecology, 12, 835–858.

Devitt, B.D.; Philogène, B.J.R.; Hinks, C.F. 1980. Effects of veratrine, berberine, nicotine and atropine on developmental characteristics and survival of the dark-sided cutworm, *Euxoa messoria* (Lepidoptera: Noctuidae). Phytoprotection, 61, 88–102.

Donskotch, R.W.; El Feraly, F.S. 1969. Isolation and characterization of (+)-sesamin and β-cyclo-pyrethrosin from pyrethrum-D flowers: chrysanthemum-cinerariaefolium-D lignan sesquiterpene. Canadian Journal of Chemistry, 47, 1139–1142.

Dowd, P.F.; Smith, C.M.; Sparks, T.C. 1983. Detoxification of plant toxins by insects: short review. Insect Biochemistry, 13, 453–458.

Iyengar, S.; Arnason, J.T.; Philogène, B.J.R.; Morand, P.; Werstiuk, N.H.; Timmins, G. 1987. Toxikokinetics of the phototoxic allelochemical α-terthienyl in three herbivorous Lepidoptera. Pesticide Biochemistry and Physiology, 29, 1–9.

Jacobson, M. 1983. Insecticides, insect repellents, and attractants from arid/semi-arid land plants. *In* Plants: the potentials for extracting protein, medicines, and other useful chemicals. Office of Technology Assessment, Washington, DC, USA. OTA-BP-F23, pp. 138-146.

Koul, O.; Isman, M.B.; Ketkar, C.M. 1990. Properties and uses of neem, *Azadirachta indica*. Canadian Journal of Botany, 68, 1–11.

Lambert, J.H.D.; Legesse Wolde-Yohannes; Makhubu, L. 1985. Endod: potential for controlling schistosomiasis. BioScience, 35(6), 365–366.

Lemma, A. 1965. A preliminary report on the molluscicidal property of endod (*Phytolacca dodecandra*). Ethiopian Medical Journal, 3, 187–190.

Lemma, A.; Goll, P.; Duncan, J.; Mazengia, B. 1978. Control of schistosomiasis with the use of endod in Adwa, Ethiopia: results of a 5-year study. *In* Proceedings of an international conference on schistosomiasis, Cairo, Egypt. Pp. 415–436.

MacRae, W.D.; Towers, G.H.N. 1984. Biological activation of lignans. Phytochemistry, 23, 1–12.

Matsumura, F. 1975. Toxicology of insecticides. Plenum Press, New York, NY, USA. 503 pp.

Miller, J.R.; Miller, T.A. 1986. Insect–plant interactions. Springer-Verlag, New York, NY, USA. 341 pp.

Morgan, E.D.; Mandava, N.B., ed. 1990. Handbook of natural pesticides: Volume VI — Insect attractants and repellents. CRC Press, Boca Raton, FL, USA. 249 pp.

Philogène, B.J.R. 1974. Les phagostimulants, les phagorépresseurs et les insectes phytophages. Annals of the Society of Entomologists, 19, 121–126.

_____ 1982. Development rate changes in insects: the importance of photoperiod. American Naturalist, 120, 269–272.

Philogène, B.J.R.; Arnason, J.T.; Berg, C.W.; Duval, F.; Champagne, D.; Taylor, R.G.; Leitch, L.C.; Morand, P. 1985. Synthesis and evaluation of the naturally occurring phototoxin, alphaterthienyl, as a control agent for larvae of *Aedes intrudens*, *Aedes atropalpus* (Diptera: Culicidae) and *Simulium verecundum* (Diptera: Simuliidae). Journal of Economic Entomology, 78, 121–126.

Philogène, B.J.R.; Arnason, J.T.; Towers, G.H.N.; Abramowski, Z.; Campos, F.; Champagne, D.; McLachlan, D. 1984. Berberine: a naturally occurring phototoxic alkaloid. Journal of Chemical Ecology, 10, 115–123.

Schmutterer, H. 1990. Properties and potential of natural pesticides from the neem tree (*Azadirachta indica*). Annual Review of Entomology, 35, 271–297.

Webbe, G.; Lambert, J.D.H. 1983. Plants that kill snails and prospects of disease control. Nature, 302, 754.

Wharten, J.D.; Morgan, E.D. 1990. Insect feeding deterrents. *In* Morgan, E.D.; Mandava, N.B., ed., Handbook of natural pesticides: Volume VI — Insect attractants and repellents. CRC Press, Boca Raton, FL, USA. Pp. 23–134.

Young, J.W. 1983. Registration requirements for pesticides. *In* Whitehead, D.L.; Bowers, W.S., ed., Natural products for innovative pest management: current themes in tropical science (vol. 2). Pergamon Press, Oxford, UK. Pp. 523–540.

Endod, a potential natural pesticide for use in developing countries

Legesse Wolde-Yohannes

Institute of Pathobiology, Addis Ababa University, Addis Ababa, Ethiopia

In recent years, economic and ecological factors have forced many developing countries to consider alternatives to synthetic pesticides, such as the natural molluscicides found in local plants. Since the early 1930s, more than 1 071 plant species have been tested: 48% contained substances toxic to snails, the intermediate hosts of schistosomiasis, some of them as potent as manufactured molluscicides. Endod is one of the most promising plant molluscicides because of its high toxicity to the snails, low toxicity to mammals, stability under various environmental conditions, biodegradability, widespread distribution in tropical Africa, and potential for large-scale cultivation. Results of the first longitudinal studies on the cultivation, yield, and molluscicidal potency of endod types 3, 17, and 44 indicate that they can be cultivated successfully and that type 44 is most suitable for schistosomiasis control.

Plant-derived molluscicides are a reasonable option for the control of schistosomiasis, particularly in developing countries with agricultural economies, where parasitic disease affects at least 200 million people (Adams 1986; Kloos and McCullough 1987; Legesse and Kloos 1990).

Since the early 1930s, 1 071 plant species have been tested for molluscicidal activity (Kloos and McCullough 1987). Few, however, have proven to have the characteristics necessary for widespread use: low toxicity, water soluble, common, easy to cultivate, and consistently high molluscicidal potency (Legesse and Kloos 1990). If other uses could be found for these few species their value would increase (Adams 1986; Legesse et al. 1986; Kloos and McCullough 1987; Kloos et al. 1987; Legesse and Kloos 1990).

Commercially available chemical molluscicides are expensive in view of the large quantities required for control programs on a regular basis and currency constraints faced by developing countries (Lambert et al. 1985). Compounds, such as copper sulfate and sodium pentachlorphinate, have been used as molluscicides for many decades in Egypt, the Sudan, and other parts of the world to control schistosomiasis.

Recently, however, the ethanolamine salt of niclosamide (bayluscid), has been found to be more effective and is currently the only molluscicide recommended by the World Health Organization (WHO) for global use. However, partly because of its high cost (26 000 USD/t in 1974), most developing countries, especially those in Africa, are not using it to control schistosomiasis. Meanwhile, many well-intentioned agricultural development and water conservation schemes are providing more sites for snails to breed, and schistosomiasis is spreading rapidly.

Endod: a potential pesticide

Endod, a natural molluscicide, is biodegradable. Its active principle decomposes rapidly, breaking down to inert and nonharmful material within a few days. Accordingly, a concerted effort to develop simple methods for the development and safe use of endod could pave the way for use of other plant pesticides in developing countries. Ideally, each country where schistosomiasis is endemic should select local molluscicide-producing species for snail control. In all cases, however, the principle and approach remains the same.

The use of endod berries as a soap, dating back hundreds of years, led to studies of its detergent, antifungal, antiprotozoan, spermicidal, and molluscicidal properties (Lemma et al. 1979). The berries have been traditionally used in Ethiopia and other African countries for washing clothes and as a medicine (Hutchinson and Dalziel 1929; Dalziel 1936; Watt and Breyer Bandwik 1962; Humbert 1971; Thiseltch-Dryer 1973; Legesse and Kloos 1990). However, the molluscicidal property of endod has given it international recognition as a potential means of control for snails that transmit schistosomiasis (Lemma et al. 1979).

Endod or soapberry plant (*Phytolacca dodecandra*) is a member of the Phytolaccaceae and is a dioecious scandent shrub or liana averaging 2–3 m in height, but as climber sometimes growing as tall as 10 m. The male flowers are light yellowish green, in long staminate racemes, and bear no fruit. The female flowers are light yellowish green in short staminate racemes, producing fruit that is five lobed and 1 cm in diameter; ripe fruits are pink or red (Hutchinson and Dalziel 1929). Under favourable climatic conditions in Ethiopia, the plant bears fruit from November to June (Legesse 1983).

The molluscicidal property of endod has been studied over the last 25 years at the Institute of Pathobiology at Addis Ababa University, the Tropical Products Institute in London, the Stanford Research Institute in California, the G.W. Hooper Foundation at the University of California, the Harvard School of Public Health in Boston, the Public Health Service Laboratory and Field Station in Puerto Rico, the US Naval Medical Research Unit in Cairo, Carleton

University in Ottawa, and in other laboratories in different parts of the world, and the results are promising (Lambert et al. 1985).

Some 50 scientific papers have been published on the subject and several patents have been secured on different aspects of processing this plant product (Lemma et al. 1979). Chemical studies to isolate and identify the active principle in endod berries led to the discovery of a new compound, oleanolic acid glucoside, which Stanford Research Institute's chemists have named lemmatoxin (Parkhurst et al. 1974).

Agronomic studies have also been carried out to select endod strains with high molluscicidal potency and high yields as well as those that are easy to cultivate and have a high resistance to insect pests (Legesse and Kloos 1990). Genotypes of endod, two to three times more potent than the original type, have been developed by selection since the 1970s. At present, out of 65 types, 10 have been selected for their high berry yield, high molluscicidal potency, and insect resistance (Legesse and Kloos 1990). Three are in production (Lugt 1981; Legesse 1983; Legesse et al. 1986; Legesse and Kloos 1990). The selected types are producing berries with the capacity to kill all snails in a sample within 24 h at a concentration of 7–10 ppm (of dry berry material) (Legesse 1983; Legesse et al. 1986; Legesse and Kloos 1990).

The detergent properties of endod are also promising. Its foaming and thermal stability and cleaning properties equal or exceed most imported or locally made detergents in developing countries. The water extract of endod can be used as an effective substitute for dodecylbenzene sulfonic acid (DDBSA), a petrochemical byproduct used as a surfactant in commercial detergent formulations for washing fine grades of cotton, linen, and wool.

From a point of view of acceptability, endod has been used locally for centuries. People in the Ethiopian highlands have used the berries as a laundering agent for the white cotton shawls (*shama*) that are a part of the Ethiopian culture.

Addis Ababa University and the National Chemical Corporation are collaborating on research into the detergent properties of endod ("Findings on the viability of utilizing endod (*Phytolacca dodecandra*) in detergent production." Unpublished report, 1984). In two pilot projects at Reppi Soap Factory, endod extract was substituted for DDBSA. The cleaning properties of the endod-based product were comparable to those of DDBSA-based detergent. The detergent preparation can also be used without further processing for snail control (Legesse and Kloos 1990). Use of endod is expected to save foreign currency now spent on importing detergent ingredients.

Investigations have also been undertaken on the agrobotanical characteristics of endod in Ethiopia and elsewhere (Table 1). The annual dry-weight yield of endod berries, types 3, 17, and 44, increased over the 4-year study period, especially between the first and third harvests. Mean berry yield per hectare was higher for type 44 (2 040 kg) than for types 17 (1 165 kg) and 3 (829 kg).

Molluscicidal tests indicated that type 44 was most active, achieving 100% kill of *Bulinus pfeifferi* at 5 ppm for newly harvested berries and 7.5 ppm for 1-year-old and 10 ppm for 4-year-old ground berries (Table 2). At a concentration of 10 ppm, types 3 and 17 killed all snails, but effectiveness decreased considerably after storage.

The molluscicidal saponin yield of endod is about 25% of the weight of dry ground berries. Thus, the amount of saponin that can be extracted from the 4th-year harvest is 688 kg/ha for type 44, 463 kg/ha for type 17, and 262 kg/ha for type 3.

Field trials are now needed to establish molluscicidal efficacy and effect on nontarget organisms in natural snail habitats in accordance with principles and methods of field evaluation of plant molluscicides described by Sturrock and Duncan (cited in Kloos and McCullough 1987). Assuming that a satisfactory molluscicidal effect can be obtained at a concentration of endod of 20 ppm

Table 1. Berry yields for endod types 3, 17, and 44 over 4 years.

| Year | Yield (kg dry weight/ha) | | |
	Type 3	Type 17	Type 44
1981-82	410	380	1 230
1982-83	850	1 130	1 480
1983-84	1 010	1 300	2 700
1984-85	1 047	1 850	2 750
Total	3 317	4 660	8 160
Mean	829	1 165	2 040

Table 2. Molluscicidal potency (% mortality) of fresh, 1-year, and 4-year crude powdered berries of endod types 3, 17, and 44 against *B. pfeifferi*.

| | Concentration (ppm) | | | | |
	10	7.5	5	4	Control
Fresh berries					
Type 3	100	60	0	0	0
Type 17	100	20	0	0	0
Type 44	100	100	100	30	0
1-year-old berries					
Type 3	100	60	0	0	0
Type 17	100	60	0	0	0
Type 44	100	100	60	20	0
4-year-old berries					
Type 3	100	0	0	0	0
Type 17	100	20	0	0	0
Type 44	100	10	0	0	0

in the field (Lugt 1981), 137 500 m^3 of water may be treated in one application using the 4th-year berry yield of 2 750 kg/ha possible with type 44 endod.

Due to differences in plant variety, location, soil, climate, method of extraction, and even seasonal changes, standardization of the plant extract is required to ensure consistent effectiveness (Legesse and Kloos 1990). Toward these objectives, recommendations for an endod standard were made at an international workshop in Swaziland (Makhubu et al. 1986).

Comprehensive and systematic plant molluscicide development and the proposed multiple uses of endod depend to a large degree on intersectoral and interinstitutional collaboration involving ministries of health, agriculture, industry, and universities to assure broad-based botanical, chemical and toxicologic, agronomic, public health, and community-action programs. At the community level, active participation of local people in project planning, production, processing, and application of plant molluscicides and the assessment of their effectiveness in snail control is essential for the development of viable and low-cost, community-based, schistosomiasis-control programs (Kloos and McCullough 1987; Legesse and Kloos 1990).

In March 1983, the Zambian National Council for Scientific Research, in collaboration with Ethiopian scientists, convened an international scientific workshop on endod. Research was reviewed, areas for future research and development were identified, and recommendations were made for specific follow-up activities (Legesse 1983). Following the recommendations, the governments of Ethiopia, Swaziland, and Zambia expressed increased interest and support for collaborative research and development of endod, both as a locally produced molluscicide for schistosomiasis control and as an important additive in commercial detergent formulations for laundry use. Since a second international workshop (Makhubu et al. 1986), there has been a rapid increase in interest in this plant and active research is being carried out by many young and talented African scientists in different parts of the continent.

An expert group, organized by the United Nations Financing System for Science and Technology for Development and the International Development Research Centre (IDRC), developed a procedure for preparing water extract of endod (endod-S) to be used as a standard material for testing in different laboratories (UNFSSTD/IDRC 1986). The group also delineated selected basic toxicological tests to be performed to clear endod for field trials.

IDRC has since commissioned studies to establish the basic regulatory requirements for the licencing of endod, including acute mammalian and ecotoxicity studies performed in Canada, the USA, and Europe under "good laboratory practice." The Institute of Pathobiology, Addis Ababa University, participated in these studies and provided clonal material of endod type 44 grown in isogenic plots at the Institute.

An internationally accepted water preparation of endod type 44 (endod-S) has been developed and independently checked by laboratories in Canada,

Switzerland, and the USA. Using extracts from this procedure, toxicologic tests at the tier 1 level (OECD) of the premarketing guidelines have been completed. According to results to date, endod is no more toxic than bayluscid (Lambert et al. 1991). On the basis of this, IDRC is supporting comprehensive community-based field trials on endod's effect on schistosomiasis transmission in Zimbabwe.

In addition, scientists from 10 African countries and Brazil have now formally established an endod technical cooperation among developing countries (TCDC) network through which they plan to coordinate their work and collaborate further with other international experts in Canada, the USA, and various European countries. The African partners in the network are Ethiopia, the Gambia, Ghana, Kenya, Nigeria, Sudan, Swaziland, Zambia, and Zimbabwe.

A third international workshop on endod was held in Addis Ababa, Ethiopia, in October 1990. The results of the tier 1 toxicology studies were reviewed and research needs, strategies, and model protocols were determined for facilitating community-based trials, including local agrobotanical studies, ecotoxicologic studies, and integrated schistosomiasis intervention studies. Reports from field studies in Ethiopia, Swaziland, Zambia, and Zimbabwe were also presented at the meeting.

Conclusion

All three types of endod can be successfully cultivated, but type 44 has superior agronomic and molluscicidal characteristics. Endod berries may be produced on a large scale with sustained yield and potency, eliminating a major barrier for plant molluscicide development. The results of this study may facilitate the selection of the most suitable endod types for schistosomiasis control in at least three other African countries where endod cultivation and adaptation trials are being carried out (Adams 1986). This study also clearly indicates that emphasis should be placed on promotion of the multiple uses of endod-derived products.

Adams, R.P. 1986. Report of the working group on agrobotany and extraction. *In* Makhubu, L.; Lemma, A.; Heyneman, D., ed., Report of the 2nd international workshop on endod. Council on International and Public Affairs, New York, NY, USA. Pp. 36–42.

Dalziel, J.M. 1936. The useful plants of west tropical Africa. Crown Agents, London, UK.

Humbert, H. 1971. Flora de Madagascar et des Comores. Musée national d'histoire naturelle, Laboratoire de phanerogamie, Paris, France.

Hutchinson, J.; Dalziel, J.M. 1929. Phytolaccaceae: flora of west tropical Africa (vol. 1, part 1). Crown Agents, London, UK.

Kloos, H.; McCullough, F.S. 1987. Plants with recognized molluscicidal activity. *In* Mott, K.E., ed., Plant molluscicides. John Wiley and Sons, Chichester, Sussex, UK. Pp. 45–108

Kloos, H.; Waithaka Thiongo, F.; Ouma, J.H.; Butterworth, A.E. 1987. Preliminary evaluation of some wild and cultivated plants for snail control in Machaka District, Kenya. Journal of Tropical Medicine and Hygiene, 90, 197–204.

Lambert, J.D.H.; Legesse Wolde-Yohannes; Makhubu, L.P. 1985. Endod: potential for controlling schistosomiasis. Bioscience, 35, 364–366.

Lambert, J.D.H.; Temmink, J.H.M.; Marquis, J.; Parkhurst, R.M.; Lugt, C.; Holtze, K.; Warner, J.E.; Schoonen, A.J.M.; Dixon, G.; Legesse Wolde-Yohannes; de Savigny, D. 1991. Endod: safety evaluation of a plant molluscicide. Regulatory Toxicology and Pharmacology, 14, 189–201.

Legesse Wolde-Yohannes. 1983. Past and ongoing agrobotanical studies of endod (*Phytolacca dodecandra*) in Ethiopia. *In* Lemma, A.; Heyneman, D.; Silangwa, S., ed., Endod (*Phytolacca dodecandra*): report of the international scientific workshop, Lusaka. Tycooli International Publishing, Dublin, Ireland. Pp. 125–129.

Legesse Wolde-Yohannes; Demeke, T.; Lambert, J.D.H. 1986. Cultivation studies of endod (*Phytolacca dodecandra*) and its implication in schistosomiasis control. Institute of Pathobiology, Addis Ababa University, Addis Ababa, Ethiopia. Publication 3.

Legesse Wolde-Yohannes; Kloos, H. 1990. Agronomic and molluscicidal studies of three types of endod (*Phytolacca dodecandra*). In press.

Lemma, A.; Heyneman, D.; Kloos, H., ed. 1979. Studies on the molluscicidal and other properties of the endod plant (*Phytolacca dodecandra*). University of California, San Francisco, CA, USA.

Lugt, C.B. 1981. *Phytolacca dodecandra* berries as means of controlling bilharzia transmitting snails. Institute of Pathobiology, Addis Ababa University, Addis Ababa, Ethiopia.

Makhubu, L.; Lemma, A.; Heyneman, D., ed. 1986. Report of the second international workshop on endod. Council of International and Public Affairs, New York, NY, USA.

Parkhurst, R.M.; Thomas, D.W.; Skinner, W.A.; Gray, L.W. 1974. Molluscicidal saponins of *Phytolacca dodecandra*. Canadian Journal of Chemistry, 52, 702–705.

Thiseltch-Dryer, W.T. 1973. Flora of tropical Africa (vol. VI). Reeve, Ashford, Kent, UK. Section 1, p. 97.

UNFSSTD/IDRC (United Nations Financing System for Science and Technology for Development and International Development Research Centre). 1986. Endod toxicology: report of the Expert Group meeting. UNFSSTD, New York, NY, USA. 37 pp.

Watt, J.M.; Breyer Brandwik, M.G. 1962. The medicinal and poisonous plants of southern and eastern Africa. Livingston, Edinburgh, UK.

Role of *Ambrosia maritima* in controlling schistosomiasis transmission

J. Duncan

Weston, Hitchin, Hartfordshire, UK

Ambrosia maritima L. (Compositae) is indigenous to Egypt. It is an annual, growing up to 1 m high by May in moist habitats, typically on the banks of irrigation watercourses. The schistosomiasis-transmission season in the Nile Delta is from May to December peaking from June to August. When whole, dried A. maritima was applied to irrigation watercourses in May at 35 mg/L, snail numbers fell to low levels and did not recover until the following year. This suggests that a correctly timed, single, annual treatment might control transmission. Village communities have been encouraged to participate in cultivating and applying A. maritima to provide a cost-effective delivery system. Monthly surveys have shown that snail control appears to be adequate at the village level. However, the mobility of the villagers indicates the need for the control program to be pursued over a larger area.

Ambrosia maritima L. (Compositae) is an annual with a tendency to perennate. Known in Egypt as damsissa, it prefers moist soils and grows to 1 m high. Its molluscicidal properties were first noted 40 years ago when malaria-eradication teams working in the Nile Delta reported the absence of snails where *Ambrosia* grew on the banks of irrigation watercourses. Preliminary field experiments in the 1970s indicated that applications at about 70 mg/L would reduce snail populations to low levels (El Sawy et al. 1981).

Over the last 16 years, research at the High Institute of Public Health, Alexandria, has been aimed at establishing a practical delivery system for reducing schistosomiasis transmission and at conducting toxicologic studies (El Sawy et al. 1983, 1984, 1987, 1989).

Work to date has included:

- Field trials to determine the application level of the plant required to provide a marked reduction in snail numbers;

- Community participation, including applications of *Ambrosia* in village irrigation systems, involving villagers in the cultivation and application of the plant, and conducting a knowledge, attitudes and practice (KAP) study;

- Parasitology to compare prevalence, incidence, and egg load in village communities in treated and control villages as a measure of transmission control; and

- Toxicology to assess the effects of the plant on nontarget organisms and conduct mutagenicity tests in bacterial systems.

Results and discussion

Field trials with have shown that dried *Ambrosia* at 35 mg/L applied in May is sufficient to reduce snail numbers to low levels. The snail population does not recover until the following year. An appropriately timed, single, annual application of *Ambrosia* might, therefore, reduce transmission of schistosomiasis. Live snails collected in the survey were crushed and examined for cercariae; none were found. The number examined was small, however, and attempts will be made to collect larger numbers to determine percentage infection. In 1989, mice were exposed to the water in one treated village with no resultant infection. A larger sample will be used in more carefully designed exposure tests in the future.

In parasitologic studies, prevalence, egg load, and incidence, after being lowered by chemotherapy, return to pretreatment levels at similar rates in both treated and control villages. More information is needed on schistosomiasis transmission after application of a plant molluscicide (i.e., determining infection rates in snails surviving the treatment and in sentinel mice) as villagers may be contracting the major part of their reinfection outside their home village. People travel to work in other villages and to the large towns and children attend school in neighbouring villages — information is to be gathered this year on the extent of these movements. To counteract the effects of population mobility, the researchers will have to work with a group of contiguous villages covering a larger area.

A large program area will be managable only if a large proportion of the villagers is involved in *Ambrosia* cultivation. This has not been achieved to date. The results of the KAP study, which should give some indication of villager attitudes, are just becoming available. It appears that knowledge of schistosomiasis is insufficient to promote wholehearted community participation in its control. Some form of public-health education will probably be necessary to improve this situation.

Results from toxicity tests on nontarget organisms are encouraging, but the data relate only to acute contact toxicity. More detailed laboratory work and field trials are required to extend knowledge in this area.

Tests for mutagenicity using light petroleum and chloroform extracts have indicated no obvious mutagenic potential of the plant material. In contrast, the methanol fraction caused a slight increase in the number of revertants. A water extract of *Ambrosia* was shown to contain large amounts of histidine. Jongen and Koeman (1983) have shown that methanol extracts of other plants containing high levels of histidine increase the number of spontaneous revertants obtained in the mutagenicity test.

The information available on the toxicology of *Ambrosia* and related plants has been reviewed with particular reference to allergenic reactions in humans and animals (Duncan 1987). The chrysanthemum, another member of the Compositae, is one of the most common causes of allergenic contact dermatitis in florists and horticulturists. Dermatitis as a result of ragweed (a common name for the genus *Ambrosia*) is a well-known problem in the United States. Some very severe, generalized reactions occurred after the accidental introduction of *Parthenium* in grain imported to India in the late 1950s. The number of people handling *Ambrosia* during the project in Egypt has been relatively small, but no allergic reactions have been reported so far. This aspect, however, must investigated thoroughly before large-scale field trials are undertaken.

Ambrosia is commonly taken in Egypt as a strong decoction for renal colic and to aid in the expulsion of kidney stones (Sherif and El Sawy 1962). The basis for the apparent success of this remedy is not known. The sesquiterpene lactones, such as ambrosin and damsin, thought to be responsible for the plant's molluscicidal action are not likely to survive the lengthy boiling involved in preparation of this remedy. They were, in any case, shown to have very little pharmacological action in a number of laboratory tests (Abu-Shady and Soine 1953).

One of the main advantages of a plant molluscicide should be its availability at schistosomiasis-transmission sites, thus avoiding the cost of acquiring and importing commercial products. Any cost saving depends on participation of local communities in the production and application of the plant molluscicide. The feasibility of field delivery systems must be established on a small scale and backed up by detailed toxicologic studies. Currently, data are emerging to support the use of *Ambrosia* by village communities and the early toxicity studies are promising.

The Egyptian study has recently been extended into a nearby area of land reclaimed from the desert, where transmission appears to be more focused. As in the original experimental area, the main aim will be to develop a simple delivery system for the plant based on community participation. Enlisting participation may prove difficult, but the potential savings using this approach are evident. Egypt may, in fact, have a remedy for a long-standing disease problem literally growing in its own backyard.

Acknowledgment — This research has been carried out with the financial support of the International Development Research Centre, Ottawa, Canada.

Abu-Shady, H.; Soine, T.O. 1953. The chemistry of *Ambrosia maritima* L. 1 — The isolation and preliminary characterization of ambrosin and damsin. Journal of American Pharmacology Association, 42, 387–395.

Duncan, J. 1987. The toxicology of plant molluscicides. *In* Webbe, G., ed., The toxicology of molluscicides. Pergamon Press, Oxford, UK. Pp. 141–162.

El Sawy, M.F.; Bassiony, H.K.; Magdoub, A.I. 1981. Biological combat of schistosomiasis: *Ambrosia maritima* (damsissa) for snail control. Journal of the Egyptian Society of Parasitology, 11, 99–117.

El Sawy, M.F.; Duncan J.; Amer, S.; Ruweini, H.; Brown, N. 1989. The molluscicidal properties of *Ambrosia maritima* L. (Compositae): 4 — Temporal and spatial distribution of *Biomphalaria alexandrina* in Egyptian village irrigation systems with reference to schistosomiasis transmission control. Tropical Medicine and Parasitology, 40, 103–106.

El Sawy, M.F.; Duncan, J.; Amer, S.; Ruweini, H.; Brown, N.; Hills, M. 1987. The molluscicidal properties of *Ambrosia maritima* L. (Compositae): 3 — A comparative trial using dry and freshly-harvested plant material. Tropical Medicine and Parasitology, 38, 101–105.

El Sawy, M.F.; Duncan, J.; Marshall, T.F. de C.; Bassiouny, H.K.; Shehata, M.A.R. 1983. The molluscicidal properties of *Ambrosia maritima* L. (Compositae): 1 — Design for a molluscicide field trial. Tropical Medicine and Parasitology, 34, 11–14.

El Sawy, M.F.; Duncan, J.; Marshall, T.F. de C.; Shehata, M.A.R.; Brown, N. 1984. The molluscicidal properties of *Ambrosia maritima* L. (Compositae): 2 — Results from a field trial using dry plant material. Tropical Medicine and Parasitology, 35, 100–104.

Jongen, W.M.F.; Koeman, J.H. 1983. Mutagenicity testing of two tropical plant materials with pesticidal potential in *Salmonella typhimurium*: *Phytolacca dodecandra* berries and oil from seeds of *Azadirachta indica*. Environmental Mutagenesis, 5, 687–694.

Sherif, A.T.; El Sawy, M.F. 1962. Molluscicidal action of an Egyptian herb: 1 — Laboratory experimentation. Alexandria Medical Journal, 8, 139–148.

Alternative methods for pest management in developing countries

K.T. MacKay

International Center for Living Aquatic Resources Management,
Makati, Metro Manila, Philippines

Pest management in developing countries is normally carried out using synthetic pesticides. On some crops (e.g., rice and vegetables), agricultural chemicals have been misused, leading to health and environmental problems and increased resistance of agricultural pests. Given that pests may account for crop losses of up to 25%, alternative solutions are urgently needed for the millions of small-scale farmers in developing countries. The International Development Research Centre has supported research that promotes safer pesticide application and develops alternatives to chemical pesticides. The alternatives to chemical pesticides have included plant breeding for resistance, biological control, microbial pesticides, botanical pesticides, and integrated pest management. This paper discusses specific examples of the use of these approaches in developing countries.

Crop losses to pests (diseases, insects, and weeds) in developing countries can be high. Estimates throughout Southeast Asia vary from 10 to 30% depending on crop and environment (Teng and Heong 1988) and losses can be much higher when a new pest appears. In Latin America and parts of Africa, leaf diseases devastate crops of bananas and plantains, threatening the food and income of smallholders. An insect pest from Latin America has invaded Africa, drastically lowering yields of cassava, a major staple crop. Locusts periodically swarm across Africa consuming everything in their paths. A psyllid pest of *Leucaena leucocephala*, a tree widely used in agroforestry and soil conservation in Asia, has moved on air currents and airplanes across the Pacific decimating leucaena and causing considerable economic and environmental damage (MacKay and Durno 1989).

Effective pest management is essential. However, conditions in developing countries are often very different from those in the developed countries. Technology, pesticides, and approaches suitable for the developed countries often do not work in the Third World. Conditions are often ideal for pest multiplication. The high temperatures and humidity in the tropics result in many generations per year, e.g., *Plutella xylostella*, the diamondback moth, produces up to 28 generations per year in Malaysia (Ho 1965). In irrigated areas, the asynchronous planting of rice has resulted in increased pest populations, particularly of specialized pests, when compared to the previous seasonal planting (Loevinsohn and Litsinger 1989).

Cropping systems in developing countries are often complex. Multiple cropping, where two, three, or more crops are rotated in the same field during a year, and intercropping, where two or more crops are grown in the same field at the same time, are common. In addition, a one-farm family may be growing different crops in different parcels of land at the same time. The agricultural landscape is, as a consequence, much more complex and diverse in both time and space than the monocultures typical of developed-country agriculture.

Land holdings are usually small (0.5 to 4 ha) and there are, therefore, a large number of farms and farmers. Typically 50–75% of a developing country's population is directly engaged in agriculture. In Thailand, for example, over 38 million people are directly involved in agricultural production.

Farmers often have little formal education, are illiterate, or in some countries speak a language or dialect different from the national language. However, they often have considerable traditional knowledge about agricultural practices that is usually ignored by agricultural researchers and extension agents. Developing-country farmers are often poor. They have no cash to purchase inputs and often little access to formal credit. Informal credit may be available, but interest rates are high. It is only for the major cash crops (vegetables, rice, wheat, etc.) that government-supported credit schemes are available for the purchase of inputs. Farmers also have little equipment. Sophisticated sprayers cannot be purchased or maintained by small farmers. The backpack sprayer is the usual method of application for chemical pesticides; however, these sprayers are often ill-maintained and leak, causing a high degree of operator exposure. Protective clothing is often not available or not used.

Research and extension efforts directed toward pest management are limited. There is no international institute or board devoted to developing appropriate pest-management techniques for developing countries and often methods and technologies from developed countries are transferred with little in-country testing. There is usually a high farmer-to-extension-agent ratio and extension workers know little about pest management.

Priority support has been given by the (former) Agriculture, Food and Nutrition Sciences Division of the International Development Research Centre (IDRC) for research that benefits the small-scale farmers of the developing countries. This has led to a concentration on the complete farming system and researchers who work closely with farmers. IDRC support for research in pest management has been in two areas: developing alternatives to chemical pesticides where current use is high; and finding low-cost alternatives for areas and subsistence crops that now use little, if any, pesticide.

Chemical pesticide solutions

Synthetic pesticides (insecticides, herbicides, fungicides, etc.) are used widely in developing countries in plantation and large-scale farming activities and by small-scale farmers for cash crops of cotton, vegetables, and fruits for local and export markets. Their use is expanding and is expected to double in the next 10 years. In the "green revolution" areas, where grain production (rice, maize, and wheat) has increased dramatically, technical assistance and credit have been available for pest management. Pesticide use has increased substantially in these areas, whereas small-scale farmers growing subsistence crops often use little or no pesticide.

The health risk to small-scale farmers from use of synthetic pesticides is much greater (and may be unacceptable) in developing countries compared to developed countries. The backpack sprayer is the type most widely used and results in much greater operator exposure than the more sophisticated and large-scale equipment found in industrialized countries. Backpack sprayers often leak and are poorly maintained. In a Malaysian rice-growing area, 58% of sprayers were corroded, 48% had dented or cracked tanks, and 25% leaked from the tank valve or hose (Loevinsohn 1987). This results in even higher operator exposure. Suitable protective clothes are not available or cannot be worn because of the heat and humidity. Even when they are used (cotton gloves and handkerchiefs or shawls as respirators), they quickly become saturated and actually increase exposure.

Although there is little information on pesticide residues in food for developing countries, particularly food intended for local consumption, export crops are analyzed for residues to protect consumers in developed nations. Vegetables and fruits appear to receive the highest pesticide doses and may contain high levels of residue. In Southeast Asia, most of the "official" information indicates that residue levels meet standards acceptable to the Food and Agriculture Organization (FAO) (Hashim and Yeoh 1988; Ramos-Ocampo et al. 1988; Soekardi 1988; Tayaputch 1988). However, less-official reports indicate vegetables may contain unacceptable levels of pesticides.

This situation is made worse by farmers' practice of spraying right up to harvest and applying fungicides during transport to market. In the Philippines, vegetable farmers increased spray applications as harvest time approached and even dipped freshly picked vegetables in formalin to maintain consumer appeal (Rola 1988). Banned or restricted organochlorines, such as dichlorodiphenyltrichloroethane (DDT), dieldrin, and aldrin, were present in most vegetables, sometimes in excess of allowable limits. A similar situation was found in Thailand where organochlorine residues (particularly DDT, dieldrin, and heptachlor) were found at levels above FAO-acceptable limits in peanut seeds purchased in the markets (Wanleelag and Tau-Tong 1986, 1987; Wanleelag et al. 1988).

Because pesticides are required in small quantities, they are often repackaged in pop bottles and paper bags with no labels. They are routinely stored near food and the containers are often reused for food. Even when labels are applied, they are often not in the local language or are of little use to farmers who cannot read. Some Philippine vegetable producers have a 50% risk of poisoning after 5 years of pesticide use (Rola 1988). In Central Luzon, Philippines, a 27% increase in deaths among the most at-risk population occurred after the introduction of pesticides (Loevinsohn 1987).

Many pesticides considered safe under developed-country conditions are unsafe under conditions found in tropical developing countries (Loevinsohn 1989). No matter how many improvements are made in applicator safety and farmer education, certain chemicals are still too dangerous to use in this environment. This relates primarily to classes of chemicals (organophosphates) that are considered to be more toxic to humans than recently developed insecticides. However, pyrethroids, which are considered to have low human toxicity, can also have considerable effect on people applying them under tropical conditions (He et al. 1989).

There is little information on the wider environmental effects of pesticides on nontarget organisms in developing countries. In Southeast Asia, there is anecdotal evidence for loss of diversity, particularly among beneficial insects, and in rice-growing areas considerable reduction in numbers of frogs and fish in paddy and irrigation waters where pesticides are used.

As many as 1 400 insect pests may be resistant to synthetic insecticides. The diamondback moth (*Plutella xylostella* L.) in Southeast Asia is resistant to all classes of synthetic insecticides and also shows cross resistance to a number of others (Cheng 1986; Rushtapakornchai and Vattanatangum 1986). In addition, it may be resistant to the new class of insect growth regulators and *Bacillus thuringiensis*. Resistance to insecticides is compounded by the farmers' practice of often mixing a cocktail of insecticides and using application rates much higher than recommended.

It is clear that synthetic chemical insecticides as currently used are not the solution to pest management in developing countries. However, answers to the problem of pest control are desperately needed.

Alternative solutions

Breeding

Plant breeding for resistance to insects and disease has been very successful. The international agricultural research centres (IARCs) have had notable successes in major crops (rice, wheat, and maize). However, this has led to the planting of large areas with few varieties (e.g., Indonesia grows only three varieties on 90% of its rice area). In these situations, when pesticide resistance occurs, the results can be severe, e.g., in Asia recently with rice blast and brown plant hopper on rice.

Much of IDRC's support over the past 17 years has been in the field of crop breeding, including resistance to pests and disease. IDRC support has often been directed toward minor crops, such as food legumes, root crops (cassava and sweet potato), oil seeds, and bananas and plantains. In some cases, IDRC has promoted germ-plasm exchange to increase genetic diversity and make resistant material available. Some of the germ plasm of bananas and plantains from Southeast Asia (Philippines and New Guinea) is resistant to the leaf diseases devastating smallholders in Latin America and Africa. However, because of the presence of viruses in Southeast Asia, it has not been possible to exchange material. Research has, therefore, focused on identifying viruses, developing screening techniques including monoclonal antibodies, and perfecting tissue-culture techniques to enable virus indexing and clean-up necessary for the exchange of banana genetic material.

A survey of IDRC crop-breeding projects in China, Indonesia, the Philippines, and Thailand has just been completed to assess the use of pesticides and the approach to breeding in these areas. Some project teams are selecting for horizontal resistance and tolerance rather than resistance. However, their use of chemical pesticides was much higher than among local farmers. Although safety precautions and level of training of people applying pesticides was inadequate, follow-up training is planned.

Other alternatives are now receiving considerable attention from researchers and are starting to have an effect in farmers' fields. They include integrated pest management (IPM) and biocontrol. These are the areas in which IDRC has been, and continues to be, heavily involved.

Classic biocontrol

Coffee berry borer

Hypothenemus hampei, a pest of African origin, is now a problem in Ecuador and Mexico. Commercial coffee has little resistance. Parasites of this pest have been collected in Kenya and Togo, reared in quarantine, then sent to Latin America. One parasite is now established in Ecuador (*Cephalonomia stephanoderis*) and two populations of *Prorops nasuta* from east and west Africa are being tested to see if they can exert control over a wide geographic range.

Cassava green mite

Monongchellus tanajoa was accidentally introduced into Africa from South America in 1971 and significantly reduced cassava yields. Since 1974, IDRC has supported control efforts. Natural enemies were surveyed, tested in field conditions in Trinidad, reared in the United Kingdom under quarantine, evaluated in field conditions in Africa, and mass-released throughout West

Africa and now East Africa. The success in West Africa is touted as one of the untold stories of biocontrol.

Leucaena psyllid

Leucaena leucocephala, a tree much used in agroforestry for fodder, fuel, green manure, and soil conservation in Asia has been decimated by the psyllid, *Heteropsylla cubana*, which moved from Latin America through Florida, Hawaii, the Pacific Islands, Australia, Indonesia, and the Philippines, finally reaching India and Sri Lanka through aircraft and high altitude currents from 1983 to 1988.

Two parasitoids and a generalized cochinellid are being evaluated for use against this pest. Some have been introduced unofficially into Southeast Asia without appropriate quarantine. There is some concern over the possibility of introducing a damaging fungus and that the generalized predator may interfere with a weed-control program on mimosa. This has raised the issue of "prior informed consent" (PIC) and governments and quarantine authorities are aware of risks and potential conflicts of interest.

Personnel on these projects work closely with the Commonwealth Institute of Biological Control (CIBC). Its role has been to assist in the survey for potential controls, taxonomy, determination of efficacy, rearing methods, and third-country quarantine.

Inundative releases

China has used *Trichogramma* wasps, which are egg parasites of many pests. *Trichogramma dendrolimi* and *T. chilonis* have been used successfully to control sugarcane borers in southern China. The estimated cost of biocontrol is 16.50 USD/ha, about 10% of the cost of the previously used chemical control. Moreover, control is better and health problems are much reduced. *Trichogramma* has also been used for control of Asian corn borer on 13 000 ha of Beijing watershed that were formerly sprayed with DDT.

IDRC support in China has helped to develop rearing techniques for large and small eggs and artificial rearing, to carry out ecological studies on other species, and to expand the use of these methods to other crop pests. A book in Chinese and English summarizing the Chinese experience is in preparation and training is planned for 1992. This project involves the collaboration of researchers at Guelph University, who are carrying out research on rearing, storage, and pheromones and assisting in training.

Microbial control

In India, a survey of local microspodia (*Nosema* sp.) has been completed; the organism has been cultured and shown to be effective against grasshopper and locust. Field trials are currently under way and this approach will be tried in Africa against locust plagues.

Emphasis has been placed on the development of the bacterium *Bacillus thuringiensis* (*Bt*) as a biocide because of its safety to mammals and the environment. At least 2–3 t/year of commercial *Bt* products are now used in developed countries to control insect pests of agricultural crops and forest and ornamental trees. A small percentage of the biocide is used in developing countries. The adaptation of the commercial products to local conditions in developing countries has limited its use there. IDRC support has concentrated on solving this problem.

Botanicals

In the tropics, a large number of plant species have potential insecticidal properties. Many plants are used traditionally for fish poisons, arrow poisons, and human medicines. Some are used by farmers for pest control. Project teams in the Philippines and Thailand are surveying promising plants, testing them for insecticidal properties against major insect pests, investigating their mode of action (specifically looking for new classes of chemicals with novel modes of action), determining chemical structure and in some cases possible synthesis, and determining the effects on nontarget organisms and mammalian toxicity. The safest and most promising insecticides will then be tested in on-farm experiments for effectiveness under field conditions. These products may then be extracted locally from cultured plants either on a village or homestead scale or produced on a larger scale by commercial enterprises.

In Thailand, researchers used a similar approach to identify a botanical fish poison that may soon be marketed. It will offer a safer, locally produced alternative to current imported poisons. In these projects, the search is for effective botanical substances that can be produced locally, are cheaper than imported products, and have minimal health and environmental effects.

Cultural and ecological controls

IDRC is supporting several projects to examine various aspects of farming systems in the hope of identifying novel approaches to pest management. The traditional practice of combining rice and fish culture is expanding rapidly in Asia. Researchers in China, Indonesia, and Thailand report that the presence of fish decreases the number of pests and diseases in rice. Also in Indonesia, researchers are trying to determine how fish can be managed within an intensive rice-production system where chemical insecticides are used.

Integrated pest management

IPM may be the best approach to pest management. It combines a judicious use of chemicals with various other control strategies. There are, however, many constraints to its adoption (Bottrell 1987).

IDRC is supporting a unique project in the Philippines that involves an interdisciplinary team of researchers working on integrated management of rice and vegetable pests (Adalla 1990). They are working closely with farmers in two villages to develop and field-test recommended IPM strategies for rice and vegetables. Their findings have been useful in identifying recommendations for other areas and are being expanded to other IPM projects in the Philippines. However, the most important results relate to constraints concerning the adoption of IPM:

- Trust in the use of chemicals and mistrust of the unknown IPM approach;

- Household decision-making with regard to pest management; and

- Pest monitoring.

The researchers found that farmers are using a number of chemical insecticides, some of which are dangerous to their health. The farmers recognize the dangers, and surveys indicate that half had experienced some pesticide-related effect. Even though pesticide-related deaths occurred, the farmers continued to use chemical insecticides and sprayed more frequently than required. To change the farmers' attitudes, the researchers carried out on-farm experiments that demonstrated that IPM is successful and allows a higher return than the use of chemicals. On the basis of this, some farmers are adopting the IPM techniques.

A detailed study of decision-making and the role of women has shown that, although women control household income, they rely on advice from men in decisions on pest control in rice crops. However, women are more intimately involved in vegetable production and make all the pest-management decisions themselves. This knowledge allows extension workers to target their advice and training programs more effectively.

Effective IPM requires a thorough knowledge of the interaction between pest and crop. This is often done by examining a portion of the plants to determine the number and types of pests present. If a critical threshold is reached, the farmers apply insecticides. In a situation where farmers claimed their eyesight was not good enough for this type of active surveillance, one project team trained school children as scouts. This system worked well as long as the project subsidized the scouts' wages; farmers were unwilling to pay for the service. The project team is now examining other labour-saving ways of monitoring pests.

Policy

In many developing countries, pesticides that have been banned or restricted in developed countries are widely used. In addition, some pesticides that are approved for use in developed countries are not safe for use in developing countries because of higher operator-exposure rates. IDRC projects are starting to address both of these issues.

A small team led by the Pesticides Trust, UK, is working closely with a consortium of nongovernmental organizations (NGOs) including the Pesticide Action Network, Malaysia. The researchers are assisting in monitoring and investigating compliance to the FAO (1986) code of conduct for pesticide use, especially the principle of prior informed consent (Pesticides Trust 1989).

Conclusion

Pest management problems will continue as new pests appear and others become resistant to existing controls. The most urgent need is to develop alternative control systems. This requires emphasis on research into biocontrol, microbials, botanicals, and cultural control, supplemented by continued efforts in breeding, particularly for horizontal resistance, and IPM. However, it is becoming increasingly clear that pest problems in some cases are the symptom of much wider problems. Pest problems often increase when plants are stressed (White 1984). The nitrogen status of a plant can also be a factor (Mattson 1980). It is, therefore, important to take a "systems" view of pest management.

New research methods will be required for investigating many of the alternative pest-management systems being developed. Traditionally, pesticide experiments adopt an intensive approach in which experiments are done at a few sites and results extrapolated to a much larger area, because the pesticides are applied at saturation levels and minimize site-to-site variation. However, an extensive approach is necessary with alternative biological control systems, because between-site variability is much greater and a large number of multi-locational trials are needed. Ultimately, the farmers must be more involved in the search for feasible alternatives. Many of the techniques used in farming systems research are applicable in this context (MacKay 1989).

Different skills are also needed by farmers. Alternative pest-management strategies require more knowledge than the chemical strategies they are designed to replace. There is a much greater need to determine not only farmers' indigenous knowledge and current practices, but, more importantly, farmers' attitudes to pests and pest management and the process of decision-making (especially gender issues). Despite knowledge of the harmful effects of chemical pesticides, farmers continue to use them (Rola 1988). This appears to be partly due to lack of knowledge of alternatives, but also to the reluctance to risk crop damage. Close involvement of extension workers with both farmers and researchers to identify farmers' knowledge and practices and to assist in the training of farmers in the new approaches is essential. Unfortunately, extension workers are often much more knowledgeable about chemical pest-control techniques than they are of the alternatives. Considerable training of extension agents is needed.

Although emphasis should be on the development and application of alternative methods, there is also a need to influence policymakers. In many countries, this will involve studies of the health and environmental consequences and costs of current chemical pesticide use to convince national policymakers of the need for regulation, enforcement, and research on alternatives. In addition, pressure must be exerted on national governments and international pesticide manufacturers to ensure that:

- Pesticides banned in developed countries are not "dumped" in developing countries;

- Pesticide-safety standards are designed for the realities of the tropics; and

- Pesticides are not promoted (through development projects or pesticide manufacturers) in developing countries where they are not needed.

This pressure will have to be exerted by NGOs in both developing and developed countries.

As chemical insecticide use will continue and indeed expand in developing countries in the near future, there is also a continuing need for programs to develop safer methods for pesticide storage and application. These should include better container design, better labeling (in local languages), safer sprayers, and suitable clothing to minimize exposure. Training in the safe use of pesticides for both extension agents and applicators will continue to be essential.

To accomplish these aims, there is an increased need for regional and international cooperation and networking. This will involve donors and international and national agricultural research centres. IDRC is currently involved with a group of other donors to explore ways to increase support for IPM. However, the developing-country researchers themselves must solve their countries' pest-management problems. It is to them that IDRC must continue to direct support for training, networking, and research.

Adalla, C. 1990. IPM, women and extension: final report for phase I submitted to the International Development Research Centre, Ottawa. Department of Entomology, University of the Philippines, Los Baños, Philippines.

Bottrell, D.G. 1987. Applications and problems of integrated pest management in the tropics. Journal of Plant Protection in the Tropics, 4(1), 1–8.

Cheng, E.Y. 1986. The resistance, cross resistance, and chemical control of diamondback moth, *Plutella xylostella*, in Taiwan. *In* Griggs, T.D., ed., Diamondback moth management: proceedings of the first international workshop,
Tainan, Taiwan, 11–15 March 1985. Asian Vegetable Research and Development
Center, Shanhua, Taiwan. Pp. 329–345.

FAO (Food and Agriculture Organization of the United Nations). 1986. International code of conduct on the distribution and use of pesticides. FAO, Rome,
Italy. 31 pp.

Hashim, B.L.; Yeoh, H.F. 1988. Pesticide residue studies in Malaysia. *In* Teng, P.S.;
Heong, K.L., ed., Pesticide management and integrated pest management in
Southeast Asia: proceedings of the Southeast Asia pesticide management and
integrated pest management workshop, 23–27 February 1987, Pattaya,
Thailand. Island Publishing House, Manila, Philippines. Pp. 349–354.

He, F.; Wang, S.; Liu, L.; Chen, S.; Zhang, Z.; Sun, J. 1989. Clinical manifestations
and diagnosis of acute pyrethroid poisoning. Archives of Toxicology, 63, 54–58.

Ho, T.H. 1965. The life history and control of the diamondback moth in Malaya.
Ministries of Agriculture and Co-operatives Bulletin, 118, 26.

Loevinsohn, M.E. 1987. Insecticide use and increased mortality in rural Central
Luzon, Philippines. Lancet, 1987 (June 13), 1 359–1 362.

_____ 1989. Pesticides in the Third World: controlling hazards where regulation
is weak. Paper presented at the annual meeting of the Agriculture Institute of
Canada/Canadian Society for Pest Management, 10–12 July 1989, Montreal,
Quebec, Canada. Agricultural Institute of Canada, Ottawa, ON, Canada.

Loevinsohn, M.E.; Litsinger, J.A. 1989. Time and the abundance of rice pests. Paper
presented at the annual meeting of the Agriculture Institute of Canada/
Canadian Society for Pest Management, 10–12 July 1989, Montreal, Quebec,
Canada. Agricultural Institute of Canada, Ottawa, ON, Canada.

MacKay, K.T. 1989. Sustainable agricultural systems: issues for farming systems
research. Paper presented at the international workshop on developments in
procedures for farming systems research, 13–17 March 1989, Bogor, Indonesia.
IDRC, Singapore. 14 pp.

MacKay, K.T.; Durno, J. 1989. Do magic bullets work? Lessons from the miracle
tree, *Leucaena leucocephala*. Paper presented at the 7th international scientific
conference, 2–5 January 1989, Ouagadougou, Burkina Faso. International Fed-
eration of Organic Agriculture Movements, Oberwil, Switzerland.

Mattson, W.J. 1980. Herbivory in relation to plant nitrogen content. Annual Review
Ecology and Systematics, 11, 119–161.

Pesticides Trust. 1989. Prior informed consent in the international code of conduct
on the distribution and use of pesticides. Pesticides Trust, London, UK.

Ramos-Ocampo, V.E.; Magallona, E.D.; Tejada, A.W. 1988. Pesticide residues in the Philippines. *In* Teng, P.S.; Heong, K.L., ed., Pesticide management and integrated pest management in Southeast Asia: proceedings of the Southeast Asia pesticide management and integrated pest management workshop, 23–27 February 1987, Pattaya, Thailand. Island Publishing House, Manila, Philippines. Pp. 355–367.

Rola, A.C. 1988. Pesticides, health risks and farm productivity: a Philippine experience. Final report (Assessing the benefits and risks of pesticide use in Philippine agriculture) submitted to the Agricultural Policy Research Program, University of the Philippines, Los Baños, Philippines. 122 pp.

Rushtapakornchai, W.; Vattanatangum, A. 1986. Present status of insecticidal control of diamondback moth, *Plutella xylostella*, in Thailand. *In* Griggs, T.D., ed., Diamondback moth management: proceedings of the first international workshop, Tainan, Taiwan, 11–15 March 1985. Asian Vegetable Research and Development Center, Shanhua, Taiwan. Pp. 307–312.

Soekardi, M. 1988. Pesticide residue control and monitoring in Indonesia. *In* Teng, P.S.; Heong, K.L., ed., Pesticide management and integrated pest management in Southeast Asia: proceedings of the Southeast Asia pesticide management and integrated pest management workshop, 23–27 February 1987, Pattaya, Thailand. Island Publishing House, Manila, Philippines. Pp. 373–378.

Tayaputch, N. 1988. Pesticide residues in Thailand. *In* Teng, P.S.; Heong, K.L., ed., Pesticide management and integrated pest management in Southeast Asia: proceedings of the Southeast Asia pesticide management and integrated pest management workshop, 23–27 February 1987, Pattaya, Thailand. Island Publishing House, Manila, Philippines. Pp. 343–347.

Teng, P.S.; Heong, K.L., ed. 1988. Pesticide management and integrated pest management in Southeast Asia: proceedings of the Southeast Asia pesticide management and integrated pest management workshop, 23–27 February 1987, Pattaya, Thailand. Island Publishing House, Manila, Philippines. 473 pp.

Wanleelag, N.; Tau-Thong, P. 1986. Insect pests and residual analysis of toxic substances in groundnut. Kasetsart University, Bangkok, Thailand. Groundnut improvement project, research reports for 1982–1985, pp. 151–161.

_____ 1987. Residual analysis of the soil. Kasetsart University, Bangkok, Thailand. Groundnut improvement project, progress report for 1986, pp. 51–60.

Wanleelag, N.; Tau-Thong, P.; Impitak, S.; Sothikul, A.; Chawengsri, V. 1988. Residues of heptachlor and heptachlor epoxide in groundnut. Kasetsart University, Bangkok, Thailand. Groundnut improvement project, progress report for 1987, pp. 36–43.

White, T.C.R. 1984. The abundance of invertebrate herbivores in relation to the availability of nitrogren in stressed food plants. Oecologia, 63, 90–105.

Alternative methods for pest management in vegetable crops in Calamba, Philippines

P.R. Hagerman

Department of Environmental Biology, University of Guelph, Guelph, Ontario, Canada

The results of preliminary investigations to find a nonchemical control method for eggplant pests are reported. Two biological control agents (the wasp Trichogramma chilonis *and the earwig* Euborellia annulata*) and two cultural control methods (manual and trapping) were compared with insecticide-treated and control (untreated) plots in an experimental field. Although crop damage was lower in all treated plots, the untreated control plots produced higher yields. The crop in the experimental field was compared to that in an adjacent pesticide-treated farmer's field. The yield from the experimental field was slightly lower (4%), but resulted in higher net income (24%) than the farmer's field. Recommendations for integrated pest management based on these experiments are now being tested by farmers and are leading to more profitable farming with lower human-health risks and lower environmental impact.*

Since the 1960s, pesticides have been the primary tool used by Filipino vegetable farmers to combat insect pests (UPLB 1990). Pesticides are frequently applied at higher-than-recommended doses and at shorter-than-recommended intervals. Yet even these treatments are not always effective in controlling insect pests, suggesting that some insects may have developed resistance to certain pesticides.

Because many Filipino farmers have experienced health problems after spraying pesticides, the practice of hiring casual labourers to spray fields is common. However, this and the other costs associated with the use of pesticides substantially reduce the net income of farmers. Moreover, regardless of who applies the pesticides, farm families still risk exposure through contamination of food and drinking water (shallow wells for domestic use are often located at the edge of vegetable fields). Studies have found high levels of pesticide residues on fresh and cooked vegetables, indicating that consumers are also exposed (UPLB 1990).

In the town of Calamba, situated in the lowlands of Laguna Province, Philippines, vegetables (tomato, eggplant, string beans, and cucurbits) are grown for market and local consumption. In response to requests from farmers, research into integrated pest management (IPM) strategies has been carried out on several farms. The research described here was carried out on eggplants, but the approach is applicable to pest management in other crops.

The most serious pest in the cultivation of eggplant is the eggplant shoot borer, *Leucinodes orbonalis* Guenée (Pyralidae), which kills young shoots and bores into fruit rendering it unmarketable. Although recorded in the Philippines as early as 1948 (Capps 1948), *Leucinodes* achieved pest status only in the 1970s. Its outbreak at that time may have been sparked by environmental changes resulting from increased pesticide use (Navasero 1983). Today, Filipino farmers rely almost exclusively on pesticides to control the eggplant shoot borer, which can damage 20–92% of a crop (Esguerra and Barroga 1982; Saavedra 1987).

Minor damage to eggplants is also caused by mites, cutworms, leaf hoppers, scarab beetles, aphids, and mealybugs. In the dry season, thrips (*Thrips palmi*) can cause loss of vigour, leaf yellowing, and eventual defoliation by sucking plant juices. A complicating factor is that pesticides used on shoot borer can actually increase the population of thrips (E.N. Bernardo, University of the Philippines, Los Baños, Philippines, personal communication).

Generally, plant diseases have not been a problem for eggplant farmers and weeds are controlled by hand. Farmers occasionally clear damaged fruit and shoots to reduce the population of *Leucinodes* in the field. Although, Saavedra (1987) has suggested such manual "cleaning" as a control method, to date there has been no thorough assessment of its efficacy.

An IPM strategy for eggplant cultivation must concentrate on insect control methods that do not cause an increase in the number of weeds or diseases. Because thrips causes only minor damage in the absence of insecticides, an inexpensive, nonchemical control package for *Leucinodes* formed the focus for this research study.

Several nonchemical alternatives for the control of *Leucinodes* have been proposed. For example, *Trichogramma chilonis* wasps, which are egg parasites, have proved to be effective biological control agents of many pests. Baltazar (1989) recommended regular releases of *T. chilonis* for control of shoot borer. Navasero (1983) identified earwigs as a promising predator of larval *Leucinodes*. Similarly, Javier (1989) showed that the earwig *Euborellia annulata* was an effective biological control agent for another pyralid, the Asian corn borer.

Both *Euborellia* and *Trichogramma* attack only one stage of a pest's life cycle; thus, employing them in a control strategy requires detailed knowledge of the population dynamics of the target pest. Population monitoring is an important component of any IPM strategy and various trapping methods have been used for this purpose. Pheromone traps have been particularly useful for monitoring and reducing the activity of Lepidopteran adults (Roelofs 1981); however, *Leucinodes* pheromone has not yet been isolated.

Methods

The research was carried out in Calamba, Laguna, from February to June 1990 as part of an ongoing project on IPM in rice and vegetable crops, funded by the International Development Research Centre (IDRC). The fields are owned by a farmer, who managed all nonresearch activities. The fields were also used to demonstrate research results to local vegetable farmers.

The study was designed to facilitate adoption of procedures by farmers. All research was carried out in a farmer's fields with the farm family fully informed of all activities and participating in most. Neighbouring farmers often inquired about the experimental results and they were invited to two demonstration meetings at the fields. During monthly meetings at a local school, researchers also explained IPM and its use to farmers. Because women play a large role in vegetable production, fieldwork, decision-making, and marketing (UPLB 1990), most of the staff were female, and extension activities were directed toward both women and men.

Plots (4×5 m) were laid out in the experimental field on which six pest-control strategies were tested: two biological agents (*Trichogramma* and *Euborellia*); two culture methods (manual sanitation and trapping); chemical spraying; and no treatment. Each treatment was replicated three times with an average of 25 plants in each plot (Fig. 1). This layout permitted minimal interference between treatments. Because of the strong prevailing winds, the experimental field had to be east of the farmer's field to avoid pesticide drift, and different treatments were separated by 12 m in a north-south direction.

The methods used for the various treatments were:

- *Trichogramma chilonis* were released at a rate of 300 per 20-m^2 plot (150 000/ha). The population of insects in the field was monitored beginning in February to time *Trichogramma* releases to occur 1 day after peak adult flights of *Leucinodes*. Releases began on 23 April.

- *Euborellia annulata* were released at a rate of 17–35 per 20-m^2 plot (8 500–17 500/ha). Releases began on 1 June and were timed to coincide with increases in the larval population of *Leucinodes*.

- Manual control involved handpicking and destroying all damaged fruits and shoots from the plots and from a 2-m border around the plots after each harvest.

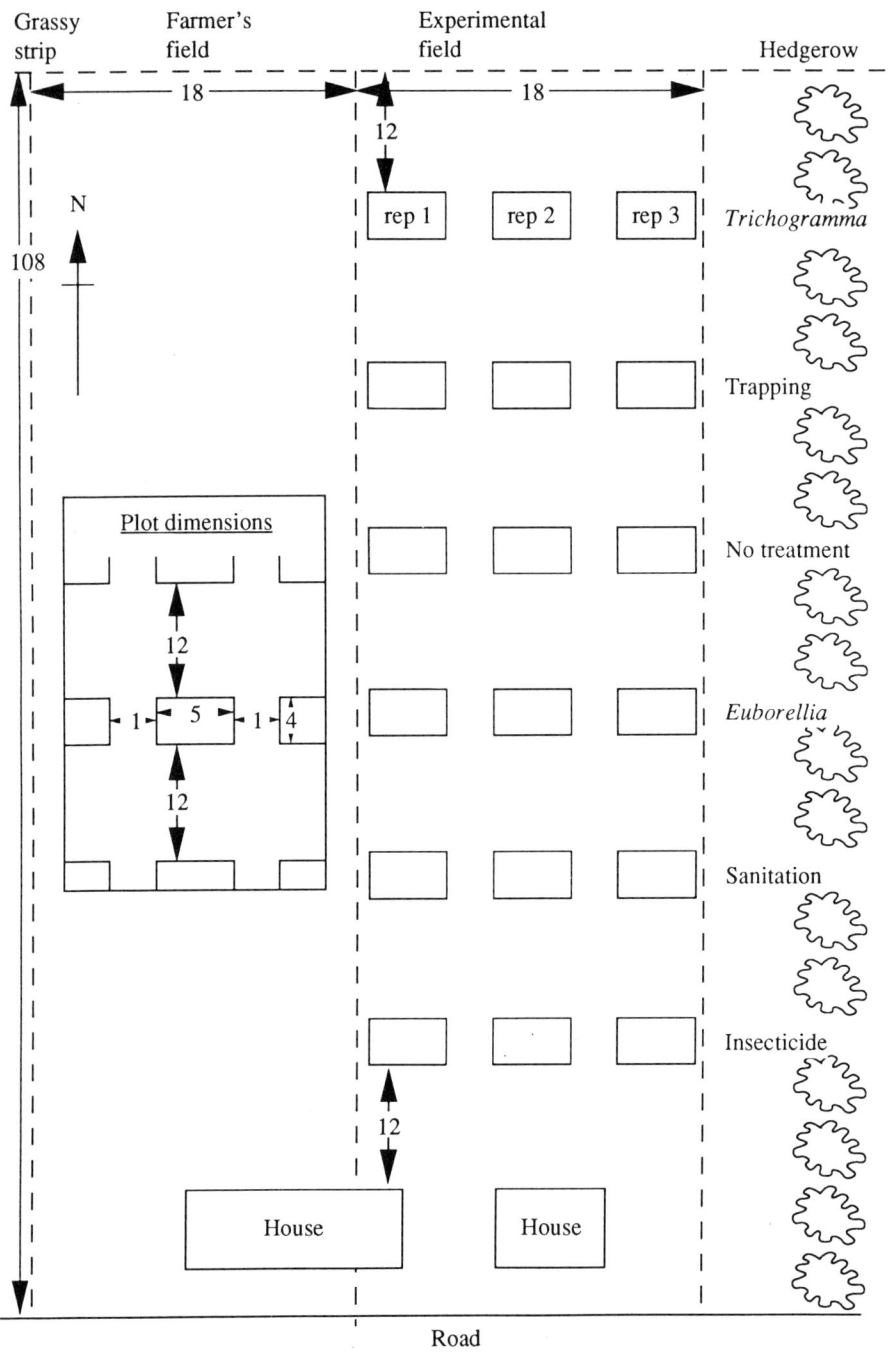

Fig. 1. Layout of plots to test methods of pest management in vegetable crops.
(All dimensions are in metres.)

- Trapping was done using three sticky traps, three pheromone traps, and a light trap (Table 1).

- Insecticide-treated plots were sprayed by the farmer using the same timing and dosage as on his own adjacent field. Between 24 February and 25 June, insecticides were applied 18 times; Hostathion (triazophos) was used most frequently and Dicarzol (formetanate hydrochloride), Decis (deltamethrin), and Azodrin (monocrotophos) less often.

- Untreated plots served as controls.

Eggplants were harvested every 3–4 days and fruit was graded as class A (large and undamaged), class B (small or slightly damaged), and rejects (damaged by insects, rot, or discoloration). Weight and market price of each grade were recorded. After harvest, five plants were selected at random in each plot and examined for fruit and shoots damaged by *Leucinodes*, an indirect but efficient method of assessing larval population. The average number of other pests and beneficial insects per leaf was also recorded.

Previous studies on *Leucinodes*, including those using an unsprayed control plot, have usually taken place in fields where most of the area was sprayed with insecticides. Such practices do not allow for the establishment of a full complement of predators and parasites in control fields. In this study, all unsprayed plots were grouped together, with large border areas, to create the largest possible block of unsprayed plants: 60 m^2 was allocated to each treatment in the midst of 1 584 m^2 of untreated plants. The overall yield of the experimental field was compared with that of the adjacent farmer's field.

Vegetables grown without pesticides often sell at higher prices in the market-place because customers are willing to pay more for the assurance of pesticide-free food. These premium prices provide an extra incentive for farmers who are considering nonchemical pest-control methods. To determine whether local consumers would pay more for unsprayed eggplant, a preliminary marketing study was conducted at the University of the Philippines at Los Baños (UPLB).

Results and discussion

Previous work (Navasero 1983) determined that the life cycle of *Leucinodes* (egg to egg) in the Philippines was 23–30 days (mean, 25 days). Monitoring larval damage revealed distinct peaks at 23- to 31-day intervals (mean, 27 days), with minima over the same period occurring on average 16 days after the peaks, indicating the presence of a nearly synchronous population. Damage was highest when most of the pest species was present as larvae. The lows occurred when larvae had left the plants to pupate.

Understanding the life cycle and population dynamics of a pest is crucial for timing the release of biological control agents that usually attack only one

Table 1. Description and efficiency of traps used to control *Leucinodes* in eggplant crops.

Type and description	Placement	No. used per night	No. nights	No. *Leucinodes* caught	Catch per trap per night
Light					
Kerosene lamp suspended over basin of water	In reps 1 and 3, trap plots	2	11	0	0
Sticky					
A. Commercial flypaper (20 × 25 cm) placed vertically	In rep 2, all treatments	6	5	0	0
B. White, yellow, or unpainted plywood (20 × 20 cm), placed vertically, with petrolatum	All	18	128	0	0
C. Green plywood (20 × 20 cm), placed horizontally, with petrolatum	All	3	116	0	0
Pheromone					
A. Caged female *Leucinodes* in centre of yellow plywood (20 × 20 cm), placed vertically, with petrolatum	All	3	1	0	0
B. Caged female *Leucinodes* suspended over basin of water	In rep 3, trap plots	1	41	1 419	34.6
C. Rubber septum with pheromone extract, suspended over basin of water	In rep 3, trap plots, and 2 sites in experimental field outside plots	3	10	17	0.6

Note: All traps were placed at canopy height and were checked each morning. Rep = replicates of experimental treatment.

stage of the life cycle. *Trichogramma* are most effective if released as the adult pest begins to lay eggs. For *Leucinodes*, this occurs, on average, 1 day after emergence of the adults or 10 days after cocoon spinning (Navasero 1983), i.e., when larval population is lowest. Field release can be timed from the last peak (add 16 days) or the last low (add 27 days).

Euborellia consume the larvae of *Leucinodes* and must be released during the 11-day larval period. This period is marked by an abrupt increase in crop damage as eggs hatch and larvae begin boring into shoots and fruit. *Euborellia* released at the beginning of this stage are most effective in controlling the pest species. Fieldwork is continuing to evaluate the level of control attained with one and two releases of *Euborellia* per *Leucinodes* life cycle.

Knowledge of the population fluctuations of *Leucinodes* may also prove useful for control methods not considered in this study. For example, the nematode *Neoplectana carpocapsae* may be used against pupal *Leucinodes* (Janardar and Bardhan 1974). On the other hand, if chemical methods prove necessary, their effectiveness can be increased by spraying during the most susceptible stage of the pest's life cycle.

Over a period of 17 weeks, none of the biological, cultural, or chemical control methods tested produced higher yields than the untreated plots in the experimental field (Table 2). However, *Leucinodes* caused less damage in four treated plots than in the control plots. Damage was lowest in plots sprayed with insecticide, followed by those cleared manually, *Trichogramma* treated, and *Euborellia* treated. These results may be due to the late start of biological control treatments and damage caused by insects other than *Leucinodes*, especially thrips and mites.

The insecticides used against shoot borers do not control thrips and mites; populations of both these pests increased noticeably in sprayed plots during the dry season. They caused leaf yellowing, reduced yield, and defoliation. In unsprayed plots, these symptoms appeared later and were less severe, probably because the pests were held in check by predators and parasites present

Table 2. Comparison of *Leucinodes* damage and yield of eggplant for various treatments.

Treatment	Damage[a]	Yield (kg)[b]
Control	347	111.0
Manual sanitation	263	106.6
Trichogramma	266	105.3
Euborellia	341	98.7
Trapping	395	97.7
Insecticide	169	93.2

[a] Cumulative number of *Leucinodes*-damaged fruit and shoots in 90 plants, monitored after every harvest.

[b] Cumulative yield of three 20-m2 plots.

Table 3. Cost (PHP) of pest management treatments.

Treatment	Applications/ month	Average cost per application per ha			Monthly cost/ha	Cost for this study[a]
		Material	Labour	Total		
Insecticide	4.2	951	96	1 047	4 397	118.72
Manual clearing	10	0	98	98	980	26.46
Trapping	Continuous	100	160	260	260	7.02
Trichogramma	1	—	—	12[b]	12	0.32
Euborellia	1	—	—	800[c]	800	21.60
Control	—	0	0	0	0	0

Note: 23 Philippine pesos (PHP) = 1 US dollar (USD).

[a] Based on 60 m^2 per treatment over 4.5 months.

[b] From biocontrol lab, UPLB, based on 150 000 individuals released.

[c] From Javier (1989), based on 18 000 individuals released.

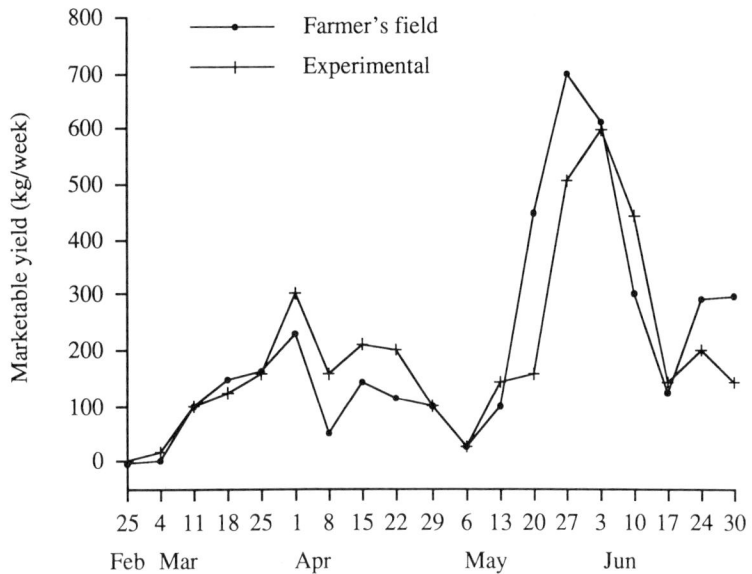

Fig. 2. Marketable yield of eggplants from the farmer's field and the experimental field.

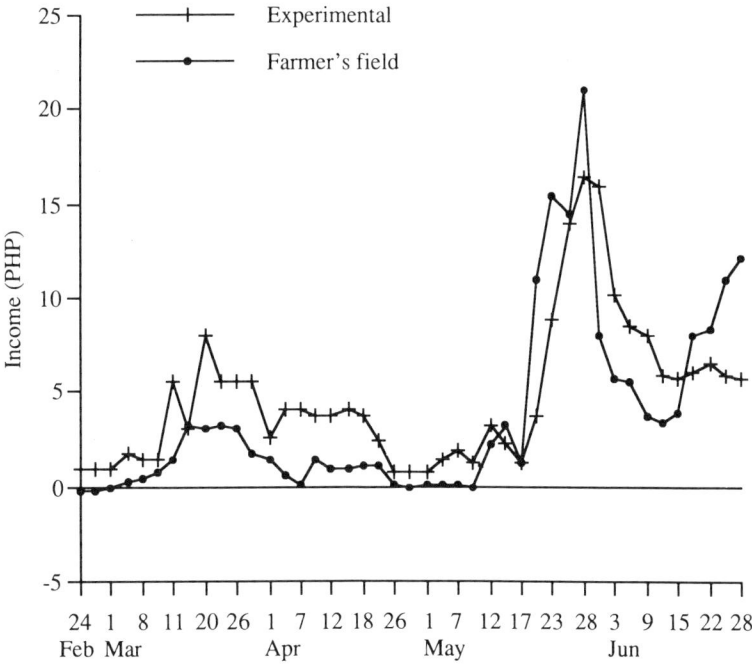

Fig. 3. Net income from eggplants from the farmer's field and the experimental field. (23 Philippine pesos (PHP) = 1 US dollar (USD).)

in the field. These pests were the main cause of lower yields during the dry season in sprayed experimental plots and in the adjacent farmer's field compared with unsprayed plots. The best control for thrips and mites seems to be avoiding the use of insecticides.

Light traps and sticky traps used in this study failed to catch any *Leucinodes*. This species may not be attracted by visual stimuli, but inadequacy of trap design may be the problem. A pheromone trap with live female bait was most effective for catching adult *Leucinodes*. During 41 nonconsecutive nights of trapping, this trap caught an average of 35 males/night, with a peak catch of 282.

Adult female *Leucinodes* are short-lived (2–3 days in a trap) and the tendency toward synchronous populations makes it difficult to rear a continuous supply. An attempt was made to extract the female sex pheromone to prolong its use in the field. Early results indicate that the extracted pheromone is not as attractive as live females, but with better understanding of *Leucinodes* mating behaviour, pheromone trapping may prove to be useful in controlling this pest. It will at least provide a valuable tool for monitoring purposes.

Earwigs were collected from the natural population in the eggplant field. In the dry season, an unidentified species was found in low numbers. In the wet season, *Euborellia annulata* was abundant. Only *E. annulata* was found inside eggplant fruit, in an empty *Leucinodes* tunnel. *E. annulata* is probably the species that preys on *Leucinodes* and was the species observed by Navasero (1983).

The total yield of class A and B fruit from February to June differed by only 4% between the experimental field and the farmer's field. The yield from the experimental field was higher on most harvest days in the dry season (until mid-May), but the yield from the farmer's field was usually higher in the wet season (Fig. 2). In the dry season, the yield from the farmer's field was reduced and plant growth was stunted by thrips and mites. Because the experimental field plants were taller, they suffered more damage from wind and rain later in the year. Fallen plants were more susceptible to plant pathogens, so marketable yield from the experimental field was reduced. Staking proved to be too labour-intensive and not very effective.

Unsprayed eggplants were sold at UPLB nine times between 14 March and 19 April. The price was set 10–50% higher than Los Baños market prices to test whether consumers placed a higher value on pesticide-free produce. Sales were promoted by posters, telephone, and word of mouth. Response was positive, with weekly sales averaging 61 kg; many customers inquired about the availability of other unsprayed vegetables.

The costs of spraying and individual treatments (Table 3) were deducted from the gross income generated from the farmer's field and experimental field. With these adjustments, together with the higher prices paid for unsprayed eggplants, the income from the experimental field was 24% higher (Fig. 3):

20% due to differences in cost of inputs and 4% due to differences in selling price. Remembering that 97% of the experimental field received no insecticide and 85% received no treatment at all, this is strong support for the effectiveness of natural control in combatting eggplant pests.

On the basis of these findings, recommendations were prepared for the integrated management of eggplant pests based on natural control augmented by the monthly release of *Trichogramma* and *Euborellia*. Plans are underway to test this strategy on another farm in Calamba.

Farmers have difficulty monitoring their fields for insects (UPLB 1990). The recommendations advocate monitoring by field inspection, but procedures could be simplified to reduce the time involved. Monitoring by inspection of harvest culls is suggested, and ongoing work on a *Leucinodes* pheromone could provide pheromone traps for continuous monitoring. Either would provide a cheap, easy, and quick method for following the shoot borer's life cycle and timing release of biological control agents.

Baltazar, C.R. 1989. Natural enemies of the major pests of cruciferous and solanaceous crops (3rd report). Department of Entomology, University of the Philippines, Los Baños, Philippines.

Capps, H.W. 1948. Status of the pyraustid moths of the genus *Leucinodes* in the new world, with descriptions of new genera and species. Proceeding of the United States National Museum, Smithsonian Institute, 98(3 223), 69.

Esguerra, N.M.; Barroga, S.F. 1982. Status of pest control in solanaceous crops. *In* The state of the art in growing vegetables. Philippine Council for Agriculture and Resources Research and Development, Los Baños, Philippines. 19 pp.

Janardar, S.; Bardhan, A.K. 1974. Effectiveness of DD-136, an entomophilic nematode against insect pests of agricultural importance. Current Science, 43(19), 622.

Javier, P.A. 1989. Natural enemy component on Asian corn borer with emphasis on mass rearing technique, field performance and selectivity to insecticides of predatory earwigs *Euborellia annulata* (Fabricus). University of the Philippines, Los Baños, Philippines. PhD thesis.

Navasero, M.V. 1983. Biology and chemical control of the eggplant fruit and shoot borer *Leucinodes orbonalis* Gueneé (Pyraustidae: Lepidoptera). Department of Entomology, University of the Philippines, Los Baños, Philippines. BSc thesis.

Roelofs, W.L. 1981. Attractive and aggregating pheromones. *In* Nordlund, D.A.; Jones, R.L.; Lewis, W.J., ed., Semiochemicals: their role in pest control. John Wiley and Sons, Toronto, ON, Canada.

Saavedra, N.T. 1987. Eggplant fruit and shoot borer *Leucinodes orbonalis*. National Crop Protection Centre Newsletter, 2(1), 3–4.

UPLB (University of the Philippines at Los Baños). 1990. Integrated pest management, extension and women (Philippines) (vol. 1). UPLB, Los Baños, Philippines. Integrated terminal report.

PART V

APPENDICES

Participants

Mahmoud M. **Amr**, Kasr El-Aini Faculty of Medicine, University of Cairo, Cairo, Egypt

Maria Elena **Arroyave**, c/o Yves Bergevin, Department of Epidemiology and Biostatistics, McGill University, 1020 Pine Avenue West, Montreal, PQ, Canada H3A 1A2

Thanawadee **Boonlue**, Faculty of Communication Arts, Chulalongkorn University, Phyathai Road, Bangkok 10500, Thailand

Borys **Brezden**, Department of Biology, Carleton University, Ottawa, ON, Canada K1S 5B6

Jacques **Brodeur**, Département de biologie, Laval University, Ste Foy, PQ, Canada G1K 7P4

Carmen **Castaneda**, Department of Pharmacology, College of Medicine, University of the Philippines at Manila, 547 Petro Gil, Ermita, PO Box 593, 1000 Manila, Philippines

Robert **Chase**, Department of Medical Education, McMaster University, 148 Wright Avenue, Hamilton, ON, Canada L8N 3Z5

Donald **Cole**, McMaster University, 148 Wright Avenue, Hamilton, ON, Canada L8N 3Z5

Guido **Condarco** (Aguilar), Instituto Nacional de Salud Ocupacional, Ministerio de Prevision Social y Salud Publica, Casilla 1832, La Paz, Bolivia

F. **Diaz-Barriga**, Facultad de Medicina, Universidad Autonoma de San Luis Potosi, Av. V. Carranza 2405, AP 142, 78210 San Luis Potosi, Mexico

Lawrence S. **Dollimore**, CHSEL/23, Shell Centre, London, SE1 7NA, United Kingdom

John **Duncan**, Goldthorn Cottage, Maiden Street, Weston, Hitchin, Hertfordshire, SG4 7AA, United Kingdom

William **Durham**, 6200 Bent Pine Place, Raleigh, NC 27615, USA

Ravindra **Fernando**, National Poisons Information Centre, General Hospital, Colombo 8, Sri Lanka

Gilles **Forget**, Health Sciences Division, International Development Research Centre, PO Box 8500, Ottawa, ON, Canada K1G 3H9

Diane **Gagnon**, Department of Biology, University of Ottawa, Ottawa, ON, Canada K1N 6N5

Paul **Hagerman**, Department of Environmental Biology, University of Guelph, Guelph, ON, Canada N1G 2W1

Fengsheng **He**, Institute of Occupational Medicine, Chinese Academy of Preventive Medicine, 29 Nan Wei Road, Beijing 100050, People's Republic of China

Jerry **Jeyaratnam**, Department of Community, Occupational and Family Medicine, National University Hospital of Singapore, Lower Kent Ridge Road, Singapore 0511

Md. **Jusoh Mamat**, Pests and Beneficial Organisms Research Unit, Malaysian Agricultural Research and Development Institute, PO Box 12301, 50774 Kuala Lumpur, Malaysia

Sam **Kacew**, Faculty of Health Sciences, University of Ottawa, Ottawa, ON, Canada K1N 6N5

Violet **Kimani**, Department of Community Health, University of Nairobi, PO Box 19676, Nairobi, Kenya

John D.H. **Lambert**, Department of Biology, Carleton University, Ottawa, ON, Canada K1S 5B6

Legesse Wolde-Yohannes, Institute of Pathobiology, Addis Ababa University, PO Box 1176, Addis Ababa, Ethiopia

Jyrki **Liesivuori**, Kuopio Regional Institute of Occupational Health, Neulaniementie 4, PO Box 93, SF-70701 Kuopio, Finland

Michael **Loevinsohn**, BP 259, Butare, Rwanda

Ken **MacKay**, Agriculture, Food and Nutrition Sciences Division, International Development Research Centre, PO Box 8500, Ottawa, ON, Canada K1G 3H9

Marco **Maroni**, International Centre for Pesticide Safety, Via Magenta 25, 20020 Busto Garolfo, Italy

Félicité Domngang **Mbiapo**, Faculté des Sciences, Université de Yaoundé, BP 812, Yaoundé, Cameroon

Michel **Mercier**, International Program on Chemical Safety, World Health Organization, 20 avenue Appia, 1211 Geneva 27, Switzerland

D.N. **Mfitimukiza**, Ministry of Labour, PO Box 4637, Kampala, Uganda

Pierre **Mineau**, Toxic Substance Evaluation and Monitoring Division, Wildlife Toxicology and Surveys Branch, Environment Canada, 351 St-Joseph Blvd, Hull, PQ, Canada K1A 0H3

O.N. **Morris**, Agriculture Canada, Research Station, 195 Dafoe Road, Winnipeg, MB, Canada R3T 2M9

Mutuku **Mwanthi**, Department of Community Health, University of Nairobi, PO Box 19676, Nairobi, Kenya

Manuel **Nasif**, Ministerio de Prevision Social y Salud Publica, Instituto Nacional de Salud Ocupacional, Casilla 1832, La Paz, Bolivia

Derek **Neal**, 1786 McMaster Avenue, Ottawa, ON, Canada K1H 6R8

A.V.F. **Ngowi**, Tropical Pesticides Research Institute, PO Box 3024, Arusha, Tanzania

Pablo **Ocampo**, c/o Sam Kacew, Faculty of Health Sciences, University of Ottawa, Ottawa, ON, Canada K1N 6N5

Grace **Ohayo**, Kenya Medical Research Institute, PO Box 20752, Nairobi, Kenya

Timo **Partanen**, Department of Epidemiology and Biostatistics, Institute of Occupational Health, Topeliuksenkatu 41 a A, SF-00250 Helsinki, Finland

Bernard **Philogène**, University of Ottawa, 550 Cumberland, Ottawa, ON, Canada K1N 6N5

Emma **Rubin de Celis**, Huayuna Instituto para el Desarollo, Paseo de la Castellana 217, Lima 33, Peru

Ramzi **Sansur**, Birzeit University, PO Box 14, Birzeit, West Bank, via Israel

Deogratias **Sekimpi**, Medical Department, Bank of Uganda, PO Box 7120, Kampala, Uganda

W.F. **Tordoir**, c/o Shell Internationale Petroleum, PO Box 162, 2501 AN The Hague, The Netherlands

H. **Versteeg**, Pest Management Advisory Board, 171 Slater Street, Room 701, Ottawa, ON, Canada K1P 5H7

Arnold **de Villiers**, Health Sciences Division, International Development Research Centre, PO Box 8500, Ottawa, ON, Canada K1G 3H9

Frank **White**, Pan-American Health Organization, Caribbean Epidemiology Centre, PO Box 164, Port-of-Spain, Trinidad and Tobago

Panduka **Wijeyaratne**, Health Sciences Division, International Development Research Centre, PO Box 8500, Ottawa, ON, Canada K1G 3H9

Xue Shou-Zheng, Box 206, School of Public Health, Shanghai Medical University, 138 Yi Xue Yuan Road, Shanghai 200032, People's Republic of China

Abbreviations and acronyms

2,4-D	2,4-dichlorophenoxyacetic acid
α-T	alpha-terthienyl
AChE	acetylcholinesterase
AIDS	acquired immune deficiency syndrome
ASEAN	Association of South East Asian Nations
BHC	γ-benzene hexachloride (lindane)
BPMC	fenobucarb, 2-sec-butyl-N-methyl carbamate
Br2A	dibromovinyl-dimethyl-cyclopropane carboxylic acid
Bti	*Bacillus thuringiensis* var. *israelensis*
CDA	control droplet applicator
CEOHS	Center for Environmental and Occupational Health Sciences
ChE	cholinesterase
CIBC	Commonwealth Institute for Biological Control
COPD	chronic obstructive pulmonary disease
CTPB	community-targeted, problem-based
DDBSA	dodecylbenzene sulfonic acid
DDE	1,1-dichloro-2,2-bis (4-dichlorophenyl) ethylene
DDT	dichlorodiphenyltrichloroethane
DDVP	2,2-dichlorovinyl dimethyl phosphate
DEET	diethyl toluamide
DNOC	dinitro-o-cresol
DOAE	Department of Agricultural Extension
EC	emulsifiable concentrate
ECG	electrocardiograph

ECHO	abdominal ultrasonograph
EEG	electroencephalograph
ELC	Environment Liaison Centre
EMG	electromyograph
EPA	US Environmental Protection Agency
EPN	o-ethyl o-4-nitrophenyl phenyl phosphonothionate
FAO	Food and Agriculture Organization of the United Nations
$FEV_{1.0}$	forced expiratory volume per second
FVC	forced vital capacity
GIFAP	Groupement international des associations nationales de fabricants de produits agrochimiques
GPT	glutamic pyruvic transaminase
GTZ	Agency for Technical Cooperation, Germany
HDPE	high-density polyethylene
HPLC	high-performance liquid chromatography
HYV	high-yielding variety
IARC	international agricultural research centre
ICOH	International Commission of Occupational Health
ICPS	International Centre for Pesticide Safety
IDRC	International Development Research Centre
ILO	International Labour Organization
INSO	Instituto Nacional de Salud Ocupacional
IOCU	International Organization of Consumers' Unions
IPCS	International Programme on Chemical Safety
IPM	integrated pest management
IRRI	International Rice Research Institute
JMPR	joint meeting on pesticide residues
KAP	knowledge, attitudes, and practices
KEMRI	Kenya Medical Research Institute
KETRI	Kenya Trypanosomiasis Research Institute
KSH	Kenyan shilling

LC$_{50}$	lethal concentration to 50% of organisms tested
LD$_{50}$	lethal dose to 50% of animals tested
LDH	lactic dehydrogenase
MACA	Malaysian Agricultural Chemicals Association
MARDI	Malaysian Agricultural Research and Development Institute
MSMA	methylarsonic acid
MYR	Malaysian ringitt
NGO	Nongovernmental organization
NIOSH	National Institute for Occupational Safety and Health
NMEO	Nepal Malaria Eradication Organization
NOAEL	"No adverse effect level"
NPIC	National Poison Information Centre
OCP	Onchocerciasis control program
OECD	Organisation for Economic Co-operation and Development
PAHO	Pan American Health Organization
Paraquat	1,1'-dimethyl-4,4'-bipyridylium dichloride
PET	polyethylene terephthalate
PHC	primary health care
PIC	prior informed consent
RV	residual volume
SIRIM	Standard and Industrial Research Institute
TCDC	technical cooperation among developing countries network
TCDD	2,3,7,8-tetrachlorodibenzo-p-dioxin
THB	Thai bhatt
TLC	total lung capacity
ULV	ultra low volume
UN	United Nations
UNEP	United Nations Environment Programme
UNFSSTD	United Nations Financing System for Science and Technology for Development
UPLB	University of the Philippines at Los Baos

URT	upper respiratory tract
USAID	US Agency for International Development
USD	United States dollar
WC	water-soluble concentrate
WHO	World Health Organization
WHOPES	WHO pesticides evaluation scheme
WP	wettable powder

The International Development Research Centre is a public corporation created by the Parliament of Canada in 1970 to support technical and policy research designed to adapt science and technology to the needs of developing countries. The Centre's five program sectors are Environment and Natural Resources, Social Sciences, Health Sciences, Information Sciences and Systems, and Corporate Affairs and Initiatives. The Centre's funds are provided by the Parliament of Canada; IDRC's policies, however, are set by an international Board of Governors. The Centre's headquarters are in Ottawa, Canada. Regional offices are located in Africa, Asia, Latin America, and the Middle East.

Head Office
IDRC, PO Box 8500, Ottawa, Ontario, Canada K1G 3H9

Regional Office for Southeast and East Asia
IDRC, Tanglin PO Box 101, Singapore 9124, Republic of Singapore

Regional Office for South Asia
IDRC, 11 Jor Bagh, New Delhi 110003, India

Regional Office for Eastern and Southern Africa
IDRC, PO Box 62084, Nairobi, Kenya

Office for South Africa
IDRC, 9th Floor Braamfontein Centre, Corner Bertha and Jorissen Streets, Braamfontein, 2001 Johannesburg, South Africa

Regional Office for Middle East and North Africa
IDRC, PO Box 14 Orman, Giza, Cairo, Egypt

Regional Office for West and Central Africa
IDRC, BP 11007, CD Annexe, Dakar, Senegal

Regional Office for Latin America and the Caribbean
IDRC, Casilla de Correos 6379, Montevideo, Uruguay

Please direct requests for information about IDRC and its activities to the IDRC office in your region.